# Recent Advances in Biophysics

# Recent Advances in Biophysics

Edited by **Betty Karasek**

R CALLISTO REFERENCE

New York

Published by Callisto Reference,
106 Park Avenue, Suite 200,
New York, NY 10016, USA
www.callistoreference.com

**Recent Advances in Biophysics**
Edited by Betty Karasek

International Standard Book Number: 978-1-63239-528-3 (Hardback)

# Contents

Preface     IX

Chapter 1 **Computational Fluid Dynamics Study of Swimmer's Hand Velocity, Orientation, and Shape: Contributions to Hydrodynamics**     1
Milda Bilinauskaite, Vishveshwar Rajendra Mantha,
Abel Ilah Rouboa, Pranas Ziliukas, and Antonio Jose Silva

Chapter 2 **Lower-Limb Joint Coordination Pattern in Obese Subjects**     15
Alberto Ranavolo, Lorenzo M. Donini, Silvia Mari, Mariano Serrao,
Alessio Silvetti, Sergio Iavicoli, Edda Cava, Rosa Asprino,
Alessandro Pinto, and Francesco Draicchio

Chapter 3 **Extremely Low-Frequency Magnetic Field Enhances the Therapeutic Efficacy of Low-Dose Cisplatin in the Treatment of Ehrlich Carcinoma**     24
Nihal S. El-Bialy and Monira M. Rageh

Chapter 4 **Determination of Poisson Ratio of Bovine Extraocular Muscle by Computed X-Ray Tomography**     31
Hansang Kim, Lawrence Yoo, Andrew Shin, and Joseph L. Demer

Chapter 5 **Repeated Bout Effect Was More Expressed in Young Adult Males Than in Elderly Males and Boys**     36
Giedrius Gorianovas, Albertas Skurvydas, Vytautas Streckis,
Marius Brazaitis, Sigitas Kamandulis, and Malachy P. McHugh

Chapter 6 **Ascorbic Acid and BSA Protein in Solution and Films: Interaction and Surface Morphological Structure**     46
Rafael R. G. Maciel, Adriele A. de Almeida, Odin G. C. Godinho,
Filipe D. S. Gorza, Graciela C. Pedro, Tarquin F. Trescher,
Josmary R. Silva, and Nara C. de Souza

Chapter 7 **Corticospinal Reorganization after Locomotor Training in a Person with Motor Incomplete Paraplegia**     53
Nupur Hajela, Chaithanya K. Mummidisetty,
Andrew C. Smith, and Maria Knikou

Chapter 8 **Cell Mechanosensitivity: Mechanical Properties and Interaction with Gravitational Field**     61
I. V. Ogneva

Chapter 9     **Production of Chemoenzymatic Catalyzed Monoepoxide**
              **Biolubricant: Optimization and Physicochemical Characteristics**         78
              Jumat Salimon, Nadia Salih, and Bashar Mudhaffar Abdullah

Chapter 10    **Development and Kinematic Verification of a Finite Element Model**
              **for the Lumbar Spine: Application to Disc Degeneration**                  89
              Elena Ibarz, Antonio Herrera, Yolanda Más, Javier Rodríguez-Vela,
              José Cegoñino, Sergio Puértolas, and Luis Gracia

Chapter 11    **Assessment of Genotoxic and Cytotoxic Hazards in Brain**
              **and Bone Marrow Cells of Newborn Rats Exposed to**
              **Extremely Low-Frequency Magnetic Field**                                 107
              Monira M. Rageh, Reem H. EL-Gebaly, and Nihal S. El-Bialy

Chapter 12    **Relationship between Anthropometric Factors, Gender, and Balance**
              **under Unstable Conditions in Young Adults**                              114
              Júlia Maria D'Andréa Greve, Mutlu Cuğ, Deniz Dülgeroğlu,
              Guilherme Carlos Brech, and Angelica Castilho Alonso

Chapter 13    **Interplay of Biomechanical, Energetic, Coordinative, and Muscular**
              **Factors in a 200 m Front Crawl Swim**                                    119
              Pedro Figueiredo, David R. Pendergast, João Paulo Vilas-Boas,
              and Ricardo J. Fernandes

Chapter 14    **Kinematic Measures during a Clinical Diagnostic Technique for Human**
              **Neck Disorder: Inter- and Intraexaminer Comparisons**                    131
              Joseph Vorro, Tamara R. Bush, Brad Rutledge, and Mingfei Li

Chapter 15    **Exploratory Study on the Methodology of Fast Imaging of Unilateral**
              **Stroke Lesions by Electrical Impedance Asymmetry in Human Heads**         140
              Jieshi Ma, Canhua Xu, Meng Dai, Fusheng You, Xuetao Shi,
              Xiuzhen Dong, and Feng Fu

Chapter 16    **Membrane Properties Involved in Calcium-Stimulated Microparticle**
              **Release from the Plasma Membranes of S49 Lymphoma Cells**                 158
              Lauryl E. Campbell, Jennifer Nelson, Elizabeth Gibbons,
              Allan M. Judd, and John D. Bell

Chapter 17    **Slow Diffusion Underlies Alternation of Fast and Slow Growth**
              **Periods of Microtubule Assembly**                                        165
              Ming Yang

Chapter 18    **Water-Protein Interactions: The Secret of Protein Dynamics**             171
              Silvia Martini, Claudia Bonechi, Alberto Foletti, and Claudio Rossi

Chapter 19    **Biophysical Insights into Cancer Transformation and Treatment**          177
              Jiří Pokorný, Alberto Foletti, Jitka Kobilková, Anna Jandová,
              Jan Vrba, Jan Vrba Jr., Martina Nedbalová,
              Aleš Čoček, Andrea Danani, and Jack A. Tuszyński

Chapter 20   *In Vivo* **Healthy Knee Kinematics during Dynamic Full Flexion**                    **188**
Satoshi Hamai, Taka-aki Moro-oka, Nicholas J. Dunbar,
Hiromasa Miura, Yukihide Iwamoto, and Scott A. Banks

**Permissions**

**List of Contributors**

# Preface

Biophysics is a field of scientific study which is premised on biology, computational biology and physics. This branch essentially circumscribes the research and study of all living organisms. The ever innovative and endlessly emerging field of biophysics, is also related to the study of agriculture and medicine. One of the crucial topics that falls under this branch is photobiology, which deals with the study of electromagnetic radiation and its impact and resourcefulness in life of different organisms, bio-elementological perspectives; laser correlation spectroscopy approach to bio-elementology problems: nutritional, ecological and toxic aspects; chlorophyll fluorescence in plant biology and Thermo-luminescence in chloroplast thylakoid. Topics on biotechnology and medical biophysics deal with molecular motors, micotools and applications of hemodynamic theory in mechanical cardiac assist devices. Also, fluorescent techniques are frequently used as a tool to characterize the structure and dynamics of bio-membranes in this field. However, a novel computer simulation method for the study of molecular dynamics (MD) of membranes for atomic-scale information using fluorescent membrane probes has now been discovered. Studies on chlorophyll fluorescence in plant biology and thermoluminescence in chloroplast thylakoid cover the aspects of techniques and applications of bio-membranes.

Each chapter includes an extensive review of literature as well as the latest trends in this field. We have been fortunate as to have an outstanding group of biophysics specialists from all over the world contributing to this publication. I'd like to thank the editorial team at the publishing house for their constant support at each and every step.

**Editor**

# Computational Fluid Dynamics Study of Swimmer's Hand Velocity, Orientation, and Shape: Contributions to Hydrodynamics

**Milda Bilinauskaite,**[1,2,3,4] **Vishveshwar Rajendra Mantha,**[3,4,5]
**Abel Ilah Rouboa,**[3,6] **Pranas Ziliukas,**[1] **and Antonio Jose Silva**[4,5]

[1] *Department of Mechanical Engineering, Kaunas University of Technology,*
*LT-44029 Kaunas, Lithuania*
[2] *Mechatronics Centre for Research, Studies and Information, Kaunas University of Technology,*
*LT-44029 Kaunas, Lithuania*
[3] *Department of Mechanical Engineering, University of Tras-os-Montes and Alto Douro,*
*5001-801 Vila Real, Portugal*
[4] *Centre of Research in Sports, Health and Human Development, CIDESD,*
*5001-801 Vila Real, Portugal*
[5] *Department of Sport Sciences, Exercise and Health, University of Tras-os-Montes and Alto Douro,*
*5001-801 Vila Real, Portugal*
[6] *Department of Mechanical Engineering and Applied Mechanics, University of Pennsylvania,*
*Philadelphia, PA 19104, USA*

Correspondence should be addressed to Vishveshwar Rajendra Mantha; vishveshwar@gmail.com

Academic Editor: Giuseppe Spinella

The aim of this paper is to determine the hydrodynamic characteristics of swimmer's scanned hand models for various combinations of both the angle of attack and the sweepback angle and shape and velocity of swimmer's hand, simulating separate underwater arm stroke phases of freestyle (front crawl) swimming. Four realistic 3D models of swimmer's hand corresponding to different combinations of separated/closed fingers positions were used to simulate different underwater front crawl phases. The fluid flow was simulated using FLUENT (ANSYS, PA, USA). Drag force and drag coefficient were calculated using (computational fluid dynamics) CFD in steady state. Results showed that the drag force and coefficient varied at the different flow velocities on all shapes of the hand and variation was observed for different hand positions corresponding to different stroke phases. The models of the hand with thumb adducted and abducted generated the highest drag forces and drag coefficients. The current study suggests that the realistic variation of both the orientation angles influenced higher values of drag, lift, and resultant coefficients and forces. To augment resultant force, which affects swimmer's propulsion, the swimmer should concentrate in effectively optimising achievable hand areas during crucial propulsive phases.

## 1. Introduction

The swimming propulsion is the result of the interaction of applied forces with water and is predominantly attributed to muscular force applied by hands and forearms. It was emphasised that the major part of about 85% to 90% of propulsion generation in water is created by the application of force by an arm [1, 2]. It was suggested that, the shoulder has the least propulsive potential, the forearm and hand approximately equal according to linear and angular velocities which are least by the shoulder and greater by the forearm and hand [3]. Miller [3] used the mathematical model of the front

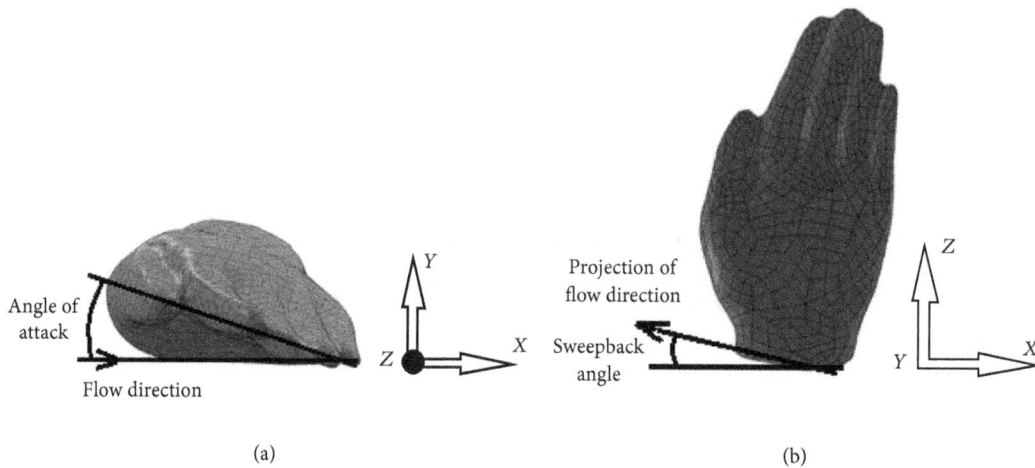

FIGURE 1: Orientations of swimmer's hand model: (a) angle of attack; (b) sweepback angle.

crawl arm pull with bending elbow and showed that the ratio of hydrodynamic forces of hand was about 2.5 bigger when compared with forearm. Toussaint and Truijens [4] studied the visualization of flow tufts around arm and hand. They showed that a strong pressure gradient along the arm occurred that induced axial flow directed from elbow to the hand. Apart from that, the biggest influence of pressure (relative to atmospheric pressure) was identified on the palm of hand when swimming at sprint speed. The authors noted that the pressure was not corrected for differences in hydrostatic pressure due to differences in the depth of the sensors.

In reality, swimmers can change the depth, orientation, shape, and velocity of hand throughout underwater front crawl cycle. All these parameters have direct influence on propulsion force. The prevailing theory of propulsion generation relates to Newton's second and third laws of motion where propulsion is the vector sum of drag ($F_D$: force opposite to the direction of fluid flow) and lift ($F_L$: forces that are perpendicular to the fluid flow) forces [4, 5]. These components depend on the density of the fluid, the velocity of the limb relative to the fluid, the projected surface area of the limb, and the coefficient of drag ($C_D$: a dimensionless constant used to show the resistance of the object in a fluid environment) and lift ($C_L$) which vary according to the shape of the limb and its orientation (i.e., the angle of attack (or pitch angle) and the sweepback angle) (Figure 1). To uncover an influence of various parameters influencing propulsive force or its components, experimental studies [6–8] as well as numerical studies were carried out in the past [5, 9–12].

One of the approachs to calculate and assess forces acting on swimmer or on separated body segments is by an application of computational fluid dynamics (CFD) method. This methodology allows the analysis of the water flow with a reduced amount of complexity and is economical alternative to an experimental method. The validity studies of CFD results are usually carried out with comparative experimental studies. Gardano and Dabnichki [7] used replica of the entire

human arm to assess drag and lift coefficients, measured in a low speed wind tunnel based experiment. The computational hand model was created from human body similar to experimental arm model. Authors showed that lift and drag coefficients obtained in wind tunnel experiments were in good correlation with CFD obtained from CFD simulations carried out using FLUENT (ANSYS, PA, USA). In similar lines, Bixler and Riewald [5] simulated the steady flow around a swimmer's hand and arm at various angles of attack in steady state by using CFD method. They improved validity of calculation and demonstrate that the force coefficients computed for the hand and arm compared well with steady-state coefficients determined experimentally by the previous studies.

In consideration of significant improvement in the methodology and validity of the CFD studies, this method is applied more often in swimming investigations. The calculations through CFD method were carried out to analyse the propulsive forces produced by the propelling segments and the drag force resisting forward motion. Rouboa et al. [11] calculated drag and lift coefficients and drag force for steady flow around a 2D swimmer's hand/forearm model placed at different angles of attack with application CFD technique. These results were compared with previous CFD studies and experimental results, which illustrate similar values under the steady-state flow conditions. It was explained that the increase in flow velocity did not have much influence on the variation of drag coefficient when the model of hand/forearm simulated is positioned at the same pitch angle. This was followed by the steady-state CFD analysis of the hydrodynamic characteristics of a true swimmer's hand model with the thumb in different positions, while the other fingers are kept close together [9]. Drag and lift coefficients were calculated for different angles of attack (the sweepback angle was equal to zero). The combination of drag and lift coefficients (resultant force coefficient) showed that the hand model positioned with the thumb fully abducted presented higher values than the other positions with the thumb partially abducted and adducted at angles of attack of 0° and 45°. However, at an angle of attack of 90°, the position with the thumb adducted presented the highest value of resultant force coefficient. Also, the lift and drag coefficients were steady despite the increase in the flow velocity, and they varied according to changes in pitch angle. Minetti et al. [12] used CFD calculation and in their short communication hypothesized that an intermediate finger spacing in the 3D hand model could increase a higher coefficient of drag providing swimmers with additional thrust. It was indicated that an optimal finger spacing (12°), roughly corresponding to the resting hand posture increases, the drag coefficient (+8.8%), which is functionally equivalent to a greater hand palm area. In the following year, Marinho et al. [10] confirmed these findings using CFD method in their study of the effect of finger spread on the propulsive force production in swimming. The steady-state CFD and 3D scanned models of the hand at the different angles of attack (the sweepback angle equal zero) were used. These results confirmed that the model with a small spread between fingers presented higher values of drag coefficient than did the models with fingers closed and fingers with a large spread. And it was concluded that the optimum finger spread could allow the hand to create more propulsive force during swimming.

Researchers proved that the hand orientation as well as hand shape contributes to the values of hydrodynamics parameters. Rouboa et al. [11] agree that both the propulsion and drag forces will fluctuate during individual phases of stroke cycle along with the respective variation of angle of attack and sweepback angle. Schleihauf (as cited in [13]) summarised that the ideal pitch angle during the underwater motion of the hand will produce an optimal combination of lift and drag forces, which in turn will generate a resultant force that could be predominantly directed in forward direction. Gardano and Dabnichki [7] noticed that, in addition to the pitch angle having the influence on drag force, at the high Reynolds numbers (which swimmers usually undergo in the competitions), two main factors affect the drag force: shape and arm orientation. In previous CFD studies of free style arm strokes, the effect of swimmer's hand orientation and its effect on propulsion force were evaluated while varying exclusively the angle of attack and full consideration of influence of both the angles were seldom considered.

This quasi-steady state (with initial data is taken from real kinematic experimental research) CFD study is based on flow simulation around swimmer's scanned hand models by using FLUENT (ANSYS, PA, USA). It is hypothesize that the changes of hand shape, velocity, and both angles of hand orientation (pitch and sweepback) complement drag force and drag coefficient in underwater freestyle stroke. Also, it is hypothesised that the realistic variation of both the orientation angles contributes to higher values of lift, drag, and resultant coefficients and forces when compared with values from previous work based on CFD method. Therefore, the aim of this study was to evaluate drag force and drag coefficient corresponding to the variation in kinematic parameters (orientation, shape, and velocity) of swimmer's hand during complete underwater front crawl stroke cycle.

## 2. Methods

In this study, the hydrodynamic components of propulsion force ($F_D$, $C_D$) using realistic models of human hand were calculated. The FLUENT (ANSYS, PA, USA) software was used to simulate the fluid flow, allowing the analysis of distribution of pressure and flow around the hand model of the swimmer.

*2.1. Hand's Geometries and Initial Parameters.* Artec L 3D scanner (Artec Group Inc., Luxembourg, Lux) was used to scan the left hand of international level swimmer. Working parameters of scanner are video frame rate of up to 15 fps and data acquisition speed up to 500000 points/s. The scanner was directly connected with computer, and all the data was transferred to image-processing program Artec Studio (Artec Group Inc., Luxembourg, Lux). Finally, using this software, cloud point nodal 3D hand models were created from scanned frames: $H_{adducted}$—with thumb adducted; $H_{abducted}$—with thumb fully abducted; $H_{spread}$—with spread fingers; $H_{adducted, spread}$—with adducted thumb and spread fingers. To optimize the models and to get good resolution by the selection of relevant parts of model and to transform the point cloud into a complete 3D polygon model, all the data from Artec Studio (Artec Group Inc, Luxembourg, Lux) was transferred to Leios 2 (EGS srl, Bologna, Italy) software. Typical models of hands with 1182, 1204, 1218, and 1142 surfaces corresponding to $H_{adducted}$, $H_{abducted}$, $H_{spread}$ and $H_{adducted, spread}$ were created (Figure 2). The Leios2 (EGS srl, Bologna, Italy) software program can export the datasets into IGES (Initial Graphics Exchange Specification) format which allows file import by CFD software FLUENT (ANSYS, PA, USA).

Initial kinematic parameters corresponding to real swimmer's single front crawl underwater stroke cycle were obtained from previous experimental study [13]. The single front crawl stroke cycle is usually divided into four phases: glide—from the entry of the hand into the water to its maximal forward displacement in the longitudinal displacement; pull—from the maximal forward displacement of the hand in the longitudinal displacement (i.e. $Y$ direction) corresponding to the stroke time when the hand is located exactly under the shoulder; push—from the end of the pull phase until the exit of the hand out of the water; recovery—from the hand's exit to its reentry into the water [14]. Each model of hand was positioned in nine different positions with the combinations of various pitch (defines angle between hand velocity vector and the plane of the hand) and sweepback angles (which define the inclination of the leading edge of hand) using SolidWorks (Dassault Systémes SolidWorks Corporation, MA, USA) software. Different values of average velocity correspond to different times (Table 1), resulting in three parts of underwater stroke time from glide phase, three from pull and three from push phases. Gourgoulis et al. [13] determined the velocity of the hand as the mean of the resultant velocities of the 2nd and the 5th metacarpophalangeal joints and transformed from the external (O; X, Y, Z) to the local reference system (O; x, y, z) of the swimmer's hand. In CFD calculations, this velocity was used as the resultant

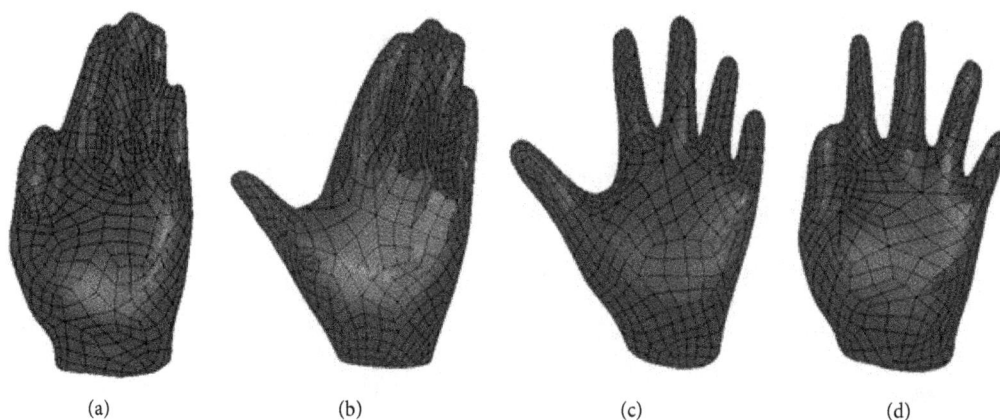

FIGURE 2: Four different models of swimmer's hand: (a) $H_{\text{adducted}}$—with thumb adducted; (b) $H_{\text{abducted}}$—with thumb abducted; (c) $H_{\text{spread}}$—with spread all fingers; (d) $H_{\text{adducted, spread}}$—with spreading fingers and thumb adducted.

TABLE 1: Initial experimental kinematic data (Gourgoulis et al. [13]) applied in steady state CFD calculations.

| Phases | Time (%) | Velocity (m·s$^{-1}$) | Angles of attack (deg) | Sweepback angle (deg) |
|---|---|---|---|---|
| | 10 | 2.15 | 4.44 | 94.75 |
| Glide | 20 | 2.15 | 12.57 | 97.45 |
| | 30 | 2.09 | 19.23 | 102.87 |
| | 55 | 1.79 | 44.37 | 127.23 |
| Pull | 65 | 1.82 | 55.46 | 192.20 |
| | 75 | 2.01 | 42.89 | 295.06 |
| | 85 | 2.20 | 36.98 | 243.63 |
| Push | 90 | 2.53 | 25.14 | 232.80 |
| | 95 | 2.75 | 20.71 | 232.80 |

average velocity of water flow while the hand model was kept stationary.

### 2.2. Computational Domain.

All hand positions of four hand models were exported to geometry modelling software DesignModeler (ANSYS, PA, USA) to generate fluid domain and later to Meshing (ANSYS, PA, USA) software to generate mesh (Figure 3(a)). The hand models were placed in the centre of 3D rectangular domain. The flow domain in this study extends from 0.53 m upstream of the hand model up to 0.88 m downstream with full dimensions: length: 1.6 m; width and height are each equal to 0.89 m. To ensure that the model would provide accurate results, the grid was refined in presumed regions of high velocity and pressure gradients.

A series of tests were carried out to estimate the independence of the results in relation to the grid resolution. $H_{\text{adducted}}$ model was used at hand speed of 1.79 m·s$^{-1}$. The pressure force was calculated for each grid (Figure 3(b)). The number of grid elements equal to 844740 was chosen in further calculations to ensure reasonably fast and accurate computations, the independence of grid used, and repeatability of the results.

### 2.3. Computational Fluid Dynamics (CFD)

*Governing Equations.* CFD method was used to simulate drag force and resultant drag coefficient of the hand models. Steady velocity is implemented by keeping hand static with the fluid flowing at constant velocity. CFD method is based on the Navier-Stokes equations which fully govern fluid flow. These equations arise from applying Newton's second law to fluid motion, together with the assumption that the fluid stress is the sum of a diffusing viscous term (proportional to the gradient of velocity), plus a pressure term. The solution of the Navier-Stokes equations is a velocity field or flow field, which is a description of the velocity of the fluid at a given point in space and time. For the incompressible fluids, the continuity equation is only function of velocity and not a function of pressure. Only the momentum equations contain pressure term. A direct method is to discretize the equations of continuity and momentum and solve them simultaneously to obtain results of pressure.

CFD methodology consists of a mathematical model that replaces the Navier-Stokes equations with discretized algebraic expressions that can be solved by algorithms on

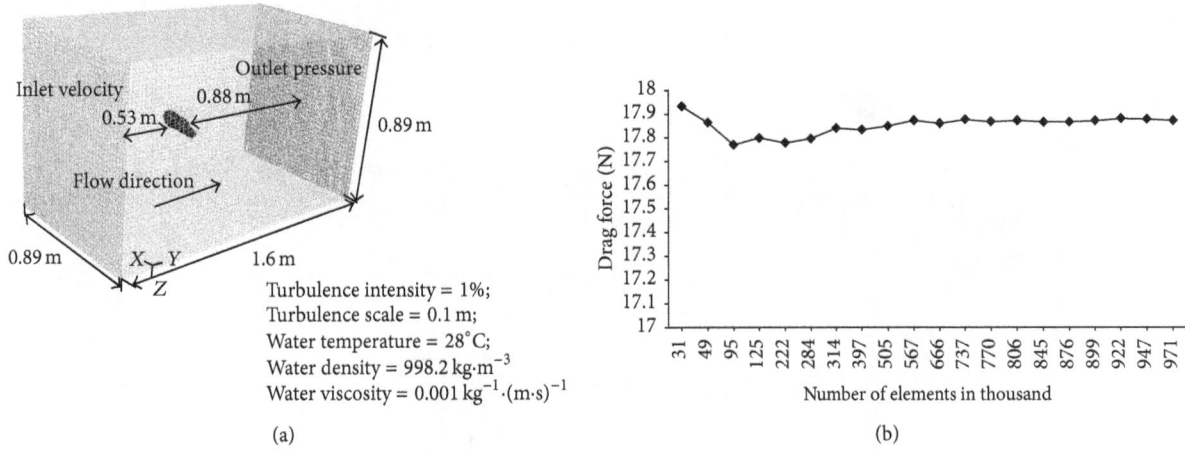

FIGURE 3: (a) Computational mesh; (b) curve of grid independence test.

the finite discretized domain consisting of volumetric mesh with the prediction of fluctuating velocities with the help of turbulent model. The problem of the turbulent modelling was solved using $k$-$\varepsilon$ model. The system of equations for solving three-dimensional, incompressible fluid flow in steady-state regime is as follows.

Continuity equation

$$\frac{\partial}{\partial x_i}\left(\overline{U_i}\right) = 0. \tag{1}$$

Navier-Stokes (momentum) equations

$$\frac{\partial}{\partial x_j}\left(\rho\overline{U_i}\overline{U_j}\right)$$

$$= -\frac{\partial\overline{p}}{\partial x_i} + \frac{\partial}{\partial x_j}\left[\left(\mu+\mu_t\right)\left(\frac{\partial}{\partial x_j}\left(\overline{U_i}\right)+\frac{\partial}{\partial x_i}\left(\overline{U_j}\right)\right) - \frac{2}{3}\delta_{ij}\rho k\right], \tag{2}$$

where $\overline{U_i}(t) \equiv \overline{U_i} + \overline{u_i}$ is the component of instantaneous velocity in $i$-direction (m·s$^{-1}$), $\overline{U_i}$ is the component of time averaged mean velocity in $i$-direction (m·s$^{-1}$), $u_i$ is the component of fluctuating velocity in $i$-direction (m·s$^{-1}$), $i, j$ are the coordinate direction vectors, $\rho$ is average fluid density (kg·m$^{-3}$), $\mu$ is dynamic viscosity of fluid (kg·(m·s)$^{-1}$), $\mu_t$ is turbulent viscosity of fluid (kg·(m·s)$^{-1}$), $\overline{p}$ is average pressure (N·m$^{-2}$), $k = (1/2)(\overline{u_i}\overline{u_j})$ is the turbulent kinetic energy per unit mass (m$^2$·s$^{-2}$), and $\delta_{ij}$ is the Kronecker delta with the condition that, $\delta_{ij} = 1$ if $i = j$ and $\delta_{ij} = 0$ if $i \neq j$.

*2.4. Boundary Conditions.* For the steady state fluid flow simulations, appropriate boundary conditions were considered. On the left side vertical surface of the domain (inlet velocity, Figure 3(a)), the horizontal component of the initial velocity was applied for all hand positions, respectively, (Table 1) and the vertical component of the velocity was assumed to be equal to zero. The pressure was set equal to zero Pascal on the right side of vertical surface (outlet pressure, Figure 3(a)).

The remaining side surfaces and bottom of the domain were considered as symmetry. Incompressible flow was assumed with turbulence intensity of 1.0% and turbulence scale of 0.10 m. The water temperature was 28°C with a density of 998.2 kg·m$^{-3}$ and viscosity of 0.001 kg$^{-1}$·(m·s)$^{-1}$ with consideration of the gravity of 9.81 m·s$^{-2}$.

*2.5. Numerical Scheme.* More accurate solution was considered with the choice of second-order numerical computational schemes. The simulations are based on finite volume method of discretization. In generic terms, the convergence of the calculation is checked by the value of the residuals of the various flow parameters. The convergence criteria in FLUENT (ANSYS, PA, USA) were set at $10^{-6}$. This criterion is assumed sufficient to ensure the convergence of the solution for the present study. The appropriate number of tetrahedral grids cells in the simulation model was arrived, which was an outcome of grid independence test carried out at the beginning of actual simulations. It was found that the difference in solutions for the drag coefficients for subsequent refinement in tetrahedral grid was less than 1%. In order to limit numerical dissipation, particularly when the geometry is complex consisting of an unstructured grid, as seen in Figure 3(a), the choice of second-order upwind discretization scheme for the convection terms in the solution equations and Pressure-Implicit with Splitting of Operators (PISO) pressure-velocity coupling scheme for the double precision pressure-based solver was chosen. The PISO pressure-velocity coupling scheme, part of the SIMPLE family of algorithms, is based on the higher degree of the approximate relation between the corrections for pressure and velocity.

# 3. Results

The calculated values of drag forces, drag coefficients, and areas of hands of all the studied phases at three different velocities (differences from initial velocity $\pm$ 0.5 m·s$^{-1}$) are presented in Tables 2 and 3, along with the corresponding data of all hand models used. The largest mean of drag force

TABLE 2: Values of the areas, drag forces, and drag coefficients at different increase of velocity in separated underwater phases varied according to the model of hand: $H_{adducted}$ and $H_{abducted}$.

| Phases | $H_{adducted}$ area ($\times 10^{-3}$ m$^2$) | Initial velocity, m·s$^{-1}$ | | Initial velocity −0.5 m·s$^{-1}$ | | Initial velocity +0.5 m·s$^{-1}$ | | $H_{abducted}$ area ($\times 10^{-3}$ m$^2$) | Initial velocity, m·s$^{-1}$ | | Initial velocity −0.5 m·s$^{-1}$ | | Initial velocity +0.5 m·s$^{-1}$ | |
|---|---|---|---|---|---|---|---|---|---|---|---|---|---|---|
| | | $F_D(N)$ | $C_D$ | $F_D(N)$ | $C_D$ | $F_D(N)$ | $C_D$ | | $F_D(N)$ | $C_D$ | $F_D(N)$ | $C_D$ | $F_D(N)$ | $C_D$ |
| Glide 1 | 4.77 | 8.33 | 0.757 | 6.22 | 0.959 | 10.57 | 0.633 | 4.97 | 8.53 | 0.744 | 6.34 | 0.939 | 10.89 | 0.626 |
| Glide 2 | 5.01 | 8.48 | 0.733 | 6.32 | 0.928 | 10.78 | 0.613 | 5.03 | 8.46 | 0.729 | 6.28 | 0.920 | 10.83 | 0.614 |
| Glide 3 | 4.90 | 8.71 | 0.816 | 6.43 | 1.040 | 11.14 | 0.680 | 5.22 | 9.07 | 0.796 | 6.69 | 1.015 | 11.66 | 0.667 |
| Mean | 4.89 | 8.51 | 0.77 | 6.32 | 0.98 | 10.83 | 0.64 | 5.07 | 8.69 | 0.76 | 6.44 | 0.96 | 11.13 | 0.64 |
| s | 0.12 | 0.19 | 0.04 | 0.10 | 0.06 | 0.29 | 0.03 | 0.13 | 0.33 | 0.04 | 0.22 | 0.05 | 0.46 | 0.03 |
| Pull 1 | 7.89 | 16.15 | 1.280 | 11.16 | 1.703 | 21.62 | 1.047 | 8.16 | 17.23 | 1.320 | 11.77 | 1.736 | 23.24 | 1.088 |
| Pull 2 | 13.51 | 41.59 | 1.862 | 27.83 | 2.369 | 57.36 | 1.580 | 13.60 | 38.06 | 1.694 | 25.50 | 2.156 | 52.35 | 1.433 |
| Pull 3 | 11.04 | 32.50 | 1.461 | 22.83 | 1.817 | 43.58 | 1.256 | 10.24 | 26.79 | 1.297 | 18.82 | 1.616 | 35.64 | 1.107 |
| Mean | 10.81 | 30.08 | 1.53 | 20.60 | 1.96 | 40.85 | 1.29 | 10.67 | 27.36 | 1.44 | 18.70 | 1.84 | 37.08 | 1.21 |
| s | 2.82 | 12.89 | 0.30 | 8.56 | 0.36 | 18.02 | 0.27 | 2.74 | 10.43 | 0.22 | 6.86 | 0.28 | 14.61 | 0.19 |
| Push 1 | 10.63 | 32.40 | 1.261 | 23.42 | 1.527 | 42.63 | 1.102 | 11.60 | 36.87 | 1.316 | 26.24 | 1.569 | 49.13 | 1.164 |
| Push 2 | 9.17 | 30.32 | 1.035 | 22.97 | 1.218 | 38.48 | 0.915 | 9.56 | 32.89 | 1.077 | 24.63 | 1.252 | 42.18 | 0.963 |
| Push 3 | 8.44 | 30.12 | 0.946 | 23.41 | 1.098 | 37.48 | 0.842 | 8.77 | 32.23 | 0.974 | 24.85 | 1.122 | 40.55 | 0.877 |
| Mean | 9.42 | 30.95 | 1.08 | 23.27 | 1.28 | 39.53 | 0.95 | 9.97 | 34.00 | 1.12 | 25.24 | 1.31 | 43.95 | 1.00 |
| s | 1.12 | 1.26 | 0.16 | 0.26 | 0.22 | 2.73 | 0.13 | 1.46 | 2.51 | 0.18 | 0.87 | 0.23 | 4.56 | 0.15 |

TABLE 3: Values of the areas, drag forces, and drag coefficients at different increase of velocity in separated underwater phases varied according to the model of hand: $H_{spread}$ and $H_{adducted,spread}$.

| Phases | $H_{spread}$ area ($\times 10^{-3}$ m$^2$) | Initial velocity, m·s$^{-1}$ | | Initial velocity, −0.5 m·s$^{-1}$ | | Initial velocity, +0.5 m·s$^{-1}$ | | $H_{adducted,spread}$ area ($\times 10^{-3}$ m$^2$) | Initial velocity, m·s$^{-1}$ | | Initial velocity, −0.5 m·s$^{-1}$ | | Initial velocity, +0.5 m·s$^{-1}$ | |
|---|---|---|---|---|---|---|---|---|---|---|---|---|---|---|
| | | $F_D(N)$ | $C_D$ | $F_D(N)$ | $C_D$ | $F_D(N)$ | $C_D$ | | $F_D(N)$ | $C_D$ | $F_D(N)$ | $C_D$ | $F_D(N)$ | $C_D$ |
| Glide 1 | 6.09 | 10.59 | 0.754 | 7.90 | 0.955 | 13.46 | 0.631 | 5.24 | 8.21 | 0.680 | 6.09 | 0.855 | 10.55 | 0.575 |
| Glide 2 | 6.50 | 10.86 | 0.725 | 8.09 | 0.916 | 13.82 | 0.607 | 5.76 | 10.72 | 0.806 | 8.35 | 1.067 | 13.94 | 0.690 |
| Glide 3 | 6.67 | 11.24 | 0.773 | 8.27 | 0.982 | 14.45 | 0.647 | 6.55 | 12.02 | 0.842 | 8.69 | 1.051 | 15.68 | 0.715 |
| Mean | 6.42 | 10.90 | 0.75 | 8.09 | 0.95 | 13.91 | 0.63 | 5.85 | 10.32 | 0.78 | 7.71 | 0.99 | 13.39 | 0.66 |
| s | 0.30 | 0.33 | 0.02 | 0.18 | 0.03 | 0.50 | 0.02 | 0.66 | 1.94 | 0.09 | 1.41 | 0.12 | 2.61 | 0.07 |
| Pull 1 | 9.91 | 19.68 | 1.242 | 13.38 | 1.626 | 26.60 | 1.026 | 9.61 | 18.86 | 1.227 | 12.90 | 1.616 | 25.46 | 1.012 |
| Pull 2 | 15.46 | 34.86 | 1.364 | 23.14 | 1.721 | 48.22 | 1.161 | 14.17 | 36.65 | 1.564 | 24.18 | 1.962 | 50.99 | 1.339 |
| Pull 3 | 13.58 | 35.22 | 1.286 | 24.56 | 1.589 | 47.23 | 1.106 | 12.45 | 34.03 | 1.356 | 23.72 | 1.674 | 45.80 | 1.170 |
| Mean | 12.98 | 29.92 | 1.30 | 20.36 | 1.65 | 40.69 | 1.10 | 12.08 | 29.85 | 1.38 | 20.26 | 1.75 | 40.75 | 1.17 |
| s | 2.82 | 8.87 | 0.06 | 6.09 | 0.07 | 12.21 | 0.07 | 2.31 | 9.60 | 0.17 | 6.39 | 0.19 | 13.49 | 0.16 |
| Push 1 | 13.63 | 40.38 | 1.226 | 28.86 | 1.467 | 53.43 | 1.077 | 11.66 | 34.03 | 1.208 | 24.56 | 1.461 | 44.71 | 1.054 |
| Push 2 | 11.77 | 39.17 | 1.042 | 29.40 | 1.215 | 50.02 | 0.928 | 10.31 | 31.85 | 0.967 | 24.16 | 1.139 | 40.30 | 0.853 |
| Push 3 | 10.78 | 39.74 | 0.977 | 30.59 | 1.124 | 49.78 | 0.876 | 9.43 | 31.81 | 0.894 | 24.72 | 1.037 | 39.52 | 0.795 |
| Mean | 12.06 | 39.76 | 1.08 | 29.62 | 1.27 | 51.08 | 0.96 | 10.47 | 32.56 | 1.02 | 24.48 | 1.21 | 41.51 | 0.90 |
| s | 1.45 | 0.61 | 0.13 | 0.89 | 0.18 | 2.04 | 0.10 | 1.12 | 1.27 | 0.16 | 0.29 | 0.22 | 2.80 | 0.14 |

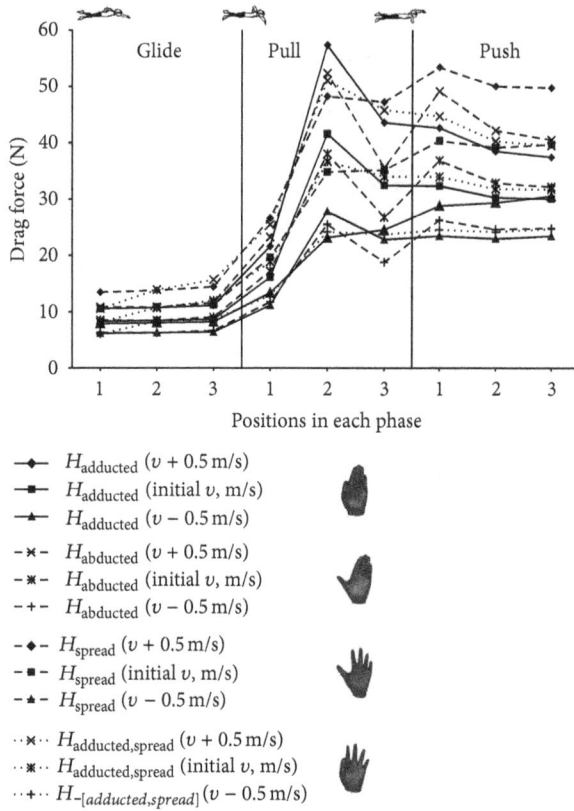

FIGURE 4: Drag force versus different hand position acting on separated swimmer's hand models corresponding to three different phases with respective orientations and increment of velocity.

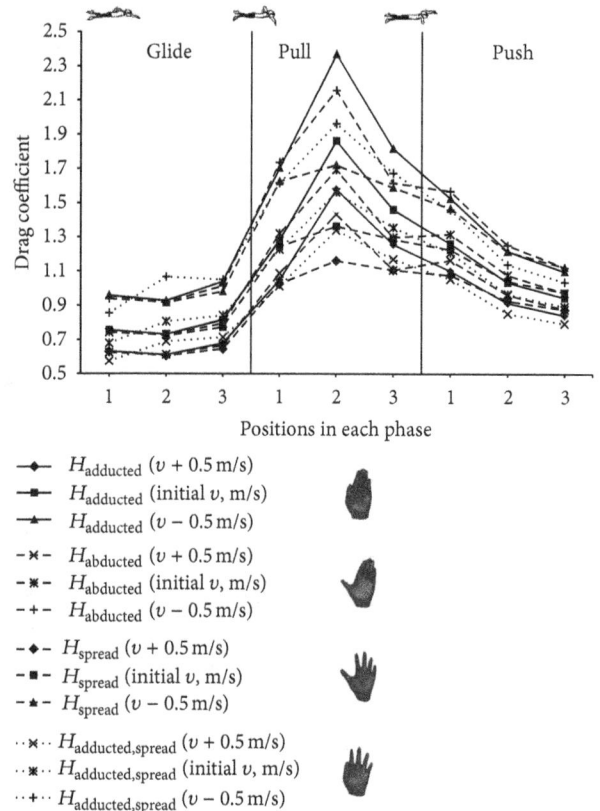

FIGURE 5: Drag coefficient versus different hand position acting on separated swimmer's hand models corresponding to three different phases with respective orientations and increment of velocity.

(40.85 N, $s$ = 18.02 N) by $H_{adducted}$ was observed during pull phase, which also presented the largest projected hand area, whereas hand models $H_{abducted}$, $H_{spread}$, $H_{adducted, spread}$ (appropriately: 43.95 N, $s$ = 4.56 N; 51.08 N, $s$ = 2.04 N; 41.51 N, $s$ = 2.80 N) presented largest mean of drag force during the push phase, when the initial mean velocity was greater than before. The lowest drag forces of all hand models were during the glide phase when initial flow velocity was decreased 0.5 m·s$^{-1}$ and projected hand area was the smallest. The maximum values of $C_D$ were calculated during the pull phase of all hand models (mean values: 1.96, $s$ = 0.36, 1.84, $s$ = 0.28, 1.65, $s$ = 0.07, and 1.75, $s$ = 0.19, accordingly $H_{adducted}$, $H_{abducted}$, $H_{spread}$, and $H_{adducted, spread}$), when flow velocity was the lowest. Minimum mean values of drag coefficients were reached during the glide phase, when initial flow velocity was increased 0.5 m·s$^{-1}$ and projected hand area was the smallest.

A variation of drag force and drag coefficient is observed due to different hand orientation, which depends on both hand angles. $F_D$ slightly varied during the glide phase (Figure 4). All drag forces rose in the beginning of pull phase and reached maximum values in the middle of this phase when projected hand area was the biggest (except $H_{spread}$ max $F_D$ during the first point of push phase). The values of the drag forces fell after peaks due to decrease in projected hand area.

$F_D$ varied more when initial velocity was increased, and there was less variation when initial velocity was decreased. Drag coefficient slightly varied in glide phase (Figure 5). The values of $C_D$ rapidly increased from the 3rd position of the glide phase to the 2nd position of the pull phase due to significant changes in hand angles which in turn affected projected hand area. Drag coefficients decreased in push phase of all hand models.

There is a marked improvement in drag force and drag coefficient changes due to different shape of hand (Figures 4 and 5). $F_D$ was not significant during the glide phase; however, the biggest means were observed in $H_{spread}$ and $H_{adducted, spread}$. $H_{adducted}$ and $H_{abducted}$ reached maximum peaks in the pull phase though the values of drag force were not the biggest during push phase. All drag forces of $H_{adducted}$, $H_{abducted}$, and $H_{adducted, spread}$ slightly decreased in push phase initially and increased initial velocities and were almost stable during decreased initial velocity. Meanwhile, $F_D$ of $H_{spread}$ was negligibly increasing in push phase during decreased velocity and slightly varied during the initial and increased initial velocities. Drag coefficient varied insignificantly during the glide phase. Significant variation was observed in push phase of all hand models. $H_{adducted}$ reached the biggest values of $C_D$ in the pull phase; however, values of $C_D$ were the biggest of $H_{abducted}$ in push phase for every variation in velocity.

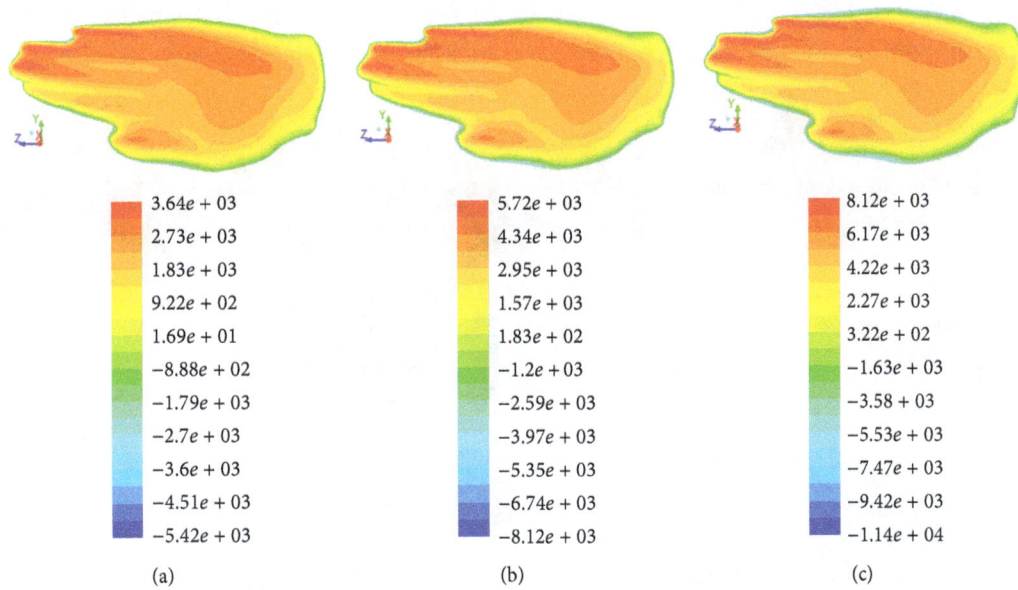

FIGURE 6: Static pressure (Pascal) acting on the same swimmer's hand model—$H_{\text{adducted}}$, at the same underwater phase with regard to different water flow velocity: (a) $1.32\,\text{m·s}^{-1}$, (b) $1.82\,\text{m·s}^{-1}$, and (c) $2.32\,\text{m·s}^{-1}$.

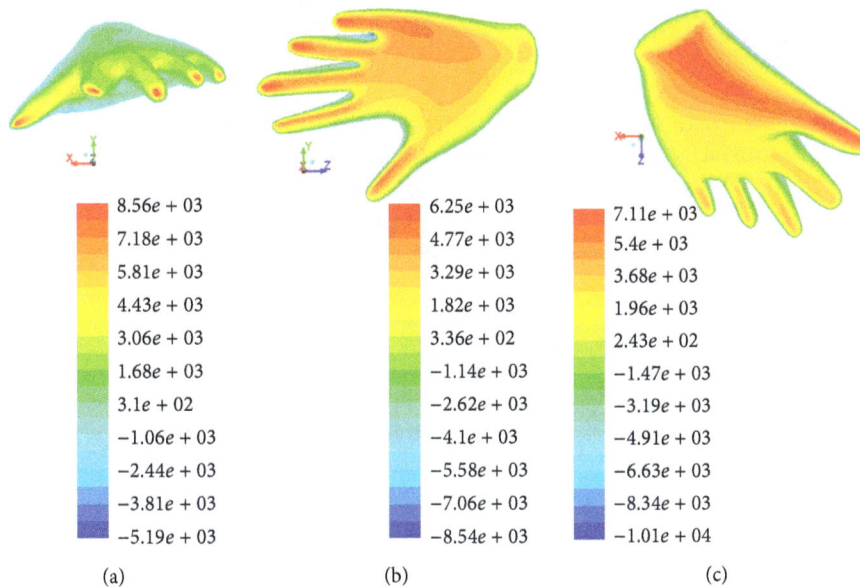

FIGURE 7: Static pressure (Pascal) acting on the same swimmer's hand model—$H_{\text{spread}}$, corresponding to different orientations at three underwater phases: (a) glide, (b) pull, and (c) push.

Pressure forces contribution to hand varied according to water flow velocity, hand orientation, and shape. Pressure visualisation considering different water flow velocity ($1.82 \pm 0.5\,\text{m·s}^{-1}$) was proposed on $H_{\text{adducted}}$ during 2nd position of pull phase (Figure 6). Pressure contour fields depending on different hand orientation of $H_{\text{spread}}$ during separated underwater freestyle phases were shown (Figure 7). The

shape of hand caused different pressure force for the same water flow velocity ($1.82\,\text{m·s}^{-1}$) during 2nd position of pull phase (Figure 8).

The vectors of flow velocity distribution on the palm of the hand are presented (Figure 9). The velocity of flow decreased when in contact with hand surface. Flow dispersed to the corners of the palm with increasing in velocity. The

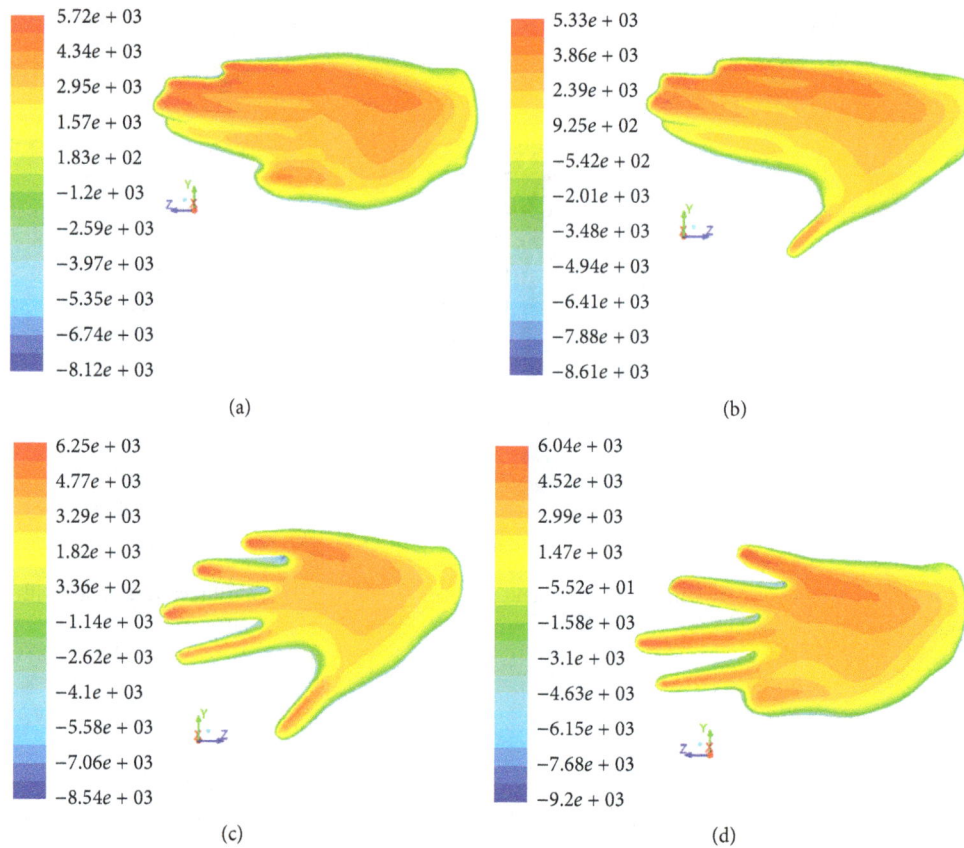

| | | | |
|---|---|---|---|
| 5.72e + 03 | | 5.33e + 03 | |
| 4.34e + 03 | | 3.86e + 03 | |
| 2.95e + 03 | | 2.39e + 03 | |
| 1.57e + 03 | | 9.25e + 02 | |
| 1.83e + 02 | | −5.42e + 02 | |
| −1.2e + 03 | | −2.01e + 03 | |
| −2.59e + 03 | | −3.48e + 03 | |
| −3.97e + 03 | | −4.94e + 03 | |
| −5.35e + 03 | | −6.41e + 03 | |
| −6.74e + 03 | | −7.88e + 03 | |
| −8.12e + 03 | | −8.61e + 03 | |
| (a) | | (b) | |
| 6.25e + 03 | | 6.04e + 03 | |
| 4.77e + 03 | | 4.52e + 03 | |
| 3.29e + 03 | | 2.99e + 03 | |
| 1.82e + 03 | | 1.47e + 03 | |
| 3.36e + 02 | | −5.52e + 01 | |
| −1.14e + 03 | | −1.58e + 03 | |
| −2.62e + 03 | | −3.1e + 03 | |
| −4.1e + 03 | | −4.63e + 03 | |
| −5.58e + 03 | | −6.15e + 03 | |
| −7.06e + 03 | | −7.68e + 03 | |
| −8.54e + 03 | | −9.2e + 03 | |
| (c) | | (d) | |

FIGURE 8: Static pressure (Pascal) acting at the same water flow velocity (1.82 m·s$^{-1}$) of 2nd pull position on separated swimmer's hand models: (a) $H_{\text{adducted}}$—with thumb adducted; (b) $H_{\text{abducted}}$—with thumb abducted; (c) $H_{\text{spread}}$—with spread all fingers; (d) $H_{\text{adducted, spread}}$—with spread fingers and thumb adducted.

biggest values of mean fluid velocity were observed near the edges of the palm of the hand and between the edges of fingers in case of hand model with spread fingers.

## 4. Discussion

The purpose of this study was to evaluate drag force and drag coefficient correspondence to variation in kinematic parameters (orientation, shape, and velocity) of swimmer's hand during complete underwater front crawl stroke cycle. Four scanned hand models with fingers in different positions were generated. The orientation of hand was set into nine different positions with accordance to average velocity. In the present study, the analysis of a particular case of front crawl underwater movement (Table 1) and initial kinematic data were taken from experimental kinematic research work [13]. Calculations were performed through FLUENT (ANSYS, PA, USA) software, which is based on computational fluid dynamics method applied to accurately solve fluid flow problems through numerical simulation. This method has

been proven to generate accurate results with repeatability of results with identical values, when performed with similar initial conditions and settings [17]. This saves time, and also the results can be accessed in detail and analysed anytime, unlike repetitive experimental tests. The outcome of this study is that the modification of three variables: hand orientation, shape, and the average water flow velocity strongly affects drag coefficient and drag force of hand and with this in some measures affects the hand propulsion.

There is clear indication of progress in the understanding of the contribution of hand shape, orientation, and velocity on drag coefficient ($C_D$) and drag force ($F_D$) in underwater hand stroke. Obvious drag force and drag coefficient dependence on velocity are shown in Figures 4 and 5. The values of $F_D$ are seen to increase and decrease with respect to the initial velocity, which was accordingly increased and decreased in 0.5 m·s$^{-1}$ intervals. A similar tendency of increase of $F_D$ with velocity was observed in a previous study performed under steady-state flow conditions of the 2D and 3D swimmer's hand model [11, 18]. Rouboa et al. [11] simulated a hand/forearm model with thumb adducted in three different

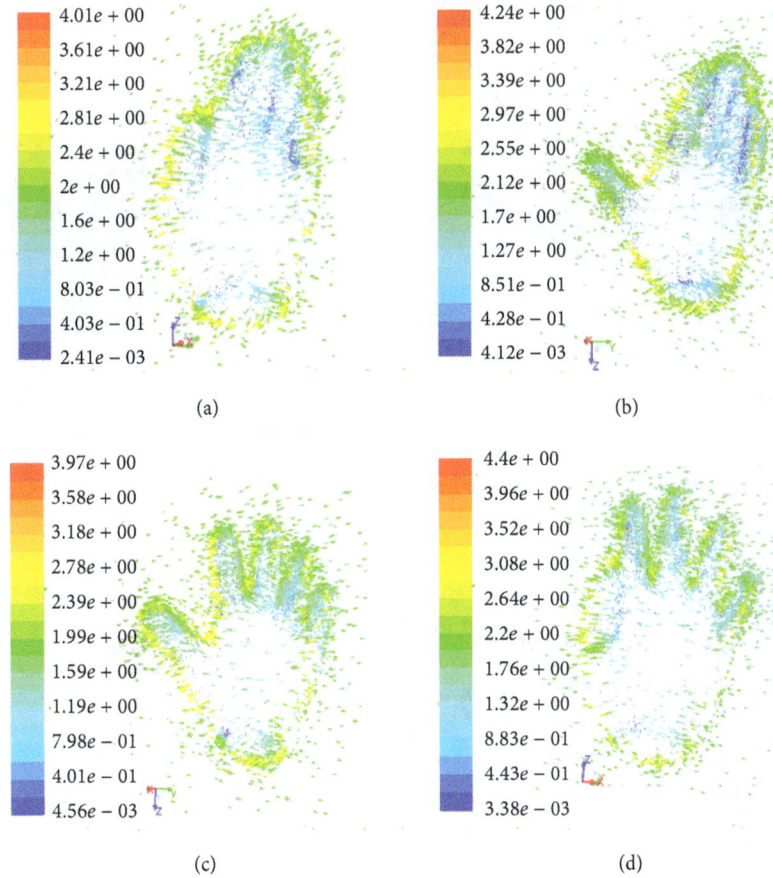

FIGURE 9: Velocity vectors on the different models of the hand: (a) $H_{\text{adducted}}$, (b) $H_{\text{abducted}}$, (c) $H_{\text{spread}}$, and (d) $H_{\text{adducted, spread}}$.

orientations where water flow velocity was increased from $0.5 \, \text{m·s}^{-1}$ to $4 \, \text{m·s}^{-1}$. Sato and Hino [18] used hand model with spread fingers in a fixed direction with water flow, and velocity was set from $0.5 \, \text{m·s}^{-1}$ to $2 \, \text{m·s}^{-1}$. However, the variation of current drag coefficient (Figure 5) and previous findings disagree. It was indicated that the drag coefficient did not vary, while there was rise in water flow velocity [9, 11, 18]. Berger et al. [6] towed model of human hand in a towing tank at the similar orientation and evaluated hydrodynamic parameters. It was shown that the $C_D$ slightly decreased within the velocity range from 0.7 to $3.0 \, \text{m·s}^{-1}$, and at the velocity lower than $0.7 \, \text{m·s}^{-1}$ $C_D$ strongly depends on velocity. Current findings indicated that the rise in velocity generated lower values of drag coefficients for all hand models during all phases. According to current findings, it can be concluded that the velocity affects values of drag force and drag coefficient during all front crawl underwater phases.

To our knowledge, there is no known work based on CFD method which compares different phases of underwater front crawl hand motion with appropriate changes of hand orientation. However, Gourgoulis et al. [19] calculated drag force of human hand by formulas while input data was taken from kinematic experiments. The drag force dependence on hand orientation appears with the variation of $F_D$ in

both current (Figure 4) and previous research. There is clear indication that, between different phases of underwater hand movement, when swimmer changes his hand orientation (i.e., pitch and sweepback angle) (Table 1), the changes directly affect the projected frontal area and hereupon the drag force. The variation of $F_D$ calculated from experimental data was similar with current computational hand model $H_{\text{abducted}}$ with similar initial flow velocity. Therefore, we could imply that the swimmer maintained his hand with thumb abducted during underwater hand stroke experimental tests.

The calculation of drag coefficient during different underwater phases of hand model allowed evaluating the influence of hand orientation on $C_D$. This study showed that small changes in pitch and sweepback angles during glide phase caused negligible variation of $C_D$ for all hand models (Figure 5). The sharp variation of hand angles during the pull phase caused major changes in values of $C_D$, and the dwindling of angles to smaller values shaped the reduction of drag coefficient during push phase. The consideration of the angle of attack and sweepback angle influenced contributions to drag coefficient during complete underwater cycle by different hand models.

The importance of hand shape variation and its contributory influence to drag force are presented (Figure 4). Different individual curve lines in the pictorial graph represent

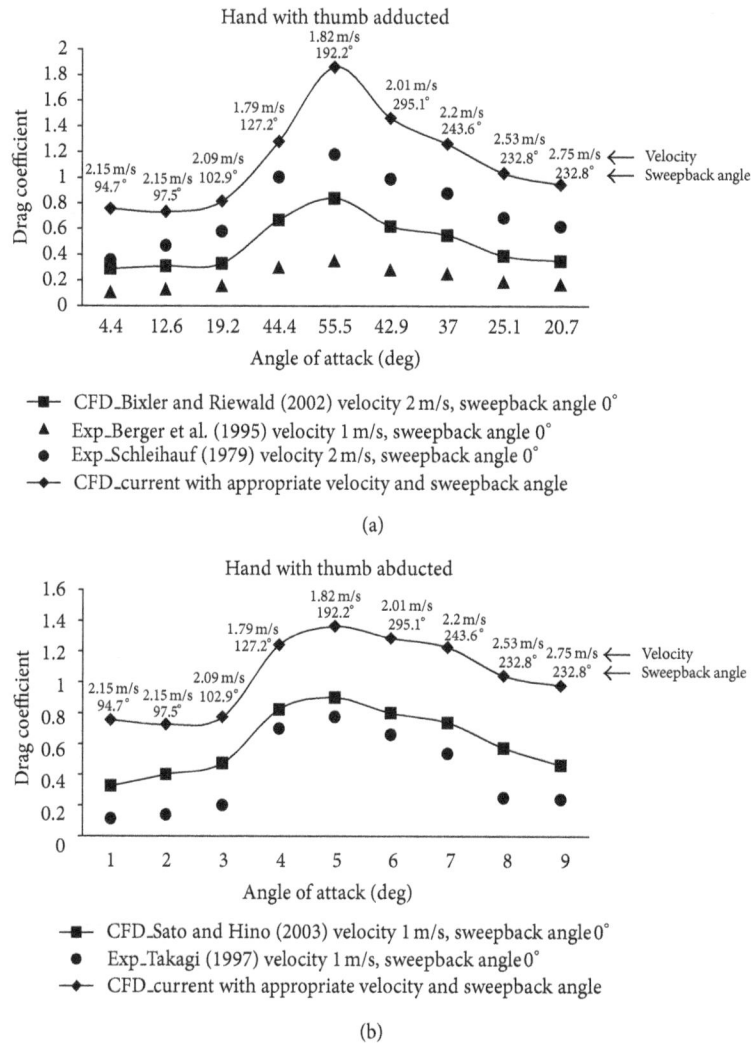

Hand with thumb adducted

- ■— CFD_Bixler and Riewald (2002) velocity 2 m/s, sweepback angle 0°
- ▲  Exp_Berger et al. (1995) velocity 1 m/s, sweepback angle 0°
- ●  Exp_Schleihauf (1979) velocity 2 m/s, sweepback angle 0°
- ◆— CFD_current with appropriate velocity and sweepback angle

(a)

Hand with thumb abducted

- ■— CFD_Sato and Hino (2003) velocity 1 m/s, sweepback angle 0°
- ●  Exp_Takagi (1997) velocity 1 m/s, sweepback angle 0°
- ◆— CFD_current with appropriate velocity and sweepback angle

(b)

FIGURE 10: Comparison of drag coefficient versus angle of attack acting on swimmer's hand: (a) current CFD results (with appropriate sweepback angle and flow velocity in each phase), Bixler and Riewald [5] CFD (velocity $2 \, \mathrm{m \cdot s^{-1}}$, sweepback angle 0°), Berger et al. [6] experimental results (velocity $1 \, \mathrm{m \cdot s^{-1}}$, sweepback angle 0°) and Schleihauf [15] experimental results (velocity $2 \, \mathrm{m \cdot s^{-1}}$, sweepback angle 0°); (b) current CFD study (with appropriate sweepback angle and flow velocity in each phase), Sato and Hino [12] CFD, and Takagi [16] (cited in [12]) experimental results where flow velocity is $1 \, \mathrm{m \cdot s^{-1}}$ and sweepback angle is 0°.

separate models of the swimmer's hand. The highest peak value of drag force was obtained from the hand model with thumb adducted and the smallest peak from the model with all fingers spread in the middle of the pull phase regardless of variation in flow velocity. Although the hand area, that directly affects the drag force, was larger in the model with fingers spread than in that with the thumb adducted, the pitch angle was the largest in this position. However, in the end of pull phase and throughout the push phase, when the hand altered its leading edge, the hand with spread fingers generated the biggest projected area of total plane area of hand and values of $F_D$, whereas the hand with the thumb adducted—the least. It means that the shape of the hand had different contributions on the drag force during separate underwater phases of front crawl stroke cycle. Likewise, it was summarised that higher drag force values could be reasonable by a hand model with optimal spacing between

fingers [10, 12]. According to the current results, it is clear that, in order to increase the drag force contribution throughout the underwater stroke cycle, the shape of the hand should be altered in separate individual phases.

The drag coefficient values (Figure 5) corresponding to different hand models were evaluated in recent CFD studies under steady state with spread fingers [10, 12, 18], with thumb in different positions [9], while current study used different values of velocity, pitch angle, and also different shapes of hand models. In previous studies, the drag coefficient was almost constant despite the changes in velocity. We observed that the drag coefficient was not constant and it varied throughout underwater front crawl cycle regardless of differences in hand model.

In the current study, we compared drag coefficients from the hand model with thumb adducted with previous studies obtained experimentally and from CFD studies, which were achieved by varying only the angle of attack in appropriate range while sweepback angle was equal to zero [5, 6]. The comparative study of current and previous $C_D$ values versus angle of attack (Figure 10(a)) showed the similar tendency of variation in $C_D$ (the hydrodynamic characteristics between all research work were similar). However, the means of current results were higher when compared with previous ones. These perceived differences could be due to differences in the usage of distinct input conditions, and, according to Berger et al. [6], material, hand size, and shape of the model might also account for the observed differences in $C_D$. Moreover, the distinct tendency of variation can also be caused by different hand orientation, which considered variation of both pitch and sweepback angles at the same time corresponding to that observed under real swimming conditions during underwater hand stroke path.

The drag coefficients of hand model with fingers spread obtained in this study were compared (Figure 10(b)) with those in Takagi's (as cited in [18]) study, obtained from experimental research, and with those in Sato and Hino's work [18] calculated through CFD method. The kinematic data from experimental work [13] and scanned human hand model were used for this study to evaluate drag force and drag coefficient; therefore, $C_D$ values and its variation present practical and further pragmatic results. Obvious differences are observed in values of the drag coefficients between present and previous CFD and experimental studies; whereas the tendency of variation was similar. Higher $C_D$ values in current work could be influenced by the choice of fluid flow model applied in the study: turbulent fluid flow was considered in this study, and laminar fluid flow was considered by Sato and Hino [18]. The values of $C_D$ were significantly higher in this study as the $C_D$ was dependent on not only the pitch angle, as considered in previous studies, but on both angles of hand orientation (pitch and sweepback). Furthermore, the consideration of experimental kinematic data of swimmer's hand motion (initial velocity, pitch, and sweepback angles) and realistic fluid flow model can affect realistic estimation of drag coefficient.

It was shown that the pressure force contribution on swimmer's hand acted differently vis-à-vis water flow velocity, hand orientation, and shape. The results of this study confirmed that the velocity of water flow directly affected pressure force on swimmer's hand (Figure 6). The contribution of pressure force to hand increased when velocity was increased. There is strong evidence that the pressure force affected different parts of hand due to altered hand sweepback angle and bigger part of hand frontal cross area came up with maximum pressure force, when pitch angle was increased (Figure 7). This maximal pressure force shifted towards hand's leading edge; this is when the swimmer altered orientation of the hand, that is, the angle of attack and sweepback angle. Different hand shapes arrived at different pressure force distributions (Figure 8). In this case, the biggest maximal value of pressure (6250 Pa) acted on hand with spread fingers (Figure 8(c)) and the least (5720 Pa)—on hand with thumb adducted (Figure 8(a)). The present study corroborates with Marinho et al. [10] and Minetti et al. [12], with reference to the maximum values of the pressure force,

suggesting that the hand with spread fingers goes through higher pressure force and in turn influences higher drag force than other forms of hand model.

## 5. Conclusions

This study, based on the CFD method, indicated that drag force and drag coefficient were significantly affected by velocity, hand frontal cross-section area, and both angles of hand orientation (pitch and sweepback) during complete underwater front crawl single cycle. Higher values of hydrodynamic parameters in this study are attributed to the consideration of real swimming conditions during simulations with the consideration of combination of both hand orientation angles observed during individual phases. To increase drag force contribution throughout the underwater stroke, the shape of the hand should be altered in separate phases.

## Acknowledgments

The authors wish to extend sincere thanks and acknowledge Professor, Ph.D., Vassilios Gourgoulis and all members of research team from the Department of Physical Education and Sport Science, Democritus University of Thrace, Komotini, Greece, for providing the experimental kinematic data of swimmers' underwater stroke which is used in the current study. The current authors have no conflict of interests. And the authors do not have any commercial or financial interests and/or professional relationships with ANSYS, PA, Artec Group Inc., Luxembourg, EGS srl, Bologna, and Dassault Systemes, SolidWorks Corporation, Italy, MA, USA.

## References

[1] J. V. Deschodt, L. M. Arsac, and A. H. Rouard, "Relative contribution of arms and legs in humans to propulsion in 25-m sprint front-crawl swimming," *European Journal of Applied Physiology and Occupational Physiology*, vol. 80, no. 3, pp. 192–199, 1999.

[2] A. P. Hollander, G. de Groot, G. J. van Ingen Schenau, R. Kahman, and H. M. Toussaint, "Contribution of the legs in front crawl swimming," in *Swimming Science V*, B. E. Ungerechts, K. Reischle, and K. Wilke, Eds., pp. 39–43, Human Kinetics Publishers, Champaign, Ill, USA, 1988.

[3] D. I. Miller, "Biomechanics of swimming," in *Exercise and Sport Sciences Reviews*, H. Willmore and J. F. Keogh, Eds., pp. 219–248, Academic Press, New York, NY, USA, 1975.

[4] H. Toussaint and M. Truijens, "Biomechanical aspects of peak performance in human swimming," *Animal Biology*, vol. 55, no. 1, pp. 17–40, 2005.

[5] B. Bixler and S. Riewald, "Analysis of a swimmer's hand and arm in steady flow conditions using computational fluid dynamics," *Journal of Biomechanics*, vol. 35, no. 5, pp. 713–717, 2002.

[6] M. A. M. Berger, G. de Groot, and A. P. Hollander, "Hydrodynamic drag and lift forces on human hand/arm models," *Journal of Biomechanics*, vol. 28, no. 2, pp. 125–133, 1995.

[7] P. Gardano and P. Dabnichki, "On hydrodynamics of drag and lift of the human arm," *Journal of Biomechanics*, vol. 39, no. 15, pp. 2767–2773, 2006.

[8] M. A. Lauder and P. Dabnichki, "Estimating propulsive forces—sink or swim?" *Journal of Biomechanics*, vol. 38, no. 10, pp. 1984–1990, 2005.

[9] D. A. Marinho, A. I. Rouboa, F. B. Alves et al., "Hydrodynamic analysis of different thumb positions in swimming," *Journal of Sports Science and Medicine*, vol. 8, no. 1, pp. 58–66, 2009.

[10] D. A. Marinho, T. M. Barbosa, V. M. Reis et al., "Swimming propulsion forces are enhanced by a small finger spread," *Journal of Applied Biomechanics*, vol. 26, no. 1, pp. 87–92, 2010.

[11] A. Rouboa, A. Silva, L. Leal, J. Rocha, and F. Alves, "The effect of swimmer's hand/forearm acceleration on propulsive forces generation using computational fluid dynamics," *Journal of Biomechanics*, vol. 39, no. 7, pp. 1239–1248, 2006.

[12] A. E. Minetti, G. Machtsiras, and J. C. Masters, "The optimum finger spacing in human swimming," *Journal of Biomechanics*, vol. 42, no. 13, pp. 2188–2190, 2009.

[13] V. Gourgoulis, P. Antoniou, N. Aggeloussis et al., "Kinematic characteristics of the stroke and orientation of the hand during front crawl resisted swimming," *Journal of Sports Sciences*, vol. 28, no. 11, pp. 1165–1173, 2010.

[14] V. Gourgoulis, N. Aggeloussis, N. Vezos, P. Antoniou, and G. Mavromatis, "Hand orientation in hand paddle swimming," *International Journal of Sports Medicine*, vol. 29, no. 5, pp. 429–434, 2008.

[15] R. E. Schleihauf, "A hydrodynamic analysis of swimming propulsion," in *Swimming III*, J. Terauds and E. W. Bedingfield, Eds., pp. 70–117, University Park Press, Baltimore, Md, USA, 1979.

[16] H. Takagi, "A quantitative analysis in a sense of catching water for competitive swimmer," *Journal of the Japan Society for Precision Engineering*, vol. 63, no. 4, pp. 495–498, 1997.

[17] D. A. Marinho, A. I. Rouboa, T. M. Barbosa, and A. J. Silva, "Modelling swimming hydrodynamics to enhance performance," *The Open Sports Sciences Journal*, vol. 3, pp. 43–46, 2010.

[18] Y. Sato and T. Hino, "Estimation of thrust of swimmer's hand using CFD," in *Proceedings of the 2nd International Symposium on Aqua Bio-Mechanisms*, Honolulu, Hawaii, USA, September 2003.

[19] V. Gourgoulis, N. Aggeloussis, N. Vezos, P. Kasimatis, P. Antoniou, and G. Mavromatis, "Estimation of hand forces and propelling efficiency during front crawl swimming with hand paddles," *Journal of Biomechanics*, vol. 41, no. 1, pp. 208–215, 2008.

# Lower-Limb Joint Coordination Pattern in Obese Subjects

**Alberto Ranavolo,**[1] **Lorenzo M. Donini,**[2,3] **Silvia Mari,**[4] **Mariano Serrao,**[5,6] **Alessio Silvetti,**[1]
**Sergio Iavicoli,**[1] **Edda Cava,**[2] **Rosa Asprino,**[2] **Alessandro Pinto,**[2] **and Francesco Draicchio**[1]

[1] Department of Occupational Medicine, INAIL, Via Fontana Candida 1, Monte Porzio Catone, 00040 Rome, Italy
[2] Department of Experimental Medicine, Medical Physiopathology, Food Science and Endocrinology Section,
Food Science and Human Nutrition Research Unit, Sapienza University of Rome, Ple Aldo Moro 5, 00185 Rome, Italy
[3] Villa delle Querce Clinical Rehabilitation Institute, Unit of Metabolic and Nutritional Rehabilitation, Via delle Vigne 19,
Nemi, 00040 Rome, Italy
[4] Fondazione Don Gnocchi, 20148 Milan, Italy
[5] Rehabilitation Centre, Policlinico Italia, Piazza del Campidano 6, 00162 Rome, Italy
[6] Department of Medical and Surgical Science and Biotechnologies, Sapienza University of Rome, Via Faggiana 34,
40100 Latina, Italy

Correspondence should be addressed to Silvia Mari; mari.silvia@virgilio.it

Academic Editor: Giuseppe Spinella

The coordinative pattern is an important feature of locomotion that has been studied in a number of pathologies. It has been observed that adaptive changes in coordination patterns are due to both external and internal constraints. Obesity is characterized by the presence of excess mass at pelvis and lower-limb areas, causing mechanical constraints that central nervous system could manage modifying the physiological interjoint coupling relationships. Since an altered coordination pattern may induce joint diseases and falls risk, the aim of this study was to analyze whether and how coordination during walking is affected by obesity. We evaluated interjoint coordination during walking in 25 obese subjects as well as in a control group. The time-distance parameters and joint kinematics were also measured. When compared with the control group, obese people displayed a substantial similarity in joint kinematic parameters and some differences in the time-distance and in the coupling parameters. Obese subjects revealed higher values in stride-to-stride intrasubjects variability in interjoint coupling parameters, whereas the coordinative mean pattern was unaltered. The increased variability in the coupling parameters is associated with an increased risk of falls and thus should be taken into account when designing treatments aimed at restoring a normal locomotion pattern.

## 1. Introduction

Obesity is a pathology with multifactorial causes that is characterized by an increase in fat body mass and is linked to a significant increase in morbidity and mortality. It is related to the interaction of erroneous eating habits, reduced energy consumption, and metabolic alterations [1]. In obese subjects, body movements are affected by the excess mass, which alters the individual's range of motions and exerts excessive joint load, thereby causing a high incidence of musculoskeletal disorders [2]. Functional tests demonstrated that obese subjects have difficulties in performing activities of daily life and experience more pain than normal-weight individuals [3].

As locomotion is one of the most important and frequent tasks in daily life, gait has been extensively analyzed in previous studies; some of which demonstrated that obesity alters the body's motor scheme, in terms of time-distance, kinematic, and kinetic parameters [4, 5]. Obese adults walk with a wider support base and a lower speed, cadence, and stride length than normal people [6, 7]. Differences emerged between obese and nonobese subjects walking at a standard gait speed in the angular kinematics of the lower-limb joints [8], and in particular reductions in the hip, knee, and ankle range of motions (ROM) on the sagittal plane [6]. By contrast, lower-limb joint kinematic parameters in obese subjects walking at a self-selected velocity were found to be similar to those of healthy subjects [9].

TABLE 1: Means, ranges, standard deviations, and $t$-test significance ($P$ values) of the personal, anthropometric, and functional characteristics of the two groups. $P$ values lower than 0.05 are shown in bold.

| Characteristics | Obese subjects | Controls | $P$ values |
|---|---|---|---|
| Sex ($n$) | | | |
| M | 8 | 8 | |
| F | 17 | 17 | |
| Age range (years) | 34–58 | 33–59 | 0.85 |
| BMI range (kg/m$^2$) | 33.8–44.0 | 19.0–27.8 | **<0.001** |
| Trunk flexibility (cm) | 14.3 ± 13 | 5.0 ± 6.1 | **0.006** |
| Waist circumference (cm) | | | |
| M | 133.5 ± 14 | 95.8 ± 8 | **<0.001** |
| F | 113 ± 11 | 78.2 ± 7 | **<0.001** |
| Fat mass (%) | | | |
| M | 37.3 ± 6 | 25.1 ± 3 | **0.002** |
| F | 43.1 ± 3 | 32.2 ± 4 | **<0.001** |
| 6MWT (m/s) | | | |
| M | 1.39 ± 0.2 | 1.38 ± 0.1 | 0.85 |
| F | 0.97 ± 0.5 | 1.26 ± 0.4 | **0.03** |
| Borg | | | |
| M | 3.6 ± 1 | 2.5 ± 0.6 | 0.16 |
| F | 4.1 ± 1 | 1.9 ± 1 | **<0.001** |
| Strength (kg) | | | |
| M | 41.6 ± 14 | 51.5 ± 6 | 0.08 |
| F | 25.1 ± 8 | 22.4 ± 9 | 0.3 |

Kinetic analyses revealed that ankle torque is higher in obese than in nonobese individuals walking at a standard speed, whereas, at a self-selected speed, obese individuals produce a gait pattern with lower knee torque and power [8]. Moreover, Malatesta et al. [10] hypothesized that obese people adopt a slower walking speed to reduce the mechanical effort exerted upon the lower extremity muscles and thus minimize energy cost during walking.

The energetic cost of walking has also been studied in relation to adding mass to the legs; it has been demonstrated that net metabolic rate during walking increases with load magnitude and more distal leg-load location, while there is a small increase in net metabolic rate with proximal loading [11].

A common difficulty of all these studies is the proper placement of passive markers in obese subjects, due to the excessive adipose tissue in the abdominal and pelvic areas. Despite efforts including manual measure of anterior superior iliac spines width [9, 12], potential errors in calculation of hip joint center are possible, as reported in a previous study [13]. Furthermore, another possible source of errors has been reduced by the use of elastic band around the waist to minimize the oscillations of adipose tissues during walking [5, 14].

Although gait pattern alterations in obese people are widely known, no information is, to our knowledge, yet available on the relationship between obesity and lower-limb joint coordination during walking.

Joint coordination is usually investigated by means of the Continuous Relative Phase (CRP) according to the dynamic system theory [15]. The CRP provides a measure of coupling or phase relationship between the actions of couples of interacting joints or segments and is frequently used to investigate lower-limb coordination in both walking and running [16–19].

Since multijoint coordination impairment accompanies numerous pathologies, the CRP has also been used to analyze the effects on gait coordination of numerous musculoskeletal [20, 21] and neurological diseases [22–25].

In this study, we hypothesised that the excess mass in the pelvic and lower-limb areas in obese subjects may represent a mechanical constraint that the central nervous system (CNS) is forced to control by means of a coordinative compensatory strategy. This strategy may be aimed at reducing the number of degrees of freedom that the CNS has to control by adopting a more in-phase coupling relationship between pairs of joints.

Since a modified coordination pattern may be a condition predisposing to joint injuries, diseases altering motor control and causing risk of falls [26–28], the aim of this study was thus to evaluate how obesity affects coordination during locomotion using the CRP method. We also analysed the time-distance parameters and lower-limb joint kinematics in the same subjects.

## 2. Materials and Methods

*2.1. Description of Obese Subjects and Controls.* Twenty-five obese subjects (BMI range: 33.8–44.0 kg/m$^2$) were enrolled in the study. Twenty-five controls (BMI range: 19.0–27.8 kg/m$^2$), gender- and age-matched volunteers, were recruited as a control group (see Table 1).

None of the controls volunteers had pathologies known to influence the normal gait pattern. Exclusion criteria were severe cardiovascular disease, neurological impairment and lower extremity trauma, lower extremity surgery, and appreciable leg discrepancy. All the participants gave their written consent. The study was approved by the local ethics committee and conformed to the Helsinki declarations. No information regarding the expected results was provided in order to avoid the results being biased, whether consciously or unconsciously.

Both controls and obese subjects underwent anthropometric and functional examinations (Table 1). Anthropometric measurements were performed and body composition was assessed. The anthropometric measurements were based on body weight and stature (SECA scale, Hamburg, Germany) for the calculation of the BMI and waist circumference (measured with an inextensible tape measure midway between the lower rib margin and the iliac crest). As regards body composition, the Siri equation was applied to estimate percentage of fat mass [29]. In addition to the afore-mentioned measurements, a "6-minute walking test" (6MWT) [30], a Borg's Perceived Exertion Scale [31], muscular strength of the forearm flexor muscles [32], and the standing trunk flexibility evaluations were performed in order to assess the functional condition. Muscular strength, expressed in kg, was evaluated by means of the Lafayette dynamometer. Trunk flexibility, defined as the distance between the fingertips and the floor, was evaluated by asking the subjects to reach down towards the floor in front of their feet as far as possible while standing with knees in an extended position.

### 2.2. Instrumental Evaluation.

*2.2. Instrumental Evaluation.* An optoelectronic motion analysis system (SMART-E System, BTS, Italy), consisting of eight infrared cameras (operating at 120 Hz), was used to detect the movements of spherical markers (15 mm diameter) covered with aluminium powder reflecting material placed over prominent bony landmarks on the skin according to Davis's protocol [33]. Meticulous attention was paid to marker placement in obese patients. If necessary, an elastic band was placed around the waist to avoid any possible movement of adipose tissue that could alter the marker trajectories [14]. All the subjects were asked to wear only underwear or shorts and a tight fitting undershirt. The use of minimal clothing was designed to ensure the correct placement of the markers over the anatomical landmarks. The calibrated walking volume consisted of a level surface that was approximately 6 m in length, with a width of 1.60 m and a height of about 2.00 m. Experiments began with a standing trial, in which rest joint angular displacements were acquired. Obese subjects and controls were then instructed to walk barefoot at a self-selected speed along the level surface. Assuming that this speed would be slower in the obese subjects, we instructed the controls to walk at low speed, too; in this way, gait characteristics could be compared between the groups without the potential velocity bias. Before formal measurements started, subjects did a practice session to familiarize themselves with the experimental procedure by walking for one hour (with some pauses to avoid fatigue). In the experimental session, which was performed the following day, twelve valid trials were acquired for each subject. A valid trial was defined as one in which marker trajectories were not lost during the subject's gait and included at least one cycle per limb. The 1st and 12th trials of each subject were discarded to reduce movement variability related to the start and end of the session. A one-minute rest period was given between groups of 3 trials to avoid fatigue.

*2.3. Data Analysis.* In the present study, a set of time-distance, kinematic, and coordination parameters were adopted to provide a thorough analysis of gait. Since walking asymmetries were not analyzed in this work, right and left leg data were considered together, yielding a set of generic lower-limb parameters.

*2.3.1. Time-Distance Data.* A stride was considered as the time between two consecutive heel-floor contacts of the same limb and was subdivided in a stance phase (from 1st initial contact to foot-off) and a swing phase (from foot-off to 2nd heel contact). The double support phase within the stride, defined as the time spent by subjects with both feet on the ground, was also considered. A step was defined as the time between a heel-floor contact of one limb and the

consecutive heel-floor contact of the other limb. As time-distance gait parameters, we evaluated stride duration, mean speed, cadence, duration of the stance, swing and double support phases within the stride (all evaluated as percentages of the stride duration), step length, and step width.

For each subject, the time-distance parameters were obtained by averaging the data of the valid strides of 10 successful trials.

*2.3.2. Kinematic Data.* Kinematic data were derived using the BTS Smart Analyzer software. Data were smoothed using a triangular four-order window filter. Joint centres of rotation were determined and joint excursions were calculated. Joint angular displacement data were normalized to the stride duration and reduced to 100 samples using a polynomial procedure, thereby defining a gait cycle. We evaluated pelvic tilt, obliquity and rotation, hip flexion-extension, abduction-adduction and rotation, knee flexion-extension, ankle dorsi-plantar flexion, and foot progression. For each angle, we calculated the range of motion (ROM), defined as the differences between the maximum and minimum values during the gait cycles.

*2.3.3. Coordination Parameters.* Inter-joint coordination on the sagittal plane was assessed by using the CRP technique [21, 34, 35]. A custom written Matlab code (version 7.0; The Mathwoks Inc., MA) was used to compute the coordination parameters. First, the so-called phase portrait of each sagittal joint motion was generated by plotting its normalized angular velocity ($\omega_N$) against its normalized angular position ($\theta_N$) [35]. The following equations were used to normalize each gait cycle:

$$\omega_{iN} = \frac{\omega_i}{\max\left[\max\left(\omega_i\right), \max\left(-\omega_i\right)\right]},$$

$$\theta_{iN} = \frac{2 * \left[\theta_i - \min\left(\theta_i\right)\right]}{\max\left(\theta_i\right) - \min\left(\theta_i\right)} - 1, \tag{1}$$

where $i$ is the percentage of the entire movement cycle, while $\theta$ and $\omega$ are, respectively, the angular displacement and angular velocity.

Once the phase portrait had been obtained, the phase angle ($\varphi_i$) of the joint motion was computed as follows:

$$\varphi_i = \tan^{-1}\frac{\omega_{iN}}{\theta_{iN}}, \tag{2}$$

where $i$ is the time point within the cycle. The CRP angle ($\phi_i$) between each pair of sagittal joint motions was computed by subtracting the value of the phase angle of the distal joint ($D_J$) from the value of the phase angle of the proximal joint ($P_J$):

$$\phi_i = \varphi_{iP_J} - \varphi_{iD_J}. \tag{3}$$

The CRP angles can range between $-360°$ and $360°$, with $0° \pm 360°$ indicating in-phase coupling, and $-180°$ and $180°$ indicating out-phase coupling. A positive CRP indicates that the proximal joint leads the distal. According to the method of Stergiou et al. [36], in order to analyze significant differences between CRP curves, two indices were used for each pair of joint relative movements. The first index is the mean absolute relative phase (MARP), which was computed by averaging the absolute values of the ensemble curve points for both the stance and the swing phases:

$$\text{MARP} = \frac{\sum_{i=1}^{p} |\overline{\phi}_i|}{p}, \tag{4}$$

where $p$ is the number of time points in each phase. The second parameter is the deviation phase (DP), which is calculated by averaging the standard deviations of the ensemble CRP curve points for both the stance and the swing phases:

$$\text{DP} = \frac{\sum_{i=1}^{p} SD_i}{p}. \tag{5}$$

DP provides a measure of stability of the organization of the neuromuscular system. A low DP value indicates a less variable intrasubjects stride-to-stride relationship between the actions of the two joints. Three couplings of joints were considered: hip-knee, knee-ankle, and hip-ankle. The $\text{MARP}_{\text{stance}}$, $\text{DP}_{\text{stance}}$, $\text{MARP}_{\text{swing}}$, and $\text{DP}_{\text{swing}}$ on the sagittal plane were calculated for each of these couplings of joints.

*2.3.4. Statistical Analysis.* The statistical analysis was performed using PASW software (PASW Statistic 17, Chicago, USA). Means and standard deviations were calculated for time-distance, kinematic, and coordination parameters. For each variable, the Shapiro-Wilk test was performed to assess the Gaussian distribution of the two samples. Then, the 2-tailed $t$-test for equality of means was applied when the parameters were found to be normally distributed. The nonparametric test (Mann-Whitney) was performed for non-Gaussian variables. We considered $P$ values of less than 0.05 as statistically significant.

To evaluate the group effect on the mean joint coupling variables we calculated the between-group coefficient of multiple correlation ($\text{CMC}_{\text{BG}}$) between the mean waveforms of the obese and the control groups. To determine the level of homogeneity within each group, we also calculated the within-group coefficient of multiple correlation ($\text{CMC}_{\text{WG}}$) between the mean waveforms of the subjects in the obese group and in the healthy group.

CMC, that is, the positive square root of the adjusted coefficient of multiple determination [37, 38], is a measure of the overall waveform similarity of a group of curves; the closer to 1 the CMC is, the more similar the waveforms are. We calculated the two CMC as follows:

$$\text{CMC} = \sqrt{1 - \frac{(1/(T(N-1)))\sum_{i=1}^{N}\sum_{t=1}^{T}(y_{it} - \overline{y}_t)^2}{(1/(TN-1))\sum_{i=1}^{N}\sum_{t=1}^{T}(y_{it} - \overline{y})^2}}, \tag{6}$$

where $T = 100$ (number of time points within the cycle), $N$ is the number of curves (2 for $\text{CMC}_{\text{BG}}$ and 25 for $\text{CMC}_{\text{WG}}$),

TABLE 2: Means, standard deviations of the time-distance parameters, and their statistical significance ($P$ values). In-bold $P$ values are lower than 0.05.

| Time-distance parameters | Obese subjects | Controls | $P$ values |
|---|---|---|---|
| Stride duration (s) | $1.20 \pm 0.11$ | $1.26 \pm 0.15$ | 0.183 |
| Mean speed (m/s) | $0.93 \pm 0.11$ | $0.92 \pm 0.19$ | 0.794 |
| Cadence (step/min) | $101.70 \pm 8.94$ | $96.86 \pm 11.09$ | 0.137 |
| Stance % | $64.06 \pm 1.76$ | $61.22 \pm 2.18$ | **<0.001** |
| Swing % | $35.94 \pm 1.76$ | $38.78 \pm 2.18$ | **<0.001** |
| Double support % | $13.77 \pm 1.74$ | $11.23 \pm 2.02$ | **<0.001** |
| Step length (m) | $0.49 \pm 0.04$ | $0.51 \pm 0.05$ | 0.199 |
| Step width (m) | $0.26 \pm 0.04$ | $0.20 \pm 0.06$ | **<0.001** |

$y_{it}$ is the value at the $t$th time point in the $i$th curve, $\overline{y}_t$ is the average at time point $t$ over $N$ curves:

$$\overline{y}_t = \frac{1}{N}\sum_{i=1}^{N} y_{it} \tag{7}$$

and $\overline{y}$ is the grand mean of all $y_{it}$:

$$\overline{y} = \frac{1}{NT}\sum_{i=1}^{N}\sum_{t=1}^{T} y_{it}. \tag{8}$$

## 3. Results

As correctly assumed, obese subjects showed a significantly reduced mean gait speed compared to controls when walking at self-selected speed ($0.93 \pm 0.11$ m/s versus $1.138 \pm 0.14$ m/s, $P = 0.001$). However, obese subject mean self-selected speed was not statistically different from that of controls when walking at low speed ($0.93 \pm 0.11$ m/s versus $0.92 \pm 0.19$ m/s, $P = 0.794$). Thus, in order to avoid the potential velocity bias, time-distance, kinematic, and coordination parameters were compared between obese subjects walking at self-selected speed and controls walking at low speed.

*3.1. Time-Distance Data.* The means and standard deviations of the time-distance parameters and $P$-values are shown in Table 2. Obese subjects spent a significantly greater percentage of the gait cycle in the stance (5% more) and double support (23% more) phases than controls, while their swing phase was shorter (a 7% reduction). Furthermore, the step width was greater (an 30% increase) in the obese group than in controls.

*3.2. Kinematic Data.* The means and the standard deviations of hip, knee, and ankle joint angular ROM on the sagittal plane are shown in Figure 1. Table 3 shows the means and standard deviations of the kinematic parameters. The table also presents the statistical analysis results. Significant reductions in knee flexion-extension (7%) and pelvic obliquity (24%) ROMs were observed in obese subjects if compared with controls. Lastly, a significant increase in pelvic tilt ROM (17%) was found in obese subjects if compared with controls.

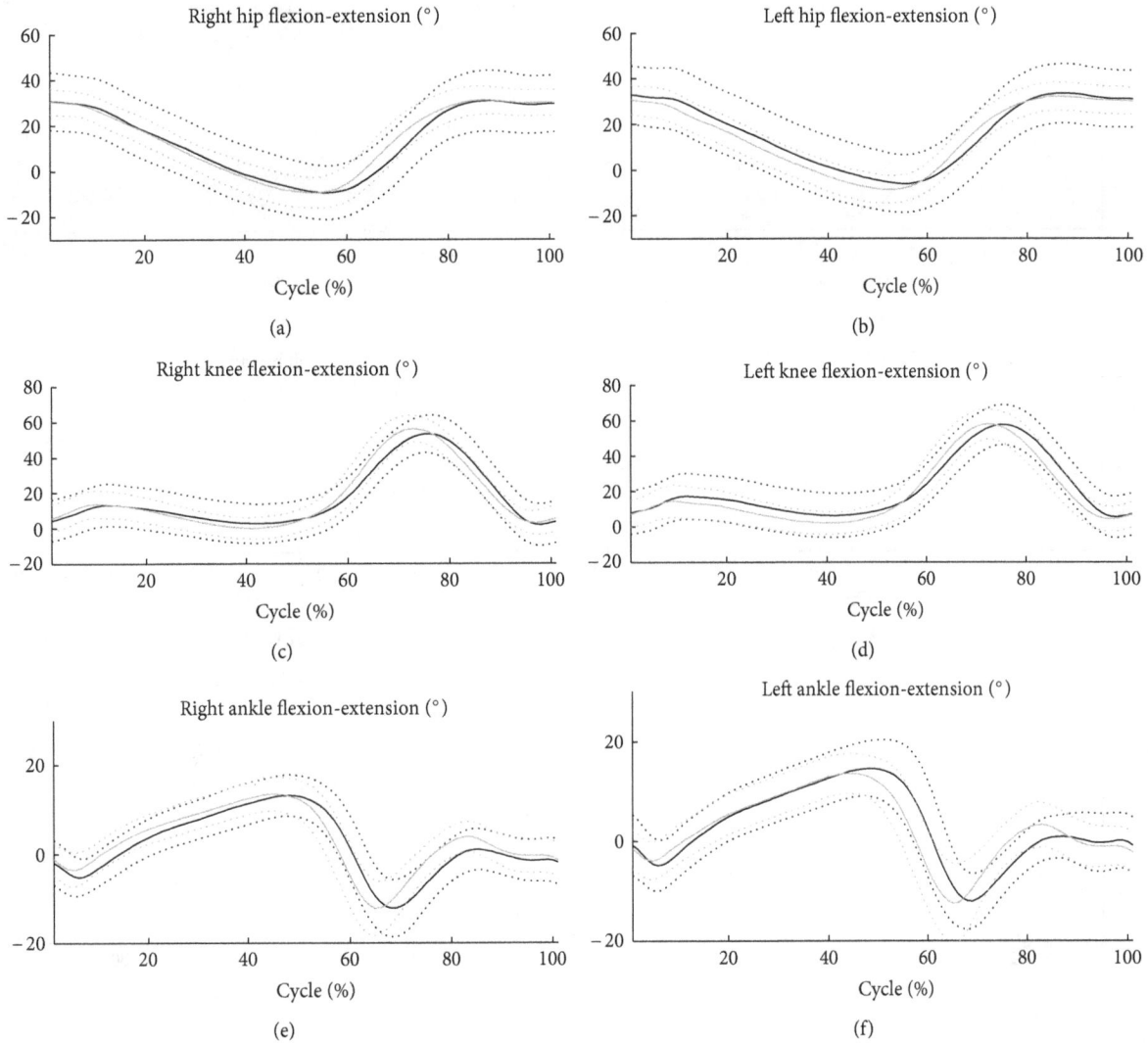

FIGURE 1: Mean and standard deviation of angular displacement of hip, knee, and ankle joint in the sagittal plane. The black curve refers to obese subjects, the gray one to controls.

TABLE 3: Means, standard deviations, and significance (*P* values) of the kinematic parameters. *P* values lower than 0.05 are shown in bold.

| ROM (°) | Obese subjects | Controls | *P* values |
|---|---|---|---|
| Ankle flexion-extension | $28.52 \pm 5.20$ | $28.35 \pm 6.21$ | 0.926 |
| Foot progression | $12.57 \pm 2.55$ | $13.95 \pm 3.49$ | 0.163 |
| Knee flexion-extension | $55.09 \pm 4.12$ | $58.98 \pm 5.48$ | **0.015** |
| Hip flexion-extension | $42.01 \pm 4.83$ | $42.31 \pm 4.73$ | 0.841 |
| Hip abduction-adduction | $14.42 \pm 2.90$ | $13.10 \pm 3.12$ | 0.174 |
| Hip rotation | $14.86 \pm 4.06$ | $14.76 \pm 3.40$ | 0.820 |
| Pelvic tilt | $3.80 \pm 0.87$ | $3.24 \pm 0.86$ | **0.021** |
| Pelvic obliquity | $5.32 \pm 1.35$ | $6.99 \pm 2.68$ | **0.018** |
| Pelvic rotation | $8.28 \pm 2.65$ | $9.78 \pm 2.93$ | 0.097 |

*3.3. Coordination Parameters.* The means and the standard deviations of the ensemble CRP curves are shown in Figure 2. The means and standard deviations of the coordination parameters and the statistical analysis results are reported

in Table 4. No significant differences in MARP mean values were observed between the two groups in both stance and swing phases.

By contrast, the joint coordination variability, as calculated by means of the DP, was always greater in obese subjects than in controls. Indeed, an increase of DP values at the hip-knee (stance 40%, swing 17%), hip-ankle (stance 35%, swing 28%), and knee-ankle (stance 40%, swing 19%) couplings were observed.

$CMC_{WG}$ and $CMC_{BG}$ values are summarized in Table 5.

## 4. Discussion

The aim of this work was to evaluate how gait coordination is influenced by obesity. To achieve this aim, we calculated time-distance data, kinematic, and inter-joint Continuous Relative Phase parameters. We observed a substantial similarity between obese subjects and controls in segmental parameters (joint ROMs) and some significant differences

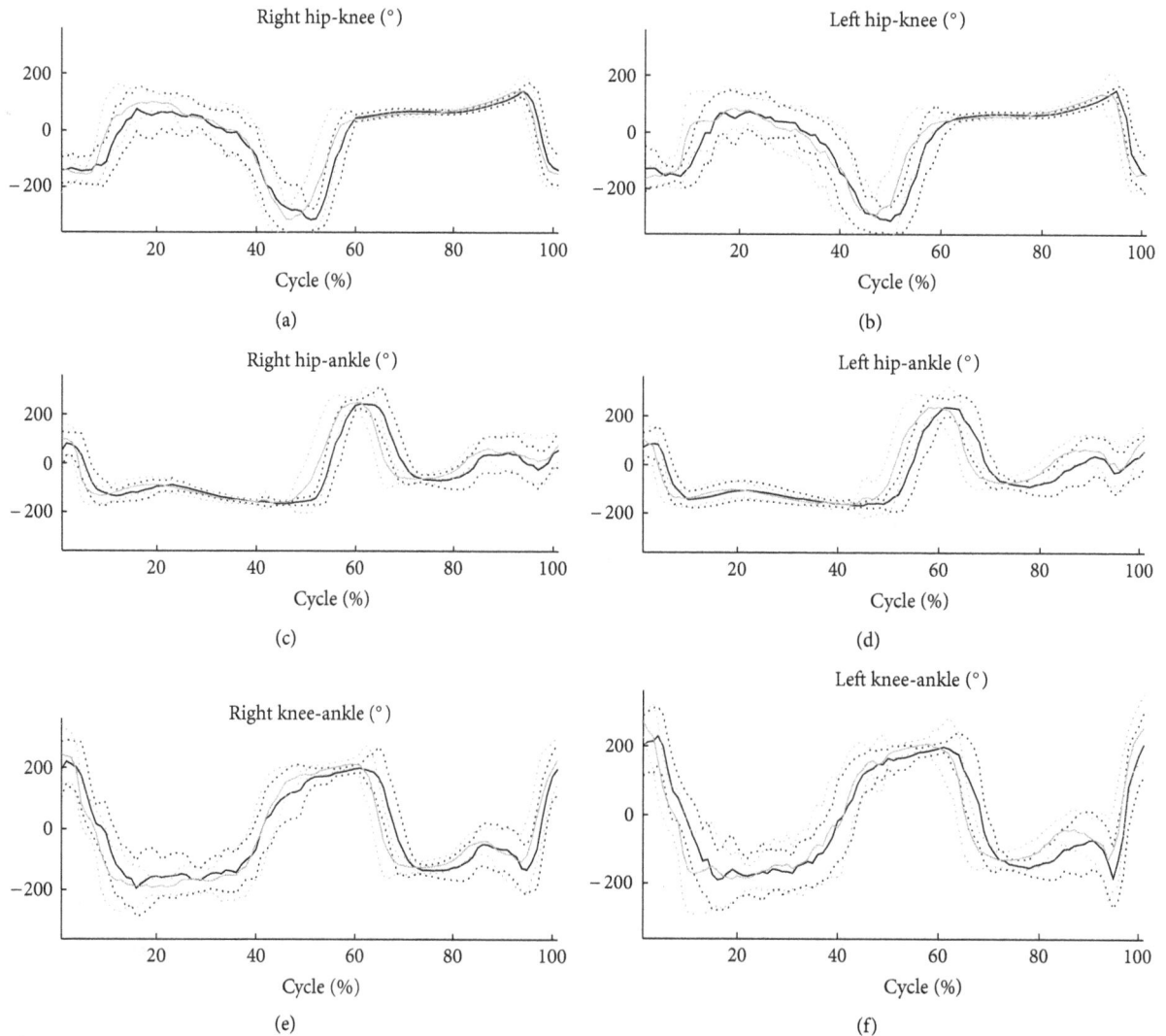

FIGURE 2: Mean and standard deviation of CRP at the hip-knee, hip-ankle, and knee-ankle joint coupling. The black curve refers to obese subjects, the gray one to controls.

between the two groups in the global (time-distance) and inter-joint coupling (CRP) parameters.

As regards the time-distance parameters, in keeping with the results of previous works [5–7, 38], our findings reveal that the gait cycle in obese subjects is characterized by a longer stance and double support phase, a shorter swing phase, and a wider base of support. This altered gait pattern is probably related to an increased need for stabilization caused by obesity. Indeed, the greater double support and stance phase is likely to provide a safer locomotion by maintaining the weight on both limbs and not overloading one limb, thereby reducing the risk of instability [9]. The increased effort made by obese individuals to maintain body balance also yields high values in step width (due, in part, also to the increased body mass encumbrance between the legs) and in a lower gait self-selected speed. Moreover, the lower gait self-selected velocity may be read as being indicative of poor physical condition in obese adults as well as an attempt to give the CNS more time to react to obstacles [39]. These altered global gait parameters are similar to the ones exhibited

by older adults who tend to fall. Indeed, according to Maki [39], changes in the time-distance parameters in obese subjects may also be associated with a preexisting fear of falling.

From a kinematic point of view, obese subjects displayed a substantial similarity in the angle curves of the pelvis and lower-limb joints within the gait cycle (see Figure 1). However, a significant decrease emerged in obese subjects in pelvic obliquity and in knee joint flexion-extension ROMs, whereas a significant increase was observed in pelvic tilt ROM. The reduced ROMs may be due to the fact that obese subjects need to keep both limbs in contact with the ground to gain stability, which increases the amount of time spent in a closed lower-limb kinematic chain condition. This reduces the degrees of freedom of the rigid lower body system and consequently exerts greater constraint on the pelvic segment and knee joint. Another possible explanation for these reduced ROMs may be the excess weight of the limbs, which represents an extra load for the muscles involved

TABLE 4: Means, standard deviations, and significance of the coordination parameters. Significant results are shown in bold. H: hip, K: knee, A: ankle.

| Coordination parameters (°) | Obese subjects | Controls | P values |
|---|---|---|---|
| Stance MARP H-K | 125.33 ± 10.76 | 123.77 ± 11.06 | 0.654 |
| Stance DP H-K | 73.90 ± 20.08 | 52.88 ± 22.87 | **0.001** |
| Stance MARP H-A | 139.83 ± 9.36 | 142.89 ± 12.32 | 0.382 |
| Stance DP H-A | 37.35 ± 10.20 | 27.76 ± 7.30 | **0.000** |
| Stance MARP K-A | 155.19 ± 12.93 | 160.48 ± 14.15 | 0.180 |
| Stance DP K-A | 72.11 ± 19.24 | 51.39 ± 21.96 | **0.001** |
| Swing MARP H-K | 88.53 ± 7.48 | 91.16 ± 7.90 | 0.288 |
| Swing DP H-K | 22.16 ± 5.66 | 18.99 ± 7.74 | **0.015** |
| Swing MARP H-A | 74.67 ± 25.34 | 72.49 ± 17.82 | 0.755 |
| Swing DP H-A | 61.57 ± 19.00 | 48.28 ± 19.45 | **0.035** |
| Swing MARP K-A | 120.74 ± 26.70 | 112.45 ± 18.35 | 0.260 |
| Swing DP K-A | 68.47 ± 15.78 | 57.46 ± 17.17 | **0.041** |

TABLE 5: $CMC_{WG}$ values for obese subjects and controls and $CMC_{BG}$ values. H: hip, K: knee, A: ankle.

| Couples of joints | $CMC_{WG}$ obese subjects | $CMC_{WG}$ controls | $CMC_{BG}$ |
|---|---|---|---|
| H-K | 0.885 | 0.845 | 0.975 |
| H-A | 0.885 | 0.870 | 0.940 |
| K-A | 0.880 | 0.875 | 0.965 |

in pelvic obliquity and knee flexion movement during the preswing subphase and swing phase.

In particular, the reduction in knee ROMs may be due to the presence of excess fat on the thigh and shank, which mechanically encumbers intersegmental rotation [40] and counteracts the antigravity action exerted by the knee flexors. According to Park et al. [40], the main cause of the reduction observed in ROMs in obese subjects is the lower level of daily physical activity, which limits both weight loss and muscle strengthening. These kinematic results are also partially in keeping with those obtained by Spyropolous et al. [6], who reported lower ROMs in obese subjects than in controls walking at a self-selected speed.

We investigated the coordinative behaviour by means of the CRP technique. The choice of such method instead of others based on principal component analysis [41] was determined by the need to quantify, sample by sample, the coupling relationship between two joints. One of the main findings of our study is that the joint coordination pattern in obese subjects was substantial similar from that observed in nonobese subjects, in both the stance and swing phases. These results indicate that the hip, knee and ankle joints play the same role in obese subjects as in controls in producing a coordinated walking pattern within the gait cycle. The fact that the mechanical constraint exerted by the lower-limb excess mass does not affect the coordinative strategy adopted by the CNS is, however, in contrast to our hypothesis.

Previous studies have shown that in obesity there is a neuromuscular adaptation, implemented by a decreasing self-selected gait speed, which results in reduced ground reaction forces, lower joint loads, and net muscle moments during gait and is aimed at reducing the risk of musculoskeletal diseases [8, 42]. We hypothesize that the afore-mentioned neuromuscular adaptations may be the result of walking control mechanisms designed to preserve a physiological inter-joint lower-limb coupling pattern. This motor behaviour, anchored to the physiological coordinative strategy, results in the high $CMC_{WG}$ and $CMC_{BG}$ values that point to a strong similarity

among obese subjects as well as between obese and control subjects in inter-joint lower-limb couplings.

A surprising finding of the present study is the wide stride-to-stride intrasubject variability (see DP values) in all the joint couplings in obese subjects if compared with controls, which may be a risk factor for falls [43]. As recently showed, gait variability gives indirect information on the control mechanism of locomotion [44]. In particular, a low variability is related to the stability of the locomotion, meaning the capacity to maintain the dynamic balance.

The higher variability we found may be due to the excess fat distributed above the pelvis and the lower-limb segments, creating a mechanical encumbrance, that has to be continuously managed stride by stride. Since the distribution of this excess fat is not the same in men and women [45], a study designed to analyze gender differences in coordinative strategies may help confirm these results.

In a recent study by Yen et al. [35], an increased variability in the coupling relationship was observed in elderly people during obstacle crossing. It was read as an age-related biomechanical change associated with a lower ability to maintain a stable body balance, which might increase the risk of falls during walking. For this reason, further studies are warranted to shed light on the relationship in obese people between the risk of falling and inter-joint coordination as well as on the relative position between the centre of pressure and the "extrapolated centre of mass" [46].

The CRP curves show that the pattern in obese subjects is topologically similar to that of controls (see Figure 2 and $CMC_{BG}$ in Table 5). The only difference that emerged is a time shift between the mean curves of obese subjects and those of the controls, which is particularly pronounced at the end of the stance phase and during the swing phase of gait. The time shift that is evident in the CRP curves of obese subjects is in keeping with the changes found in the time-distance parameters concerning the increase in the stance phase and the decrease in the swing phase.

The accuracy of the marker placement in obese subjects is one possible limitation of this work. Indeed, although we adopted strategies to minimize the likelihood of errors, marker trajectory anomalies, especially in pelvis markers, which are subject to movement-induced oscillations due to the excess fat at the waist, cannot be ruled out. For this reason, in the kinematic results, the small differences observed in pelvic ROMs between obese subjects and controls could be considered as not relevant, especially taking into account that the pelvic angular excursions are very small.

In view of the above, the reduction of the variability associated to the lower-limb joint coupling relationships analyzed in the present study may prove useful as indexes to design and assess the effectiveness of dietary treatment, of physical exercise and of rehabilitative protocols adopted.

## 5. Conclusions

In this study we performed an exhaustive gait analysis to investigate the time-distance, kinematic, and coordinative alterations that occur in obese subjects. The results of this work shed light on the motor strategy adopted by obese individuals, which is aimed at maintaining body balance and at preserving a physiological inter-joint lower-limb coupling pattern. Since obesity was found to be related to an increased intrasubject stride-to-stride coordination variability, the study of joint coordination and specific rehabilitation interventions in walking performance may help to improve activities of daily life in these subjects.

## References

[1] National Institutes oh Health, "Practical Guide to the identification, evaluation and treatment of overweight and obesity in adults," Bethesda, Md, USA, National Insitutes of Health, 2000.

[2] P. Capodaglio, G. Castelnuovo, A. Brunani, L. Vismara, V. Villa, and E. M. Capodaglio, "Functional limitations and occupational issues in obesity: a review," International Journal of Occupational Safety and Ergonomics, vol. 16, no. 4, pp. 507–523, 2010.

[3] U. E. Larsson and E. Mattsson, "Functional limitations linked to high body mass index, age and current pain in obese women," International Journal of Obesity, vol. 25, no. 6, pp. 893–899, 2001.

[4] D. L. Gushue, J. Houck, and A. L. Lerner, "Effects of childhood obesity on three-dimensional knee joint biomechanics during walking," Journal of Pediatric Orthopaedics, vol. 25, no. 6, pp. 763–768, 2005.

[5] P. P. K. Lai, A. K. L. Leung, A. N. M. Li, and M. Zhang, "Three-dimensional gait analysis of obese adults," Clinical Biomechanics, vol. 23, no. 1, supplement, pp. S2–S6, 2008.

[6] P. Spyropoulos, J. C. Pisciotta, K. N. Pavlou, M. A. Cairns, and S. R. Simon, "Biomechanical gait analysis in obese men," Archives of Physical Medicine and Rehabilitation, vol. 72, no. 13, pp. 1065–1070, 1991.

[7] S. A. F. De Souza, J. Faintuch, A. C. Valezi et al., "Gait cinematic analysis in morbidly obese patients," Obesity Surgery, vol. 15, no. 9, pp. 1238–1242, 2005.

[8] P. DeVita and T. Hortobágyi, "Obesity is not associated with increased knee joint torque and power during level walking," Journal of Biomechanics, vol. 36, no. 9, pp. 1355–1362, 2003.

[9] L. Vismara, M. Romei, M. Galli et al., "Clinical implications of gait analysis in the rehabilitation of adult patients with "prader-Willi" Syndrome: a cross-sectional comparative study ("Prader-Willi" Syndrome versus matched obese patients and healthy subjects)," Journal of Neuroengineering and Rehabilitation, vol. 4, article 14, 2007.

[10] D. Malatesta, L. Vismara, F. Menegoni, M. Galli, M. Romei, and P. Capodaglio, "Mechanical external work and recovery at preferred walking speed in obese subjects," Medicine and Science in Sports and Exercise, vol. 41, no. 2, pp. 426–434, 2009.

[11] R. C. Browning, J. R. Modica, R. Kram, and A. Goswami, "The effects of adding mass to the legs on the energetics and biomechanics of walking," Medicine and Science in Sports and Exercise, vol. 39, no. 3, pp. 515–525, 2007.

[12] V. Cimolin, L. Vismara, M. Galli, F. Zaina, and S. Negrini, "Effects of obesity and chronic low back pain on gait," Journal of Neuroengineering and Rehabilitation, vol. 26, no. 8, pp. 55–61, 2011.

[13] A. Cereatti, M. Donati, V. Camomilla, F. Margheritini, and A. Cappozzo, "Hip joint centre location: an ex vivo study," Journal of Biomechanics, vol. 42, no. 7, pp. 818–823, 2009.

[14] J. Nantel, M. Brochu, and F. Prince, "Locomotor strategies in obese and non-obese children," Obesity, vol. 14, no. 10, pp. 1789–1794, 2006.

[15] H. Haken, J. A. S. Kelso, and H. Bunz, "A theoretical model of phase transitions in human hand movements," Biological Cybernetics, vol. 51, no. 5, pp. 347–356, 1985.

[16] L. Li, E. C. H. Van Den Bogert, G. E. Caldwell, R. E. A. Van Emmerik, and J. Hamill, "Coordination patterns of walking and running at similar speed and stride frequency," Human Movement Science, vol. 18, no. 1, pp. 67–85, 1999.

[17] M. J. Kurz and N. Stergiou, "Does footwear affect ankle coordination strategies?" Journal of the American Podiatric Medical Association, vol. 94, no. 1, pp. 53–58, 2004.

[18] A. T. DeLeo, T. A. Dierks, R. Ferber, and I. S. Davis, "Lower extremity joint coupling during running: a current update," Clinical Biomechanics, vol. 19, no. 10, pp. 983–991, 2004.

[19] M. J. Kurz, N. Stergiou, U. H. Buzzi, and A. D. Georgoulis, "The effect of anterior cruciate ligament reconstruction on lower extremity relative phase dynamics during walking and running," Knee Surgery, Sports Traumatology, Arthroscopy, vol. 13, no. 2, pp. 107–115, 2005.

[20] L. K. Drewes, P. O. McKeon, G. Paolini et al., "Altered ankle kinematics and shank-rear-foot coupling in those with chronic ankle instability," Journal of Sport Rehabilitation, vol. 18, no. 3, pp. 375–388, 2009.

[21] S. L. Chiu, T. W. Lu, and L. S. Chou, "Altered inter-joint coordination during walking in patients with total hip arthroplasty," Gait and Posture, vol. 32, no. 4, pp. 656–660, 2010.

[22] A. Winogrodzka, R. C. Wagenaar, J. Booij, and E. C. Wolters, "Rigidity and bradykinesia reduce interlimb coordination in Parkinsonian gait," Archives of Physical Medicine and Rehabilitation, vol. 86, no. 2, pp. 183–189, 2005.

[23] G. Kwakkel and R. C. Wagenaar, "Effect of duration of upper- and lower-extremity rehabilitation sessions and walking speed on recovery of interlimb coordination in hemiplegic gait," Physical Therapy, vol. 82, no. 5, pp. 432–448, 2002.

[24] E. Hutin, D. Pradon, F. Barbier, J. M. Gracies, B. Bussel, and N. Roche, "Lower limb coordination patterns in hemiparetic gait: factors of knee flexion impairment," Clinical Biomechanics, vol. 26, no. 3, pp. 304–311, 2011.

[25] E. G. Fowler and E. J. Goldberg, "The effect of lower extremity selective voluntary motor control on interjoint coordination during gait in children with spastic diplegic cerebral palsy," Gait and Posture, vol. 29, no. 1, pp. 102–107, 2009.

[26] C. J. T. Van Uden, J. K. C. Bloo, J. G. M. Kooloos, A. Van Kampen, J. De Witte, and R. C. Wagenaar, "Coordination and stability of one-legged hopping patterns in patients with anterior cruciate ligament reconstruction: preliminary results," Clinical Biomechanics, vol. 18, no. 1, pp. 84–87, 2003.

[27] R. G. Crowther, W. L. Spinks, A. S. Leicht, F. Quigley, and J. Golledge, "Intralimb coordination variability in peripheral arterial disease," *Clinical Biomechanics*, vol. 23, no. 3, pp. 357–364, 2008.

[28] R. H. Miller, S. A. Meardon, T. R. Derrick, and J. C. Gillette, "Continuous relative phase variability during an exhaustive run in runners with a history of iliotibial band syndrome," *Journal of Applied Biomechanics*, vol. 24, no. 3, pp. 262–270, 2008.

[29] W. E. Siri, "Body composition from fluid spaces and density: analysis of methods," in *Techniques for Measuring Body Composition*, pp. 223–244, National Academy of Science, National Research Council, Washington, DC, USA, 1963.

[30] American Thoracic Society, "ATS statement: guidelines for the 6-minute walk test," *American Journal of Respiratory and Critical Care Medicine*, vol. 166, pp. 111–117, 2002.

[31] G. Borg, "Psychophysical scaling with applications in physical work and the perception of exertion," *Scandinavian Journal of Work, Environment and Health*, vol. 16, no. 1, pp. 55–58, 1990.

[32] A. W. Andrews, M. W. Thomas, and R. W. Bohannon, "Normative values for isometric muscle force measurements obtained with hand-held dynamometers," *Physical Therapy*, vol. 76, no. 3, pp. 248–259, 1996.

[33] R. B. Davis, S. Õunpuu, D. Tyburski, and J. R. Gage, "A gait analysis data collection and reduction technique," *Human Movement Science*, vol. 10, no. 5, pp. 575–587, 1991.

[34] T. W. Lu, H. C. Yen, and H. L. Chen, "Comparisons of the inter-joint coordination between leading and trailing limbs when crossing obstacles of different heights," *Gait and Posture*, vol. 27, no. 2, pp. 309–315, 2008.

[35] H. C. Yen, H. L. Chen, M. W. Liu, H. C. Liu, and T. W. Lu, "Age effects on the inter-joint coordination during obstacle-crossing," *Journal of Biomechanics*, vol. 42, no. 15, pp. 2501–2506, 2009.

[36] N. Stergiou, J. L. Jensen, B. T. Bates, S. D. Scholten, and G. Tzetzis, "A dynamical systems investigation of lower extremity coordination during running over obstacles," *Clinical Biomechanics*, vol. 16, no. 3, pp. 213–221, 2001.

[37] M. P. Kadaba, H. K. Ramakrishnan, M. E. Wootten, J. Gainey, G. Gorton, and G. V. B. Cochran, "Repeatability of kinematic, kinetic, and electromyographic data in normal adult gait," *Journal of Orthopaedic Research*, vol. 7, no. 6, pp. 849–860, 1989.

[38] G. Steinwender, V. Saraph, S. Scheiber, E. B. Zwick, C. Uitz, and K. Hackl, "Intrasubject repeatability of gait analysis data in normal and spastic children," *Clinical Biomechanics*, vol. 15, no. 2, pp. 134–139, 2000.

[39] B. E. Maki, "Gait changes in older adults: predictors of falls or indicators of fear?" *Journal of the American Geriatrics Society*, vol. 45, no. 3, pp. 313–320, 1997.

[40] W. Park, J. Ramachandran, P. Weisman, and E. S. Jung, "Obesity effect on male active joint range of motion," *Ergonomics*, vol. 53, no. 1, pp. 102–108, 2010.

[41] N. A. Borghese, L. Bianchi, and F. Lacquaniti, "Kinematic determinants of human locomotion," *The Journal of Physiology*, vol. 494, no. 3, pp. 863–879, 1996.

[42] R. C. Browning and R. Kram, "Effects of obesity on the biomechanics of walking at different speeds," *Medicine and Science in Sports and Exercise*, vol. 39, no. 9, pp. 1632–1641, 2007.

[43] Y. Barak, R. C. Wagenaar, and K. G. Holt, "Gait characteristics of elderly people with a history of falls: a dynamic approach," *Physical Therapy*, vol. 86, no. 11, pp. 1501–1510, 2006.

[44] J. M. Hausdorff and F. Danion, "Stride variability: beyond length and frequency (multiple letters)," *Gait and Posture*, vol. 20, no. 3, pp. 304–305, 2004.

[45] J. C. Lovejoy and A. Sainsbury, "Sex differences in obesity and the regulation of energy homeostasis," *Obesity Reviews*, vol. 10, no. 2, pp. 154–167, 2009.

[46] A. L. Hof, M. G. J. Gazendam, and W. E. Sinke, "The condition for dynamic stability," *Journal of Biomechanics*, vol. 38, no. 1, pp. 1–8, 2005.

# 3

# Extremely Low-Frequency Magnetic Field Enhances the Therapeutic Efficacy of Low-Dose Cisplatin in the Treatment of Ehrlich Carcinoma

## Nihal S. El-Bialy and Monira M. Rageh

*Biophysics Department, Faculty of Science, Cairo University, Al Gammaa Street, Giza 12613, Egypt*

Correspondence should be addressed to Monira Mahmoud Rageh; monirarageh@yahoo.com

Academic Editor: Kazim Husain

The present study examines the therapeutic efficacy of the administration of low-dose cisplatin (*cis*) followed by exposure to extremely low-frequency magnetic field (ELF-MF), with an average intensity of 10 mT, on Ehrlich carcinoma in vivo. The cytotoxic and genotoxic actions of this combination were studied using comet assay, mitotic index (MI), and the induction of micronucleus (MN). Moreover, the inhibition of tumor growth was also measured. Treatment with cisplatin and ELF-MF (group A) increased the number of damaged cells by 54% compared with 41% for mice treated with cisplatin alone (group B), 20% for mice treated by exposure to ELF-MF (group C), and 9% for the control group (group D). Also the mitotic index decreased significantly for all treated groups ($P < 0.001$). The decrement percent for the treated groups (A, B, and C) were 70%, 65%, and 22%, respectively, compared with the control group (D). Additionally, the rate of tumor growth at day 12 was suppressed significantly ($P < 0.001$) for groups A, B, and C with respect to group (D). These results suggest that ELF-MF enhanced the cytotoxic activity of cisplatin and potentiate the benefit of using a combination of low-dose cisplatin and ELF-MF in the treatment of Ehrlich carcinoma.

## 1. Introduction

Platinum-based chemotherapeutic regimens have been widely used against many human cancers including oral, lung, head and neck cancer, metastatic tumors of testis and ovaries and many other solid tumors [1, 2]. The anticancer activity of cisplatin comes from its interactions with DNA. The drug binds with N7 of purine bases forming monoadducts which are later transformed into inter and intrastrand cross links by reaction of second reactive site of the drug with the second nucleobase. Such cisplatin-DNA adducts can inhibit fundamental cellular processes including replication, transcription, translation, and DNA repair [3].

Cisplatin must be used with a very high dose to maximize its antineoplastic effect. Such dose has been impeded by its sever toxicities, including nephrotoxicity, gastrointestinal toxicity, peripheral neuropathy, and ototoxicity [4–6]. The impairment of kidney function is considered as the main side effect of cisplatin, which is able to generate reactive

oxygen species, such as superoxide anion and hydroxyl radical [7, 8]. Also nephrotoxicity is closely associated with an increase in lipid peroxidation in the kidney tissues [9]. Additionally, cisplatin-based chemotherapy induces a fall in patient plasma concentrations of various antioxidants [10]. This may lead to failure of the antioxidative defense mechanism against free-radical-mediated organ damage and genotoxicity. Accordingly, the significant risk of cisplatin frequently hinders its use with such effective dose. To address this problem, attention has been focused on finding a novel combination of anticancer agents with nonoverlapping mechanisms of action to achieve enhanced efficacy with decreased side effects.

Consequently, [11] reported the possible synergism between ELF-MF and chemotherapy, where a low dose of cisplatin was administered followed by exposure to ELF-MF in order to reduce the drug side effects while keeping its therapeutic efficiency. The study hypothesized that static

Extremely Low-Frequency Magnetic Field Enhances the Therapeutic Efficacy of Low-Dose Cisplatin in the Treatment of Ehrlich Carcinoma

25

and extremely low frequency magnetic fields (ELF-MF) selectively act on cell signaling through their effects on charged matter motion.

The influence of static and ELF-MF on tumor growth, apoptosis, and P53 immunohistochemical expression have been studied in a series of independent reports. Their results indicated that simultaneous use of static and extremely low frequency magnetic fields with an average intensity higher than 3.59 mT, significantly inhibited tumor growth, decreased tumor cell mitotic index, and lowered the proliferative activity. Moreover, an increase in apoptosis and a corresponding reduction of immunoreactive P53 expression were also observed [12–15].

Therefore, the aim of the present work is to investigate the effectiveness of administration of low-dose cisplatin followed by exposure to ELF-MF, with an average intensity of 10 mT, on the growth of Ehrlich Carcinoma by studying cytotoxicity and DNA damage in tumor cells.

## 2. Materials and Methods

*2.1. Cell Culture and Tumor Inoculation.* Ehrlich ascites carcinoma cells (obtained from National Cancer Institute "NCI", Cairo University) containing $1 \times 10^6$ cells were intraperitoneally (i.p.) injected into female mice. Ascites fluid was collected on the 7th day after injection. The Ehrlich cells were washed twice and then resuspended in 0.09 saline ($5 \times 10^6$ viable cells). Female BALB mice (obtained from the animal house of NCI, with a body weight 22–25 g, 7-8 weeks old) were injected subcutaneously in their right flanks where the tumor was developed in a single and solid form. Tumor growth was monitored postinoculation until the desired volume was about 0.3 to 0.6 cm$^3$. All animal procedures and care were performed using guidelines for the Care and Use of Laboratory Animals [16] and approved by animal Ethics Committee at Cairo University.

*2.2. Treatment Protocols.* The experiment was run on a total of 40 mice. Ten days after tumor cell inoculation, mice were randomly assigned to experimental groups. Mice of group (A) were treated three times on experimental days 1, 4, and 7 with 0.1 mL cisplatin (3 mg/kg i.p.) followed by exposure to 50 Hz, 10 mT ELF-MF, 1 hr daily for 2 weeks. Mice of group (B) were treated three times on experimental days 1, 4, and 7 with 0.1 mL cisplatin (3 mg/kg i.p.). Mice of group (C) were injected with 0.1 mL saline (instead of cisplatin) three times on experimental days 1, 4, and 7 followed by exposure to 50 Hz, 10 mT ELF-MF, 1 hr daily for 2 weeks. Mice of group (D) were neither injected with cisplatin nor exposed to ELF-MF. During the treatment protocol, the tumor growth was monitored every three days over a period of 12 days for all the experimental groups A, B, C, and D. At the end of the treatment protocol, the mice of each group were divided so that 5 mice were sacrificed for the assessment of both comet and micronucleus and the other 5 mice were used to evaluate mitotic index.

*2.3. Magnetic Field Exposure.* The exposure was performed by a magnet with a fixed magnetic field value of 10 mT ±0.025. The magnetic field was generated by a solenoid carrying current of 18 A (ampere) at 50 Hz from the main supply (220–230 Volt) via a Variac (made in Yugoslavia). The magnet consisted of a coil with 320 turns made of electrically insulated 0.8 mm copper wire. The coil was wounded around a copper cylinder of 2 mm thickness, 40 cm diameter, and 40 cm length. The cylinder wall was earthed to eliminate the electric field. The magnetic field was measured at different locations to find out the most homogenous zone inside the solenoid core. This was done using Gauss/Tesla meter model 4048 with probe T-4048 manufactured by Bell Technologies Inc. (Orlando-Florida USA). Plastic cages containing groups (A) and (C) were placed in the middle of the exposure chamber prior to ELF-MF exposure.

*2.4. Comet Assay (Single Cell Gel Electrophoresis).* Comet assay (single cell gel electrophoresis) is considered as a rapid, simple, visual, and sensitive technique to assess DNA fragmentation typical for toxic DNA damage and early stage of apoptosis [17, 18]. The comet assay was performed under alkaline conditions (pH > 13) according to the method developed by Singh et al. [19] and Tice et al. [20]. Briefly, a small piece of tumor tissues ($n$ = 5) from each group were placed in 1ml cold HBSS containing 20 mM EDTA (ethylenediaminetetraacetic acid)/10% DMSO (dimethylsulfoxide, Qualigens, CPW59). The tissues were minced into fine pieces and let settled. 5 $\mu$L of aliquot was mixed with 70 $\mu$L of 0.7% low melting point (LMP) agarose (Sigma, A9414). This agarose was prepared in $Ca^{2+}$, $Mg^{2+}$ free PBS (phosphate buffered saline, HiMedia, TS1006) at 37°C and placed on a microscope slide, which was already covered with a thin layer of 0.5% normal melting point (NMP) agarose (HiMedia.RM273). After cooling at 4°C for 5 min, slides were covered with a third layer of LMP agarose. After solidification at 4°C for 5 min, slides were immersed in freshly prepared cold lysis solution (2.5 M NaCl, 1 mM $Na_2$EDTA, 10 mM tris base, pH 10, with 1% Triton X-100 and 10% DMSO added just before use) at 4°C for at least 1 h. Following lyses, slides were placed in a horizontal gel electrophoresis unit and incubated in fresh alkaline electrophoresis buffer (1 mM $Na_2$EDTA, 300 mM NaOH, pH 13). Electrophoresis was conducted for 30 min at 24 V (~0.74 V/cm) and 300 mA at 4°C. Then, the slides were immersed in neutralized buffer (0.4 M Tris-HCl, pH 7.5) and gently washed three times for 5 min at 4°C. All the above procedures were performed under dimmed light to prevent the occurrence of additional DNA damage. Comets were visualized by 80 $\mu$L, 1X ethidium bromide staining (SigmaE-8751) and examined at 400 x magnification using a fluorescent microscope. Comet 5 image analysis software developed by Kinetic Imaging, Ltd. (Liverpool, UK) linked to a CCD camera was used to assess the quantitative and qualitative extent of DNA damage in the cells by measuring the length of DNA migration and the percentage of migrated DNA. Finally, the program calculates tail moment and Olive tail moment. In all the samples, 100 cells were analyzed and classified into 5 types (0–4) depending on their tail moment.

Type 0 represents the cells without visible damage, while cells of type 4 have total degradation of DNA (long, broad tail, poorly visible head of the comet). Types 1, 2, and 3 represent the symptoms of increasing DNA damage. To calculate the extent of DNA damage, three types of the comet: numbers 2, 3, and 4 were selected.

*2.5. Micronucleus Test.* Bone marrow slides for micronucleus assay from 5 mice of each group were prepared and stained according to the method described by Schmidt [21] using the modifications of Agarwal and Chauhan [22]. The bone marrow was flushed out from tibias using 1mL fetal calf serum and centrifuged at 2000 xg for 10 min. The supernatant was discarded. Evenly spread bone marrow smears were stained using the May-Grunwald and Giemsa protocol. Slides were scored at a magnification of 1000x using a light microscope. 1000 polychromatic erythrocytes per animal were scored, and the number of micronucleated polychromatic erythrocytes (MNPCE) was determined. In addition, the number of polychromatic erythrocytes (PCE) was counted in fields that contained 100 cells (mature and immature) to determine the score of PCE and normochromatic erythrocytes (NCE).

*2.6. Mitotic Index Determination.* Chromosomes were prepared according to the method described by Adler [23] with some modification. Briefly, 5 mice from control and treated groups were injected i.p with colchicine (2 mg/kg) 2 hours prior to tissue sampling. Bone marrow cells were collected from the tibia by flashing in KCl (0.075 M, at 37°C) and incubated at 37°C for 25 min. Material was centrifuged at 2000xg for 10 min, fixed in aceto-methanol (acetic-acid: ethanol. 1 : 3, v/v). Centrifugation and fixation (in the cold) were repeated five times at an interval of 20 min. The material was resuspended in a small volume of the fixative, dropped onto chilled slides, flame-dried, and stained the following day in 5% buffered Giemsa (pH 6.8). Slides were scored at a magnification of 1000x using a light microscope. At least 1000 cells were examined in each mouse and the number of dividing cells including late prophase and metaphase was determined. The mitotic activity is expressed by the mitotic index (MI), which is the number of dividing cells in 1000 cells per mouse.

*2.7. Tumor Size Measurements.* Due to the high growth rate in Ehrlich tumor model, change in tumor volume ($\Delta V$) was monitored over a period of 12 days for the four groups A, B, C, and D. Ellipsoidal tumor volume ($V$) was assessed and calculated using the formula $V = (\pi/6)(d)^2(D)$, where $D$ and $d$ are the long and short axes, respectively, measured with a digital caliper (accuracy 0.01 mm). Each data point was the average of 10 measurements taken every three days.

*2.8. Statistical Analysis.* Data were expressed as mean ± standard error. Statistical analysis was performed by one-way variance analysis ANOVA using SPSS (version 17.0). Difference were considered significant when $P < 0.05$.

## 3. Results

The levels of DNA damage in cells of Ehrlich tumor showed a significant increase in treated groups (A, B, and C) compared to control group (D) (Figures 1 and 2). For type (0), the data revealed that about 70% of Ehrlich tumor cells did not exhibit any DNA damage in control group (D) compared to 19, 28 and 57% in treated groups A, B, and C, respectively. Meanwhile in type (4) about 16, 9 and 4% of Ehrilch tumor cells showed complete DNA damage in treated groups A, B, and C, respectively, relative to 1% for control group (D). The total percent of DNA damage in Ehrlich tumor cells represented by types (2, 3, and 4) showed five-, four-and twicefold increases for treated groups A, B, and C, respectively, with respect to control group (D). Also Figure 3 showed a significant increase ($P < 0.019$) in Olive tail moment for all treated groups compared to the control one.

Table 1 shows the frequencies of MNPCEs, PCEs, and NCEs in bone marrow cells of tumor bearing mice for both control group (D) and treated groups (A, B, and C). The results showed a significant increase in the formation of PCE for treated groups compared with that of control one. Also, MNPCEs induction showed an about 50% increase for treated groups (A and B) compared with the control group (D).

The results of mitotic index (used to evaluate cell cycle kinetics) are summarized in (Figure 4. MI of bone marrow cells showed a significant decrease in the treated groups (A, B and C) ($P < 0.001$). The percent of decrement for the treated groups A, B, and C was about 75, 60, and 25%, respectively, in comparison with the control group (D).

Figure 5 shows the average change in tumor volume measured for mice of control group (D) and that of treated groups (A, B and C) over a period of 12 days. Under our experimental conditions, after 3 days, a significant decrease ($P < 0.001$) in tumor growth rate was observed in mice of groups A and B, while group (C) showed a slight delay in tumor growth rate ($P < 0.049$) compared with control group (D). The control group (D) showed a marked increase in tumor volume (growth) throughout the experimental time and the same behavior was observed for group (C), but with lower rate which is probably due to the existence of few viable tumor cells (Figure 5). The average tumor growth at day 12 for treated groups (A and B) was significantly less than that observed in control group (D) ($P < 0.001$).

Table 2 shows correlation coefficients between DNA damage, evaluated by comet assay, and cytogenic damage, measured by MN test. Both types of damage assessment are in good correlation, but the comet parameter, % DNA in tail, has a lower correlation with the MN values.

Table 3 shows correlation coefficients between the average change in Ehrlich tumor volume ($\Delta V$) and MI. Both parameters are highly correlated.

## 4. Discussion

Cisplatin is one of the most widely used anticancer drug for the treatment of various cancers and solid tumors [24]. However, its major side effects are the main limiting factors of its clinical use for long-term treatment [9, 25]. Various

Extremely Low-Frequency Magnetic Field Enhances the Therapeutic Efficacy of Low-Dose Cisplatin in the Treatment of
Ehrlich Carcinoma

27

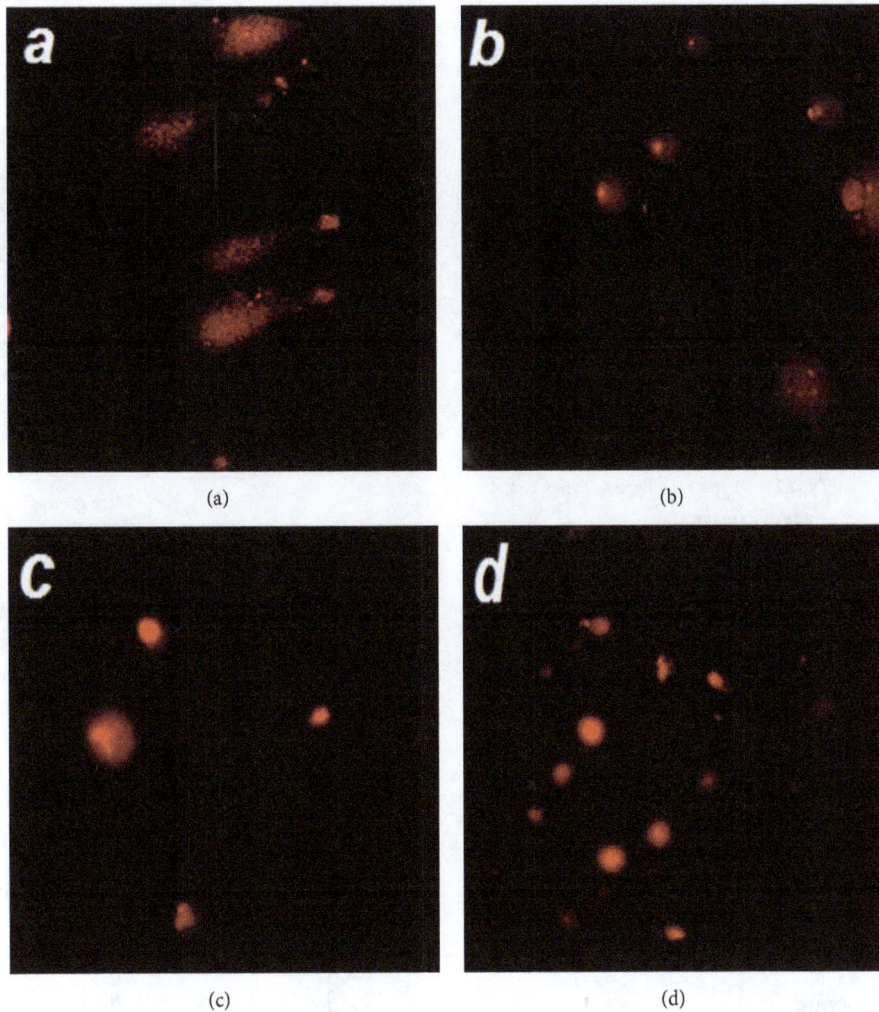

FIGURE 1: Typical comet images of Ehrlich carcinoma cells for (a) mice group (A) treated with cisplatin followed by exposure to ELF-MF, (b) mice group (B) treated with cisplatin, (c) mice group (C) treated by exposure to ELF-MF, and (d) mice group (D) the control one.

treatment strategies and curing agents have been tried and used to monitor or control its side effects.

Many anticancer agents exert their cellular toxicity through DNA damage [26], mainly DNA double-strand breaks. It is well known that DNA is the major target of cisplatin either as a result of its direct or indirect action through the generation of reactive oxygen species [10, 27–29].

The current study had revealed that administration of low dose of cisplatin followed by ELF-MF exposure disrupts the integrity and the amount of intact DNA. Comet results emphasized the increase in the number of cells with damaged DNA types (2, 3, and 4) (Figures 1, 2, and 3) in treated group (A), this damage might be due to the involvement of free radicals even if their concentration has not yet been measured. Such observed DNA damage is in agreement with previous studies [30–32] who reported that cisplatin forms covalent platinum DNA adducts and also acts as a DNA alkylator. In addition, cisplatin generates reactive

oxygen species, which trigger the opening of the mitochondrial permeability transition pore that permits the release of cytochrome c from mitochondria to cytosol and hence activates the mitochondria-dependent pathway leading to apoptosis [33, 34]. Also Tofani et al. [35] explained the synergistic activity observed between ELF-MF exposure and cisplatin by hypothesizing its ability to influence free radical chemistry exerted by the ELF-MF treatment.

Micronucleus test is a very reliable, widely used assay to measure not only DNA damage but also chromosomal instability and cell death [36]. Our results showed that ELF-MF alone did not cause MN induction in bone marrow cells of tumor bearing mice, while treatment by both cisplatin combined with ELF-MF (group A) and cisplatin alone (group B) increased the induction of MN by about 50% compared to the control (group D) (Table 1). These results are in consistent with previous work by Miyakoshi et al. [37]. Moreover, correlation coefficients between MN test and the three measured parameters, determined by comet assay,

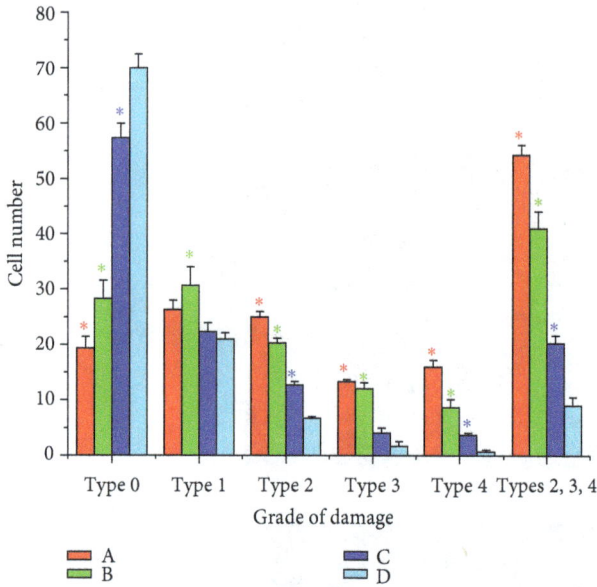

FIGURE 2: The level of DNA damage in Ehrlich tumor cells for mice group (A) treated with cisplatin followed by exposure to ELF-MF, mice group (B) treated with cisplatin, mice group (C) treated by exposure to ELF-MF, and mice group (D) the control one assessed by comet assay. Each value represents the mean ± SE ($n = 5$, $^*P < 0.001$)

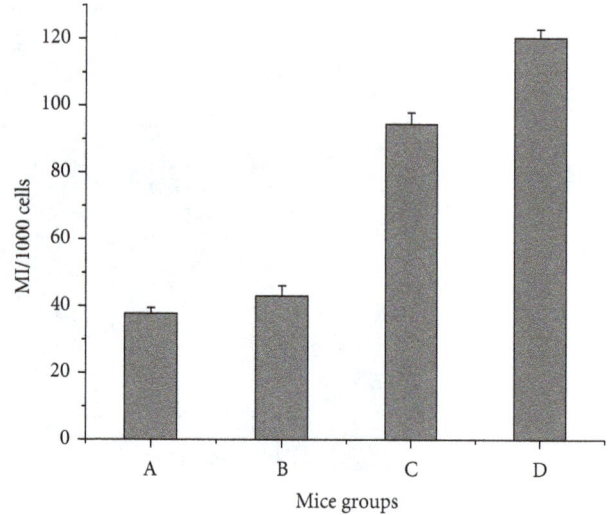

FIGURE 4: The values of mitotic index in bone marrow cells for mice group (A) treated with cisplatin followed by exposure to ELF-MF, mice group (B) treated with cisplatin, mice group (C) treated by exposure to ELF-MF, and mice group (D) the control one. Each value represents the mean ± S.E. ($P < 0.001$).

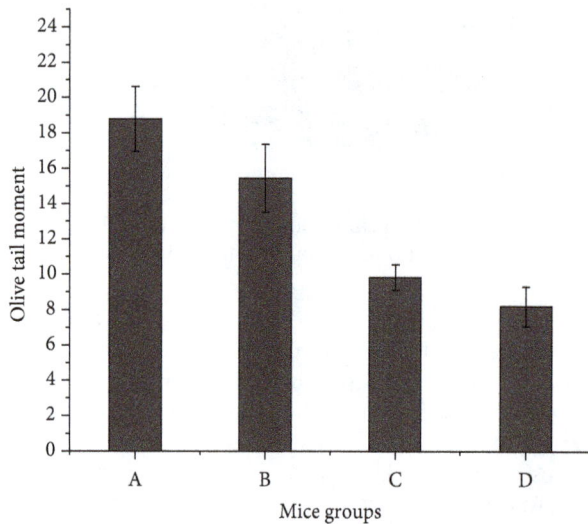

FIGURE 3: The values of olive tail moment assessed by comet assay for mice group (A) treated with cisplatin followed by exposure to ELF-MF, mice group (B) treated with cisplatin, mice group (C) treated by exposure to ELF-MF, and mice group (D) the control one. Each value represents the mean ± S.E. ($n = 5$, $P < 0.019$).

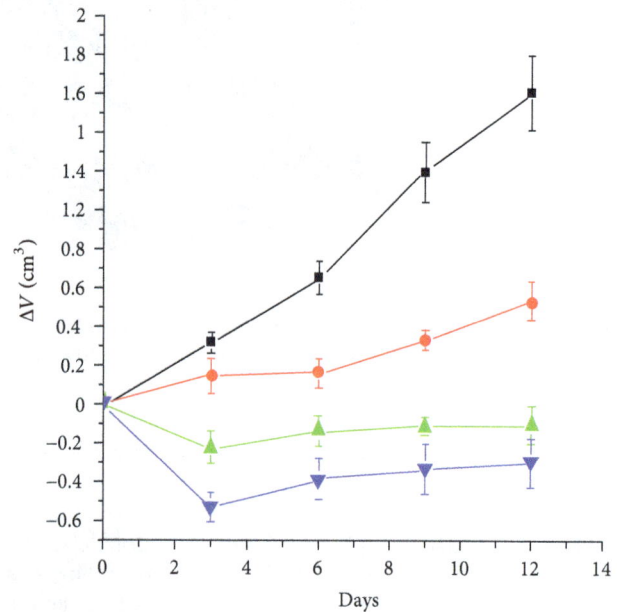

FIGURE 5: Average changes in Ehrlich tumor volume throughout a period of 12 days for mice group A (blue, (▼)) treated with cisplatin followed by exposure to ELF-MF, mice group B (green, (▲)) treated with cisplatin, mice group C (red, (●)) treated by exposure to ELF-MF, and mice group D (black, (■)) the control one. Each value represents the mean ± S.E. ($n = 10$).

pointed to a good relationship between DNA damage and MN induction induced by cisplatin and ELF-MF (Table 2).

The observed inhibition of mitotic index (Figure 4) in tumor-bearing mice bone marrow indicated that ELF-MF enhanced the cytotoxicity and genotoxicity of low dose cisplatin. These results were in good agreement with previous report on the genotoxic and cytotoxic potential of ELF-MF

[13]. Moreover, the tumor growth suppression observed in treated groups A and B (Figure 5) and the high correlation between tumor growth inhibition and mitotic index (Table 3), emphasized that the treatment protocol used in this work is therapeutically beneficial as most likely enhanced the effectiveness of low dose cisplatin. The improvement in treated group A was superior to treatment with the same

TABLE 1: PCEs, NCEs, MNPCEs induction in bone marrow cells of tumor bearing mice for control and treated groups.

| Groups | NCEs/100 | PCEs/100 | MNPCEs/1000 |
|---|---|---|---|
| A | $54 \pm 0.57^*$ | $46 \pm 0.57^*$ | $11 \pm 0.97^*$ |
| B | $57.3 \pm 0.33^*$ | $42.7 \pm 0.33^*$ | $11.0 \pm 1.15^*$ |
| C | $65.3 \pm 0.88^*$ | $34.7 \pm 0.88$ | $7 \pm 1$ |
| D | $65.3 \pm 0.88$ | $34.7 \pm 0.88$ | $7.3 \pm 0.33$ |

Mice group (A) treated with cisplatin followed by exposure to ELF-MF, mice group (B) treated with cisplatin, mice group (C) treated by exposure to ELF-MF and mice group (D) the control one. Each value represents the mean $\pm$ S.E. ($n = 5$, $^* P < 0.01$).

TABLE 2: Correlation coefficients between DNA and MN.

| | % DNA in Tail | Tail Length ($\mu$m) | Tail Moment | Olive Moment |
|---|---|---|---|---|
| MN | 0.446 | $0.649^*$ | $0.670^*$ | $0.738^{**}$ |

$^* P < 0.05$, Person's correlation, $^{**} P < 0.01$, Person's correlation.

TABLE 3: Correlation coefficients between MI and average change in Ehrlich tumor ($\Delta V$).

| | Average change in Ehrlich tumor ($\Delta V$) |
|---|---|
| MI | 0.808 |

$P < 0.001$, Person's correlation.

dose of cisplatin administrated to group B, which might be attributed to the increase in cell/tumor permeability induced by ELF- MF.

Consequently, our results indicated that increased damage of DNA by administrating low dose of cisplatin followed by ELF-MF exposure enhanced cell cytotoxicity as observed by the significant increase in micronucleus induction, in addition to a significant inhibition in both mitotic index and tumor growth. These results are in accordance with the commonly accepted assumption that extremely low frequency magnetic field enhanced the chemotherapeutic efficiency of cisplatin by increasing the production of oxygen species that caused more oxidative DNA damage.

## 5. Conclusion

The data presented here seem to indicate that exposure to ELF-MF may be a useful adjunct to chemotherapy. However, further investigations are needed to optimize ELF-MF physical parameters, chemotherapy schedule, and combination of both.

## References

[1] M. Gottfried, R. Ramlau, M. Krzakowski et al., "Cisplatin-based three drugs combination (NIP) as induction and adjuvant treatment in locally advanced non-small cell lung cancer: final results," *Journal of Thoracic Oncology*, vol. 3, no. 2, pp. 152–157, 2008.

[2] G. Türk, A. Ateşşahin, M. Sönmez, A. O. Çeribaşi, and A. Yüce, "Improvement of cisplatin-induced injuries to sperm quality, the oxidant-antioxidant system, and the histologic structure of the rat testis by ellagic acid," *Fertility and Sterility*, vol. 89, no. 5, pp. 1474–1481, 2008.

[3] Z. Suo, S. J. Lippard, and K. A. Johnson, "Single d(GpG)/cis-diammineplatinum(II) adduct-induced inhibition of DNA polymerization," *Biochemistry*, vol. 38, no. 2, pp. 715–726, 1999.

[4] I. Arany and R. L. Safirstein, "Cisplatin nephrotoxicity molecular mechanisms," *Cancer Therapy*, vol. 1, pp. 47–61, 2003.

[5] S. W. Thompson, L. E. Davis, M. Kornfeld, R. D. Hilgers, and J. Standefer, "Cisplatin neuropathy. Clinical, electrophysiologic, morphologic, and toxicologic studies," *Cancer*, vol. 54, no. 7, pp. 1269–1275, 1984.

[6] F. P. T. Hamers, W. H. Gispen, and J. P. Neijt, "Neurotoxic side-effects of cisplatin," *European Journal of Cancer*, vol. 27, no. 3, pp. 372–376, 1991.

[7] H. Masuda, T. Tanaka, and U. Takahama, "Cisplatin generates superoxide anion by interaction with DNA in a cell-free system," *Biochemical and Biophysical Research Communications*, vol. 203, no. 2, pp. 1175–1180, 1994.

[8] K. Wozniak, A. Czechowska, and J. Blasiak, "Cisplatin-evoked DNA fragmentation in normal and cancer cells and its modulation by free radical scavengers and the tyrosine kinase inhibitor ST1571," *Chemico-Biological Interactions*, vol. 147, no. 3, pp. 309–318, 2004.

[9] L. M. G. Antunes, J. D. C. Darin, and M. D. L. P. Bianchi, "Protective effects of vitamin C against cisplatin-induced nephrotoxicity and lipid peroxidation in adult rats: a dose-dependent study," *Pharmacological Research*, vol. 41, no. 4, pp. 405–411, 2000.

[10] N. I. Weijl, T. J. Elsendoorn, E. G. Lentjes et al., "Supplementation with antioxidant micronutrients and chemotherapy-induced toxicity in ancer patients treated with cisplatin-based chemotherapy: a randomized, double-blind, placebo-controlled study," *European Journal of Cancer*, vol. 40, no. 11, pp. 1713–1723, 2004.

[11] S. Tofani, "Physics may help chemistry to improve medicine: a possible mechanism for anticancer activity of static and ELF magnetic fields," *Physica Medica*, vol. 15, no. 4, pp. 291–294, 1999.

[12] S. Tofani, D. Barone, M. Cintorino et al., "Tumor growth inhibition, apoptosis and loss of P53expression induced in vivo and in vitro by magnetic fields," *Proceedings of the American Association for Cancer Research*, vol. 40, p. 488, 1999.

[13] S. Tofani, D. Barone, M. Cintorino et al., "Static and ELF magnetic fields induce tumor growth inhibition and apoptosis," *Bioelectromagnetics*, vol. 22, no. 6, pp. 419–428, 2001.

[14] S. Tofani, M. Cintorino, D. Barone et al., "Increased mouse survival, tumor growth inhibition and decreased immunoreactive P53 after exposure to magnetic fields," *Bioelectromagnetics*, vol. 23, no. 3, pp. 230–238, 2002.

[15] S. Tofani, D. Barone, S. Peano, P. Ossola, F. Ronchetto, and M. Cintorino, "Anticancer activity by magnetic fields: inhibition of metastatic spread and growth in a breast cancer model," *IEEE Transactions on Plasma Science*, vol. 30, no. 4, pp. 1552–1557, 2002.

[16] National Research Council, *Guide For the Care and Use of Laboratory Animals*, National Academy Press, Washington, DC, USA, 1996.

[17] W. M. Awara, S. H. El-Nabi, and M. El-Gohary, "Assessment of vinyl chloride-induced DNA damage in lymphocytes of plastic industry workers using a single-cell gel electrophoresis technique," *Toxicology*, vol. 128, no. 1, pp. 9–16, 1998.

[18] P. Moller, L. E. Knudsen, S. Loft, and H. Wallin, "The comet assay as a rapid test in biomonitoring occupational exposure to DNA-damaging agents and effect of confounding factors," *Cancer Epidemiology Biomarkers and Prevention*, vol. 9, no. 10, pp. 1005–1015, 2000.

[19] N. P. Singh, M. T. McCoy, R. R. Tice, and E. L. Schneider, "A simple technique for quantitation of low levels of DNA damage in individual cells," *Experimental Cell Research*, vol. 175, no. 1, pp. 184–191, 1988.

[20] R. R. Tice, E. Agurell, D. Anderson et al., "Single cell gel/comet assay: guidelines for in vitro and in vivo genetic toxicology testing," *Environmental and Molecular Mutagenesis*, vol. 35, no. 3, pp. 206–221, 2000.

[21] W. Schmidt, "The micronucleus test for cytogenetic analysis," in *Chemical Mutagens, Principles and Methods For Their Detection*, A. Hollaender, Ed., pp. 31–53, Plenum Press, New York, NY, USA, 1976.

[22] D. K. Agarwal and L. K. S. Chauhan, "An improved chemical substitute for fetal calf serum for the micronucleus test," *Biotechnic and Histochemistry*, vol. 68, no. 4, pp. 187–188, 1993.

[23] I. D. Adler, "Cytogenetic tests in mammals," in *Mutagenicity Testing, A Practical Approach*, S. Venitt and J. M. Parry, Eds., pp. 275–306, IRL Press, Oxoford, UK, 1984.

[24] S. C. Sweetman, *Antineoplastic and Immunosuppressant. The Complete Drug Reference*, Pharmaceutical Press, London, UK, 33rd edition, 2002.

[25] A. Zicca, S. Cafaggi, M. A. Mariggio et al., "Reduction of cisplatin hepatotoxicity by procainamide hydrochloride in rats," *European Journal of Pharmacology*, vol. 442, no. 3, pp. 265–272, 2002.

[26] E. C. Friedberg, G. C. Walker, and W. Siede, "Cross-linking agents," in *DNA, Repair and Mutagenesis*, pp. 33–42, ASM Press, Washington, DC, USA, 1995.

[27] R. C. Choudhury and M. B. Jagdale, "Vitamin E protection from/potentiation of the cytogenetic toxicity of cisplatin in swiss mice," *Journal of Chemotherapy*, vol. 14, no. 4, pp. 397–405, 2002.

[28] M. Yoshida, A. Fakuda, M. Hara, A. Terada, Y. Kitanaka, and S. Owada, "Melatonin prevents the increase in hydroxyl radical-spin trap adduct formation caused by the addition of cisplatin in vitro," *Life Sciences*, vol. 72, no. 15, pp. 1773–1780, 2003.

[29] R. Zhang, Y. Niu, and Y. Zhou, "Increased the cisplatin cytotoxicity and cisplatin-induced DNA damage in HepG2 cells by XXRCCI abrogation related mechanisms," *Toxicology Letters*, vol. 192, no. 2, pp. 108–114, 2010.

[30] B. S. De Martinis and M. D. L. P. Bianchi, "Effect of vitamin C supplementation against cisplatin-induced toxicity and oxidative DNA damage in rats," *Pharmacological Research*, vol. 44, no. 4, pp. 317–320, 2001.

[31] B. J. Chang, M. Nishikawa, E. Sato, K. Utsumi, and M. Inoue, "L-Carnitine inhibits cisplatin-induced injury of the kidney and small intestine," *Archives of Biochemistry and Biophysics*, vol. 405, no. 1, pp. 55–64, 2002.

[32] M. Satoh, N. Kashihara, S. Fujimoto et al., "A novel free radical scavenger, edarabone, protects against cisplatin-induced acute renal damage in vitro and in vivo," *Journal of Pharmacology and Experimental Therapeutics*, vol. 305, no. 3, pp. 1183–1190, 2003.

[33] J. S. Kim, L. He, and J. J. Lemasters, "Mitochondrial permeability transition: a common pathway to necrosis and apoptosis," *Biochemical and Biophysical Research Communications*, vol. 304, no. 3, pp. 463–470, 2003.

[34] G. Kroemer and J. C. Reed, "Mitochondrial control of cell death," *Nature Medicine*, vol. 6, no. 5, pp. 513–519, 2000.

[35] S. Tofani, D. Barone, M. Berardelli et al., "Static and ELF magnetic fields enhance the in vivo anti-tumor efficacy of cis-platin against lewis lung carcinoma, but not of cyclophosphamide against B16 melanotic melanoma," *Pharmacological Research*, vol. 48, no. 1, pp. 83–90, 2003.

[36] P. Thomas, N. Holland, C. Bolognesi et al., "Buccal micronucleus cytome assay," *Nature Protocols*, vol. 4, no. 6, pp. 825–837, 2009.

[37] Y. Miyakoshi, H. Yoshioka, Y. Toyama, Y. Suzuki, and H. Shimizu, "The frequencies of micronuclei induced by cisplatin in newborn rat astrocytes are increased by 50-Hz, 7.5- and 10-mT electromagnetic fields," *Environmental Health and Preventive Medicine*, vol. 10, no. 3, pp. 138–143, 2005.

# 4

# Determination of Poisson Ratio of Bovine Extraocular Muscle by Computed X-Ray Tomography

**Hansang Kim,**[1] **Lawrence Yoo,**[2, 3] **Andrew Shin,**[2, 3] **and Joseph L. Demer**[2, 4, 5, 6]

[1] *Department of Mechanical and Automotive Engineering, Gachon University, Seongnam-Si, Gyeonggi-do 461-701, Republic of Korea*
[2] *Department of Ophthalmology, Jules Stein Eye Institute, University of California, Los Angeles, CA 90095-7002, USA*
[3] *Department of Mechanical Engineering, University of California, Los Angeles, CA, USA*
[4] *Biomedical Engineering Interdepartmental Program, University of California, Los Angeles, CA, USA*
[5] *Neuroscience Interdepartmental Program, University of California, Los Angeles, CA, USA*
[6] *Department of Neurology, University of California, Los Angeles, CA, USA*

Correspondence should be addressed to Joseph L. Demer; jld@ucla.edu

Academic Editor: José M. Vilar

The Poisson ratio (PR) is a fundamental mechanical parameter that approximates the ratio of relative change in cross sectional area to tensile elongation. However, the PR of extraocular muscle (EOM) is almost never measured because of experimental constraints. The problem was overcome by determining changes in EOM dimensions using computed X-ray tomography (CT) at microscopic resolution during tensile elongation to determine transverse strain indicated by the change in cross-section. Fresh bovine EOM specimens were prepared. Specimens were clamped in a tensile fixture within a CT scanner (SkyScan, Belgium) with temperature and humidity control and stretched up to 35% of initial length. Sets of 500–800 contiguous CT images were obtained at 10-micron resolution before and after tensile loading. Digital 3D models were then built and discretized into 6–8-micron-thick elements. Changes in longitudinal thickness of each microscopic element were determined to calculate strain. Green's theorem was used to calculate areal strain in transverse directions orthogonal to the stretching direction. The mean PR from discretized 3D models for every microscopic element in 14 EOM specimens averaged 0.457 ± 0.004 (SD). The measured PR of bovine EOM is thus near the limit of incompressibility.

## 1. Introduction

Since extraocular muscles (EOMs) are manipulated mechanically during strabismus surgery to correct binocular misalignment, the mechanical properties of the EOMs should be understood in order to optimize surgical results. With increasing demand for accuracy in simulation of orbital mechanics, finite element analysis (FEA) is becoming increasingly attractive. However, casual estimation of mechanical parameters in FEA can lead to serious errors in simulation. In order to accurately determine the biomechanical properties of orbital tissues, scientists in the field have employed a variety of experimental techniques. While conventional tensile elongation tests have been performed to investigate the uniaxial force and length relationship for EOMs [1–5], micro/nano indentation has permitted measurement of the compressive modulus of other orbital tissues [6, 7]. Despite such efforts [1, 5–11], many material parameters of orbital tissues have yet to be defined. Experimental technique and the theoretical constitutive framework should be appropriate for each tested orbital tissue. For instance, triborheometry, which treats a solid from rheometric perspective, was employed for characterizing amorphous specimens such as orbital connective and fatty tissue [10]. A variety of other biomechanical methods [12] have been employed to characterize more comprehensive constitutive models for EOMs that capture their time-dependent relationships between stress and strain [1, 5, 8].

The Poisson ratio (PR) is a critical mechanical parameter required to define comprehensively the elastic behavior of a material. The PR is the ratio of the transverse contraction strain to the axial extension strain, which can be obtained during simple tensile elongation. Normally, the PR of a material ranges between 0 and 0.5, depending on the material's compressibility. However, the PR can be higher than 0.5 or lower than 0 for materials having complex matrices and inner structures [13, 14]. Most soft tissues are considered to be elastomeric materials with high bulk modulus relative to Young's modulus, so the PR is expected to approximate 0.5 [15].

In general, the PR can be measured by static or dynamic methods. Static methods, such as classical tensile or compressive testing, are most widely used in solid mechanics [15, 16]. In static determinations, the PR is calculated from transverse and axial deformations due to uniaxial stress. For dynamic determination, the PR is determined from the natural frequency of the transverse and axial waves in the material [17–19], most commonly elicited by ultrasound perturbation.

The PR has been typically assumed to be between 0.35 and 0.49 for soft tissues [20–22]. However, it has been estimated that a 20% error in the PR would result in errors of 3.8% and 4.4% in the biaxial flexural strength and the indentation modulus of a material, respectively [15]. Such errors could propagate and compound during the iterative computations in FEA. Clearly, there is a need to minimize errors by accurate experimental determination of the PR. By employing a novel X-ray computed tomographic (CT) imaging method for precise determination of strain, we aimed to extract accurate static PR for EOMs.

## 2. Materials and Methods

*2.1. Specimen Preparation.* Fresh heads of adult cows were obtained from a nearby abattoir. In the laboratory, orbits were carefully dissected for extraction of EOM and connective tissue. Transport time from abattoir to laboratory was approximately 30 min; the additional time elapsed to dissect the EOMs averaged 3 hrs. After extraction, EOMs were maintained in lactated Ringer's solution at 37°C. To minimize axial damage to EOM fibers, each specimen was initially prepared in the shape of an approximately 7 mm long prism with a 4 mm × 2 mm cross-section. For consistency, samples were prepared from the transverse center of each EOM. Given that clamping of both ends was necessary, the actual tested length was the 10 mm of middle portion of each specimen and in every case avoided the terminal tendon. Specimen preparation time was approximately 45 min.

*2.2. Experiment.* A high-resolution micro-CT scanner (Model 1172, SkyScan, Belgium) incorporating a tensile loading fixture was used to image deformed and undeformed states of 14 freshly prepared bovine EOM specimens. Scanning was accomplished by revolving and longitudinally translating the specimen and tensile fixture between a fixed, collimated X-ray source and a fixed detector (Figure 1). Once loaded in the tensile fixture, the undeformed specimen was first imaged at 10 micron spatial resolution. After the specimen was elongated 30%–35%, which is well within the linear elastic region [1], the deformed specimen was again

imaged. In order to prevent dehydration, corn oil was applied on each specimen before placement into the tensile fixture. Figure 1 shows schematics of undeformed and deformed states of the specimen in the tensile fixture.

*2.3. 3D Reconstruction of EOMs.* After EOM specimens were scanned, 500 cross-sectional area (CSA) images for undeformed and 800 CSA images for deformed states were used to create 3D reconstruction of the entire length of each EOM specimen using Matlab (Version R2010a, The MathWorks, Inc., Massachusetts) image processing tools and SolidWorks CAD software (version 2011, Dassault Systèmes SolidWorks Corp., Massachusetts). More deformed than undeformed image planes were required since the deformed specimen was elongated. The CSA in each image plane was connected to that in the adjacent planes in order to generate 3D reconstruction by using the loft feature in Solidworks.

*2.4. Poisson Ratio.* As shown in (1), the PR $v$, for materials undergoing deformations exceeding 1% [23], is expressed as the negative ratio of transverse to axial true strain [24, 25]:

$$v = -\frac{\ln\left(1 + \varepsilon_T\right)}{\ln\left(1 + \varepsilon_A\right)}, \tag{1}$$

where $\varepsilon_T$ and $\varepsilon_A$ are the transverse and axial engineering strains, respectively. Equation (1) can be rearranged and can be expressed as shown in (2):

$$\ln\left(1 + \varepsilon_A\right)^{-v} = \ln\left(1 + \varepsilon_T\right). \tag{2}$$

Recognizing the fact that an infinitesimal element in the CSA undergoes a plane deformation when the EOM is subjected to loading, the transverse strain in (2) can be shown as $\ln(\delta x/\delta x_o)$ and (2) can be rearranged as (3) with $\delta x_o$ and $\delta x$ being the length of side of the element in undeformed and deformed configuration:

$$\delta x = \delta x_o (1 + \varepsilon_A)^{-v}. \tag{3}$$

Equation (3) is valid under the assumption of isotropy or transverse isotropy, which is appropriate for an EOM [25]. Hence the CSA of the square in the deformed configuration can be expressed as (4) where $\delta A_0$ is the initial CSA:

$$\delta A = \delta x_o{}^2 (1 + \varepsilon_A)^{-2v} = \delta A_o (1 + \varepsilon_A)^{-2v}. \tag{4}$$

As reported by Vergari et al. [25], the instantaneous CSA can be expressed as (5) by summation of all the elements in the EOM CSA:

$$A = A_o (1 + \varepsilon_A)^{-2v}, \tag{5}$$

where $A$ and $A_o$ are the instantaneous and initial EOM CSA values, respectively. The CT scanner employed was specifically designed to image CSAs suitable for 3D reconstruction. After 3D reconstruction for both deformed and undeformed states, each model was then uniformly discretized into

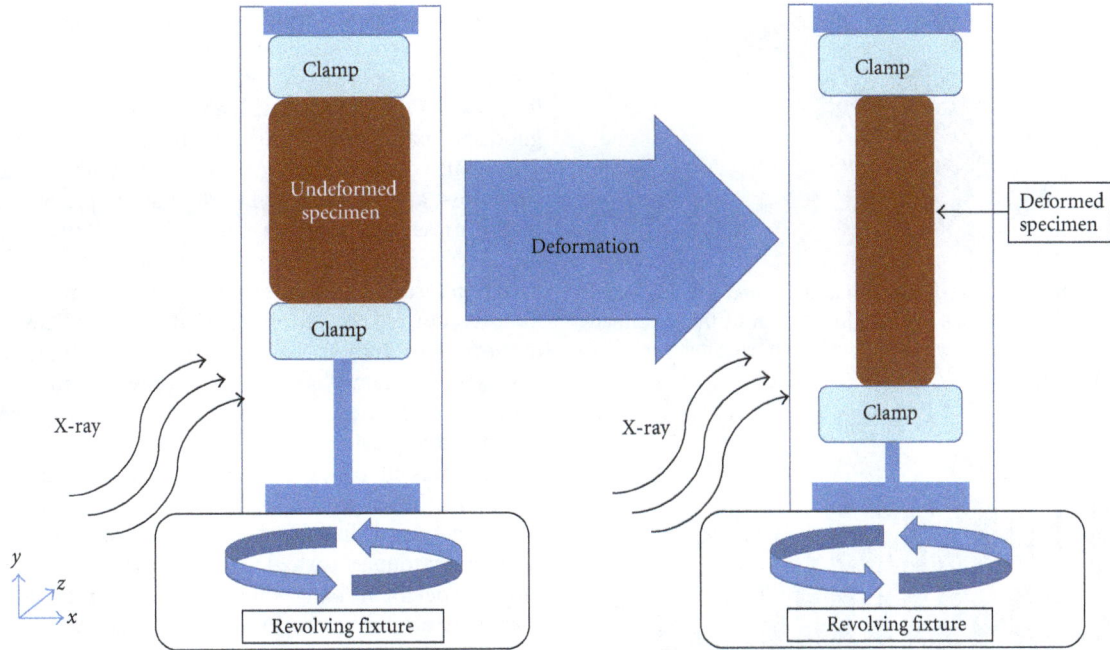

FIGURE 1: Both undeformed and deformed states of EOM specimens were imaged. As shown above, the X-ray direction was orthogonal to the revolving specimen fixture.

(a)                                                      (b)

FIGURE 2: (a) Decimated set of CSA images for an undeformed EOM specimen. (b) Finished 3D reconstruction from 500 CSAs.

8 micron thick elements. The CSA for each element was then computed using Green's theorem as shown in (6):

$$A = \oint_C x\,dy = -\oint_C y\,dx = \frac{1}{2}\int_C \left(-y\,dx + x\,dy\right). \quad (6)$$

Finally, the PR for each discretized element was calculated using (7):

$$\nu = -0.5 \frac{\ln\left(A/A_0\right)}{\ln\left(1 + \varepsilon_A\right)}. \quad (7)$$

For more precise evaluation of PR, the CAD surface reconstruction excluded regions near clamping plates that are influenced by the clamping forces. After the PR was computed for all the elements within each specimen using (7), the average PR was calculated for each of the 14 specimens tested.

## 3. Results

*3.1. 3D Reconstruction.* Figure 2 shows the 3D reconstruction from 500 CSAs (Figure 2(a)) used to build the undeformed model of an EOM specimen (Figure 2(b)).

Once 3D reconstruction was completed for each specimen for both undeformed and deformed states, the models were discretized into elements with uniform thickness of 6 microns and 8.1 microns, respectively. During preliminary experiments it was verified that all 6 anatomical EOMs exhibited similar PR values. Thus PR values for 6 EOMs were not differentiated by anatomical EOM. Figure 3 contrasts undeformed and deformed states of the same EOM specimen.

Assuming uniform stretch throughout each specimen, the PR for each element within each EOM specimen was

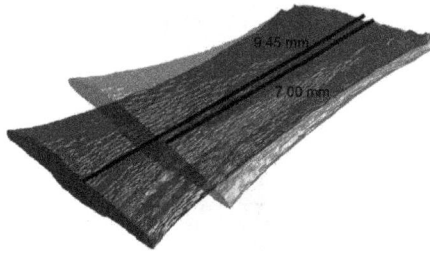

FIGURE 3: Finished 3D models for both undeformed and deformed states of an EOM specimen. The original length of the specimen was 7 mm, and the final length of the specimen after the 35% deformation was 9.45 mm.

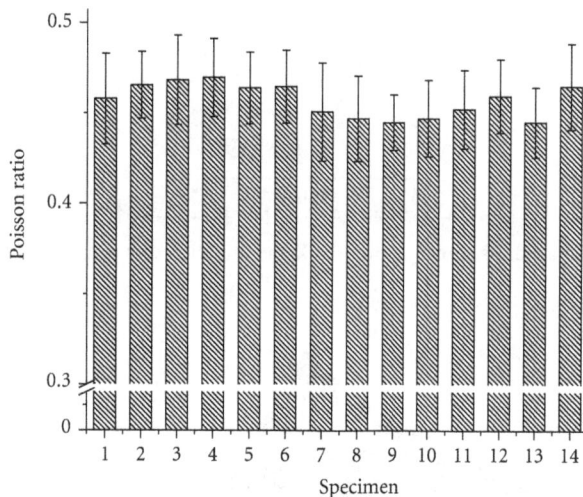

FIGURE 4: PR values for all 14 specimens. Error bars indicate SD.

computed from the CSAs and actual elongated length. Figure 4 shows the average PR values for each of the 14 specimens.

The PR over all 14 specimens averaged $0.457 \pm 0.004$ (standard deviation (SD)) was highly significantly different by Student's $t$-test from the ideal incompressible value 0.5 ($P < 10^{-9}$).

## 4. Discussion

Micro-CT imaging of EOM specimens effectively characterized the PR for bovine EOM during tensile elongation, which is a critical mechanical parameter. The present investigation is the first to evaluate bovine EOM PR directly from CSA measurements by noncontact imaging during large deformation. Prior studies were based on measurement of small, linear transverse deformations [26, 27]. In the present investigation, mean PR of 14 bovine EOM was computed to be $0.457 \pm 0.004$ (SD). While isotropic media cannot have PRs exceeding 0.5, orthotropic or transversely isotropic materials (such as tendons) sometimes have PRs exceeding 1 [25]. A PR value approximating 0.5 has interesting implications for tendon behavior: assuming conservation of mass, a PR value smaller than 0.5 implies an increase in volume during loading leading to a decrease in density.

On the other hand, a PR exceeding 0.5 implies a volume reduction and an increase in density. Still, the mean PR for all EOM specimens tested in the present study was 0.457, so EOM under axial loading can be considered to a good approximation to behave as an incompressible material. The small volume variations measured might be caused by water loss, as suggested by Lynch et al. [27]. However, since in the present experiment corn oil was employed to coat the specimen to avoid dehydration, measured small volume variations are probably due to internal rearrangements of the fiber structures [25]. The precise PR value for EOM reported in the present study should facilitate quantitative modeling of ocular motor biomechanics and is represented in a theoretical framework practical for graphical simulation of quasistatic ocular motility using FEM.

The present paper introduces micro-CT imaging as a noncontacting approach to compute the PR from specimen geometry. As it has been presented, micro-CT imaging technology coupled with 3D reconstruction based on the volumetric specimen changes allows more accurate evaluation of Poisson's ratio for bovine EOM specimens.

## 5. Conclusion

The current investigation describes a method to determine the PR for bovine EOM based upon CSAs from discretized deformed and undeformed specimens obtained from micro-CT imaging during quasistatic loading. The study also demonstrated that 3D reconstruction of micro CT imaging can be successfully performed for both deformed and undeformed EOM specimens. In the present study, the PR, a critical parameter for quantitative mechanical characterization of soft tissues that is necessary for modeling and simulation, was determined for bovine EOM specimens to be near the criterion for incompressibility.

## Conflict of Interests

The authors declare that none of them has any financial or personal relationships with people of organization that can inappropriately influence his work or the conclusions drawn from this investigation.

## Acknowledgments

The authors acknowledge Manning Beef, LLC, Pico Rivera, CA, for their generous contribution of bovine specimens. They also thank Jose Martinez, Claudia Tamayo, and Ramiro Carlos of Manning Beef for assistance with specimen preparation. This paper is supported by U.S. Public Health Service, National Eye Institute: Grants EY08313 and EY00331, and Research to Prevent Blindness. J. L. Demer is Leonard Apt Professor of ophthalmology.

## References

[1] L. Yoo, H. Kim, V. Gupta, and J. L. Demer, "Quasilinear viscoelastic behavior of bovine extraocular muscle tissue,"

*Investigative Ophthalmology and Visual Science*, vol. 50, no. 8, pp. 3721–3728, 2009.

[2] D. A. Robinson, D. M. O'Meara, A. B. Scott, and C. C. Collins, "Mechanical components of human eye movements," *Journal of Applied Physiology*, vol. 26, no. 5, pp. 548–553, 1969.

[3] C. C. Collins, M. R. Carlson, A. B. Scott, and A. Jampolsky, "Extraocular muscle forces in normal human subjects," *Investigative Ophthalmology and Visual Science*, vol. 20, no. 5, pp. 652–664, 1981.

[4] H. J. Simonsz, "Force-length recording of eye muscles during local anesthesia surgery in 32 strabismus patients," *Strabismus*, vol. 2, pp. 197–218, 1994.

[5] C. Quaia, H. S. Ying, A. M. Nichols, and L. M. Optican, "The viscoelastic properties of passive eye muscle in primates: I: static and step responses," *PLoS ONE*, vol. 4, no. 4, Article ID e4850, 2009.

[6] L. Yoo, J. Reed, J. K. Gimzewski, and J. L. Demer, "Mechanical interferometry imaging for creep modeling of the cornea," *Investigative Ophthalmology and Visual Science*, vol. 52, no. 11, pp. 8420–8424, 2011.

[7] L. Yoo, J. Reed, A. Shin et al., "Characterization of ocular tissues using microindentation and Hertzian viscoelastic models," *Investigative Ophthalmology and Visual Science*, vol. 52, no. 6, pp. 3475–3482, 2011.

[8] C. Quaia, H. S. Ying, and L. M. Optican, "The viscoelastic properties of passive eye muscle in primates. II: testing the quasi-linear theory," *PLoS ONE*, vol. 4, no. 8, Article ID e6480, 2009.

[9] L. Yoo, H. Kim, A. Shin, V. Gupta, and J. L. Demer, "Creep behavior of passive bovine extraocular muscle," *Journal of Biomedicine and Biotechnology*, vol. 2011, Article ID 526705, 2011.

[10] L. Yoo, V. Gupta, C. Lee, P. Kavehpore, and J. L. Demer, "Viscoelastic properties of bovine orbital connective tissue and fat: constitutive models," *Biomechanics and Modeling in Mechanobiology*, vol. 10, no. 6, pp. 901–914, 2011.

[11] B. L. Boyce, R. E. Jones, T. D. Nguyen, and J. M. Grazier, "Stress-controlled viscoelastic tensile response of bovine cornea," *Journal of Biomechanics*, vol. 40, no. 11, pp. 2367–2376, 2007.

[12] Y. C. Fung, *Biomechanics: Mechanical Properties of Living Tissues*, Springer, New York, NY, USA, 1993.

[13] R. Lakes, "Advances in negative poisson's ratio materials," *Advanced Materials*, vol. 5, no. 4, pp. 293–296, 1993.

[14] U. D. Larsen, O. Sigmund, and S. Bouwstra, "Design and fabrication of compliant micromechanisms and structures with negative Poisson's ratio," in *Proceedings of the 9th Annual International Workshop on Micro Electro Mechanical Systems (MEMS '96)*, An Investigation of Micro Structures, Sensors, Actuators, Machines and Systems, pp. 365–371, February 1996.

[15] S. M. Chung, A. U. J. Yap, W. K. Koh, K. T. Tsai, and C. T. Lim, "Measurement of Poisson's ratio of dental composite restorative materials," *Biomaterials*, vol. 25, no. 13, pp. 2455–2460, 2004.

[16] W. Wu, K. Sadeghipour, K. Boberick, and G. Baran, "Predictive modeling of elastic properties of particulate-reinforced composites," *Materials Science and Engineering A*, vol. 332, no. 1-2, pp. 362–370, 2002.

[17] M. Pamenius and N. G. Ohlson, "The determination of elastic constants by dynamic experiments," *Dental Materials*, vol. 2, no. 6, pp. 246–250, 1986.

[18] S. A. M. Spinner, "Elastic moduli of glasses at elevated temperatures by a dynamic method," *Journal of the American Ceramic Society*, vol. 39, pp. 113–118, 1956.

[19] M. P. D'Evelyn and T. Taniguchi, "Elastic properties of translucent polycrystalline cubic boron nitride as characterized by the dynamic resonance method," *Diamond and Related Materials*, vol. 8, no. 8-9, pp. 1522–1526, 1999.

[20] M. Zhang, Y. P. Zheng, and A. F. T. Mak, "Estimating the effective Young's modulus of soft tissues from indentation tests—nonlinear finite element analysis of effects of friction and large deformation," *Medical Engineering and Physics*, vol. 19, no. 6, pp. 512–517, 1997.

[21] S. P. W. Van Den Bedem, S. Schutte, F. C. T. Van Der Helm, and H. J. Simonsz, "Mechanical properties and functional importance of pulley bands or 'faisseaux tendineux'," *Vision Research*, vol. 45, no. 20, pp. 2710–2714, 2005.

[22] C. Sumi, A. Suzuki, and K. Nakayama, "Estimation of shear modulus distribution in soft tissue from strain distribution," *IEEE Transactions on Biomedical Engineering*, vol. 42, no. 2, pp. 193–202, 1995.

[23] S. P. Reese, S. A. Maas, and J. A. Weiss, "Micromechanical models of helical superstructures in ligament and tendon fibers predict large Poisson's ratios," *Journal of Biomechanics*, vol. 43, no. 7, pp. 1394–1400, 2010.

[24] C. W. Smith, R. J. Wootton, and K. E. Evans, "Interpretation of experimental data for Poisson's ratio of highly nonlinear materials," *Experimental Mechanics*, vol. 39, no. 4, pp. 356–362, 1999.

[25] C. Vergari, P. Pourcelot, L. Holden et al., "True stress and Poisson's ratio of tendons during loading," *Journal of Biomechanics*, vol. 44, no. 4, pp. 719–724, 2011.

[26] V. W. T. Cheng and H. R. C. Screen, "The micro-structural strain response of tendon," *Journal of Materials Science*, vol. 42, no. 21, pp. 8957–8965, 2007.

[27] H. A. Lynch, W. Johannessen, J. P. Wu, A. Jawa, and D. M. Elliott, "Effect of fiber orientation and strain rate on the nonlinear uniaxial tensile material Properties of Tendon," *Journal of Biomechanical Engineering*, vol. 125, no. 5, pp. 726–731, 2003.

# 5

# Repeated Bout Effect Was More Expressed in Young Adult Males Than in Elderly Males and Boys

Giedrius Gorianovas,[1] Albertas Skurvydas,[1] Vytautas Streckis,[1] Marius Brazaitis,[1] Sigitas Kamandulis,[1] and Malachy P. McHugh[2]

[1] Department of Applied Biology and Physiotherapy, Research Centre for Fundamental and Clinical Movement Sciences, Lithuanian Sports University, Sporto 6, 4422 Kaunas, Lithuania
[2] Nicholas Institute of Sports Medicine and Athletic Trauma, Lenox Hill Hospital, 130 East 77th Street, New York, NY 10075, USA

Correspondence should be addressed to Marius Brazaitis; marius_brazaitis@yahoo.com

Academic Editor: Giuseppe Spinella

This study investigated possible differences using the same stretch-shortening exercise (SSE) protocol on generally accepted monitoring markers (dependent variables: changes in creatine kinase, muscle soreness, and voluntary and electrically evoked torque) in males across three lifespan stages (childhood versus adulthood versus old age). The protocol consisted of 100 intermittent (30 s interval between jumps) drop jumps to determine the repeated bout effect (RBE) (first and second bouts performed at a 2-week interval). The results showed that indirect symptoms of exercise-induced muscle damage after SSE were more expressed in adult males than in boys and elderly males, suggesting that the muscles of boys and elderly males are more resistant to exercise-induced damage than those of adult males. RBE was more pronounced in adult males than in boys and elderly males, suggesting that the muscles of boys and elderly males are less adaptive to exercise-induced muscle damage than those of adult males.

## 1. Introduction

Although a large amount of research relates to exercise-induced muscle damage in adult subjects [1–7], very little research has assessed the differences in fatigue between children, adults, and elderly persons when doing the same exercise protocol—one that causes exercise-induced muscle damage [8–13]. In addition, there is a lack of research regarding the effect of age on repeated bout effect (RBE) [11, 12]. We found no study in which the effects of age (boys versus adults versus elderly) on RBE were assessed using the same exercise protocol. Also, we did not come across any special studies devoted to analysis of the effect of age on various indirect symptoms of muscle damage (i.e., creatine kinase (CK) levels, muscle soreness, and especially both voluntary (including muscle voluntary activation index) and electrically induced muscle force at different frequencies after exercise that cause mechanical muscle damage). Therefore, the effect of the same exercise-induced muscle damage protocol on RBE indicators in males across three lifespan stages

(i.e., childhood, adulthood, and old age) remains unclear, especially as regards accepted monitoring markers (i.e., CK, muscle soreness, decrease in electrically induced torque, and maximal voluntary contraction force of quadriceps muscle (MVC)). However, it has been well documented that the muscles of adults contain relatively more fast twitch muscle fibers (type II) than do the muscles of children and elderly males [14–16]; therefore, we hypothesize that muscles of children and elderly males might be more resistant to exercise-induced muscle damage than those of adult males and that RBE might be more expressed in subjects whose muscles are more sensitive to muscle damage, that is, in young adult males than in boys and elderly males. Therefore, the main purpose of this study was to test the hypothesis.

## 2. Materials and Methods

2.1. Participants. Healthy untrained boys (age 11.8 ± 0.9 years, body mass 40.2 ± 7.9 kg, height 149.3 ± 8.3 cm, and body mass index (BMI) $17.7 \pm 1.6$, $n = 11$), young adults (age $20.8 \pm 1.9$ years, body mass $82.6 \pm 10.9$ kg, height $181.9 \pm 5.9$, BMI $25.0 \pm 2.6$, $n = 11$), and elderly males (age $63.2 \pm 3.6$ years, body mass $78.4 \pm 13.9$ kg, height $176.0 \pm 6.5$, BMI $23.4 \pm 3.9$, $n = 11$) gave their informed consent to take part in the experiment within the study. The subjects were physically active but did not take part in any formal physical exercise or sport program. Each subject read and signed written informed consent form consistent with the principles outlined in the Declaration of Helsinki. This study was approved by the Ethics Committee of Kaunas University of Medicine.

*2.2. Experimental Protocol.* Three to five days before the experiment, subjects were familiarized with electrical stimulation and different tasks of voluntary performance. On the day of the experiment, after measurement of CK in the blood, the subjects completed warm-up exercises that consisted of 5 min of running on the spot at an intensity that corresponded to heart rates of 110–130 beats·min$^{-1}$. This was followed by 10 squat-stands. After their warm-up, a subject was seated in the experimental chair and after 5 min muscle contractile properties were recorded in the following sequence: P20, P100, and MVC. Quadriceps muscle voluntary activation index was registered during MVC. About 3 min later, the SSE was undertaken. Two to five min and 48 h after SSE, recordings were made of the same contractile properties, both voluntary and electrostimulation-evoked muscle contraction properties. At 24 h and 48 h after SSE, muscle soreness and CK activity were determined. The experimental protocol was performed twice, first under control conditions (*bout* 1) and then 2 weeks later (*bout* 2).

*2.3. Muscle Damaging Stretch-Shortening Exercise (SSE).* The subjects performed 100 intermittent (30 s interval between the jumps) drop jumps (DJs) from the height of 0.5 m with countermovement to $90 \pm 3$ degrees angle in the knee and immediate maximal rebound. During the jumps hands of the subjects were on the waist. The subject stepped on 0.5 m high platform with his left leg, that is, the leg in which muscle contraction force was not tested. After each jump the subjects were informed of the height and knee joint angle of the jump and were motivated to perform each jump as high as possible. Knee joint angle during drop jumps was controlled using electronic goniometer (Biometrics Ltd. Model SG150, UK) connected to Angle Display Unit device (Biometrics Ltd. Model ADU301, UK). Before SSE and during recovery control drop jumps were performed with the same techniques as during SSE. Height of the DJ was calculated by an earlier technique applying the following formula: $h$ (cm) $= g \times t^2/8$, where $h$ = height of the drop jump, $g$ = acceleration of gravity ($9.81$ m $\times$ s$^{-2}$), and $t$ = flight time (s) [17]. SSE was performed making use of the multicomponent Kistler force plate (type 9286A, USA). A similar research protocol was applied in previous research [6, 18, 19].

*2.4. Isometric Torque and Electrical Stimulation.* The isometric torque of knee extensor muscles was measured using an isokinetic dynamometer (System 3; Biodex Medical Systems, Shimley, New York). The sensitivity of the Biodex System 3 in torque measurements is $\pm 1.36$ Nm. The subjects sat upright in the dynamometer chair with the knee joint positioned at 120 degrees angle (180 degrees—knee full extension). The equipment and procedure for electrical stimulation were essentially the same as previously described [4–6]. Direct muscle stimulation was applied using two carbonized rubber electrodes, covered with a thin layer of electrode gel (ECG-EEG Gel; Medigel, Modi'in, Israel). One of the electrodes ($6 \times 11$ cm) was placed transversely across the width of the proximal portion of the quadriceps femoris. Another electrode ($6 \times 20$ cm) covered the distal portion of the muscle above the patella. A standard electrical stimulator (MG 440; Medicor, Budapest, Hungary) was used. The electrical stimulation was delivered in square-wave pulses, 0.5 ms in duration. The tolerance of volunteers to electrical stimulation was assessed on a separate occasion. All participants showed good compliance with the procedure and were recruited for the study. The intensity of electrical stimulation was selected individually by applying single stimuli to the muscles tested. During this procedure, the current was increased until no increment in single twitch torque could be detected by an additional 10% increase in current strength. The output from the force transducer was also displayed on a voltmeter in front of the subject.

The following data were measured: the torque at 20 Hz (P20) and 100 Hz (P100) of electrical stimulation using 1 strains of stimuli at each frequency, respectively. MVC at the knee angles of 120 degrees, peak of the MVC was, reached and maintained some 5 seconds before relaxation, twice. The rest interval between muscle electrostimulations was 10 s and between MVC measurements it was 2 min. The change in the ratio of P20/P100 after exercise was used for the evaluation of low frequency fatigue (LFF) [4, 7, 20].

*2.5. Voluntary Activation Index (VA) Measurement.* The volunteer was positioned in the dynamometer chair, and the stimulating electrodes were placed on the right leg. After a 5 min rest, two 5 s MVC readings separated by a 2 min rest interval were obtained. After ~3 s of MVC, a 250 ms test train of stimuli at 100 Hz (TT-100 Hz) was superimposed on the voluntary contraction. The TT-100 Hz was repeated 1-2 s after the MVC. TT-100 Hz contractions were used to assess voluntary activation of knee extensors. The amplitude of the superimposed tetani was calculated relative to the baseline, which was defined as the average torque over a period of 1 s immediately before stimulation. The superimposed TT-100 Hz produced measurable torque increments in all subjects. For the VA, the TT-100 Hz torque of the relaxed muscles was used as the control torque, and the following formula was applied: VA (%) = 1 – (superimposed TT-100 Hz torque/control TT-100 Hz torque) × 100 per cent [4, 19, 21].

*2.6. Plasma Creatine Kinase (CK) Activity.* Approximately 5 mL of blood was drawn from *vena cubiti media* of the arm at each measurement time point (before exercise as well as 24 h and 48 h after exercise). Plasma samples were

TABLE 1: Creatine kinase activity (CK) changes after *bout* 1 and *bout* 2 in boys, young adult males, and elderly males (mean ± SD).

| CK, IU/L | Before | 24 h after exercise | 48 h after exercise |
|---|---|---|---|
| Boys | | | |
| *bout* 1 | 78.8 ± 50.7 | 422.5 ± 211.5*# | 311.8 ± 236.2*# |
| *bout* 2 | 75.5 ± 46.1 | 132.8 ± 68.2* | 102.8 ± 55.4 |
| Young adult | | | |
| *bout* 1 | 106.0 ± 38.9 | 1102.2 ± 640.9*#$ | 1090.2 ± 784.7*#$ |
| *bout* 2 | 90.4 ± 40.8 | 401.9 ± 182.5*$ | 194.2 ± 76.7* |
| Elderly | | | |
| *bout* 1 | 82.7 ± 24.2 | 224.6 ± 113.6* | 340.8 ± 149.8*# |
| *bout* 2 | 92.5 ± 47.5 | 164.7 ± 81.5* | 144.5 ± 55.3* |

$^*P < 0.05$, compared with before; $^#P < 0.05$, compared with *bout* 1; $^$P < 0.05$, compared with elderly males and boys.

pipetted into microcentrifuge tubes and stored in a $-20°C$ freezer until analysis. Plasma creatine kinase activity (IU/L) was determined by using automatic biochemical analyzer "Monarch" (Instrumentation Laboratory SpA, USA-Italy). The normal reference range for men for CK using this method is between 24 and $195 \, IU \cdot L^{-1}$ according to the manual provided with the analyzer.

*2.7. Muscle Soreness.* Muscle soreness was reported subjectively using a visual analogue scale from 0 to 10 points. Each number on the scale has descriptive words for soreness: 0 (none), 1 (very slight), 2 (slight), 3 (mild), 4 (less than moderate), 5 (moderate), 6 (more than moderate), 7 (intense), 8 (very intense), 9 (barely tolerable), and 10 (intolerably intense). The participants were required to indicate the severity of soreness in their quadriceps during 2-3 squats at 24 and 48 h after *bouts* 1 and 2 [4–6].

*2.8. Statistical Analyses.* The Kolmogorov-Smirnov test confirmed that all data were normally distributed. Three-way analysis of variance (ANOVA) for repeated measures was used to determine the effect of age (three groups), time (before, 2–5 min after and 48 h after exercise), and RBE (*bout* 1 and *bout* 2) on the variables. Significant differences were subjected to post hoc testing using paired $t$-tests with Bonferroni correction for multiple comparisons. Descriptive data are presented as the mean ± SD. The level of significance was set at 0.05. The statistical powers (SP) for all mechanical indicators were calculated using an alpha level of 0.05, a sample size of 11, the standard deviation, and the average value before and after SSE.

# 3. Results

*3.1. Stretch-Shortening Exercise.* The control (before SSE) $H$ of DJs was significantly greater in young adults than in boys and elderly males ($P < 0.001$; SP > 80%); however, in boys it was significantly greater than in elderly males ($P < 0.05$; SP = 45.4%) (Figure 1). Average $H$ of DJs during 100 DJs in boys, young adult males, and elderly males was 23.4 ± 3.4 cm, 31.9 ± 4.2 cm, and 19.1 ± 3.3 cm, respectively, during *bout* 1 ($P < 0.01$, compared between groups, SP > 80%), and 24.3 ± 3.9 cm, 32.5 ± 4.2 cm, and 19.9 ± 1.5 cm during

*bout* 2 ($P < 0.01$, compared between groups, SP > 80%; $P > 0.05$ compared between *bout* 1 and *bout* 2, SP > 30%). The average contact time of 100 DJs in boys, young adult males, and elderly males was 0.58 ± 0.06 s, 0.63 ± 0.09 s and 0.86 ± 0.09 s, respectively, during *bout* 1 ($P < 0.05$, compared between groups, SP > 30%), and 0.57 ± 0.06 s, 0.64 ± 0.09 s, and 0.82 ± 0.1 during *bout* 2 ($P < 0.05$, compared between groups, SP > 30%; $P > 0.05$ compared between *bout* 1 and *bout* 2, SP > 35%). Thus, there was no significant difference in SSE loads between *bout* 1 and *bout* 2 in all groups.

*3.2. Voluntary and Electrically Induced Muscle Performance before bout 1 and bout 2.* There were no significant differences for all registered parameters before *bout* 1 and *bout* 2 in all groups (Figure 2). Preexercise electrically induced (P20 and P100) and voluntary (MVC) knee torques were significantly greater in young adults than in elderly males and boys ($P < 0.001$; SP > 80%); however, in elderly males they were significantly greater than in boys ($P < 0.01$) (Table 1). The VA index and P20/P100 of boys were significantly lower than in elderly males ($P < 0.01$; SP = 35.4%).

*3.3. Differences in RBE between Boys, Young Adult Males, and Elderly Males: Immediately after SSE.* RBE manifested immediately after SSE in all voluntary and electrically induced muscle performance in young adult males, while in boys and elderly males RBE was evident only in $H$ of DJs (Figure 2). Also, the contact time of DJs did not change significantly after SSE in all groups. Electrically induced muscle force P20 ($P < 0.001$; SP > 80%), P100 ($P < 0.001$; SP > 80%), P20/P100 ($P < 0.001$; SP > 80%), and MVC ($P < 0.001$; SP > 80%) decreased significantly in all groups after *bout* 1 and *bout* 2, whereas $H$ of DJs decreased significantly ($P < 0.05$; SP = 49.5%) after *bout* 1 in all groups and VA ($P < 0.05$) only in young adults after *bout* 1. It is of interest that changes in P20/P100 were significantly less in elderly males compared with young adults and boys ($P < 0.05$). Thus, RBE in changes of voluntary and electrically induced muscle force immediately after SSE was more pronounced in young adults.

*3.4. Differences in RBE between Boys, Young Adult Males, and Elderly Males: 48 h after SSE.* Electrically (P20, P100, and P20/P100) and voluntary induced (MVC, $H$ of DJs) muscle

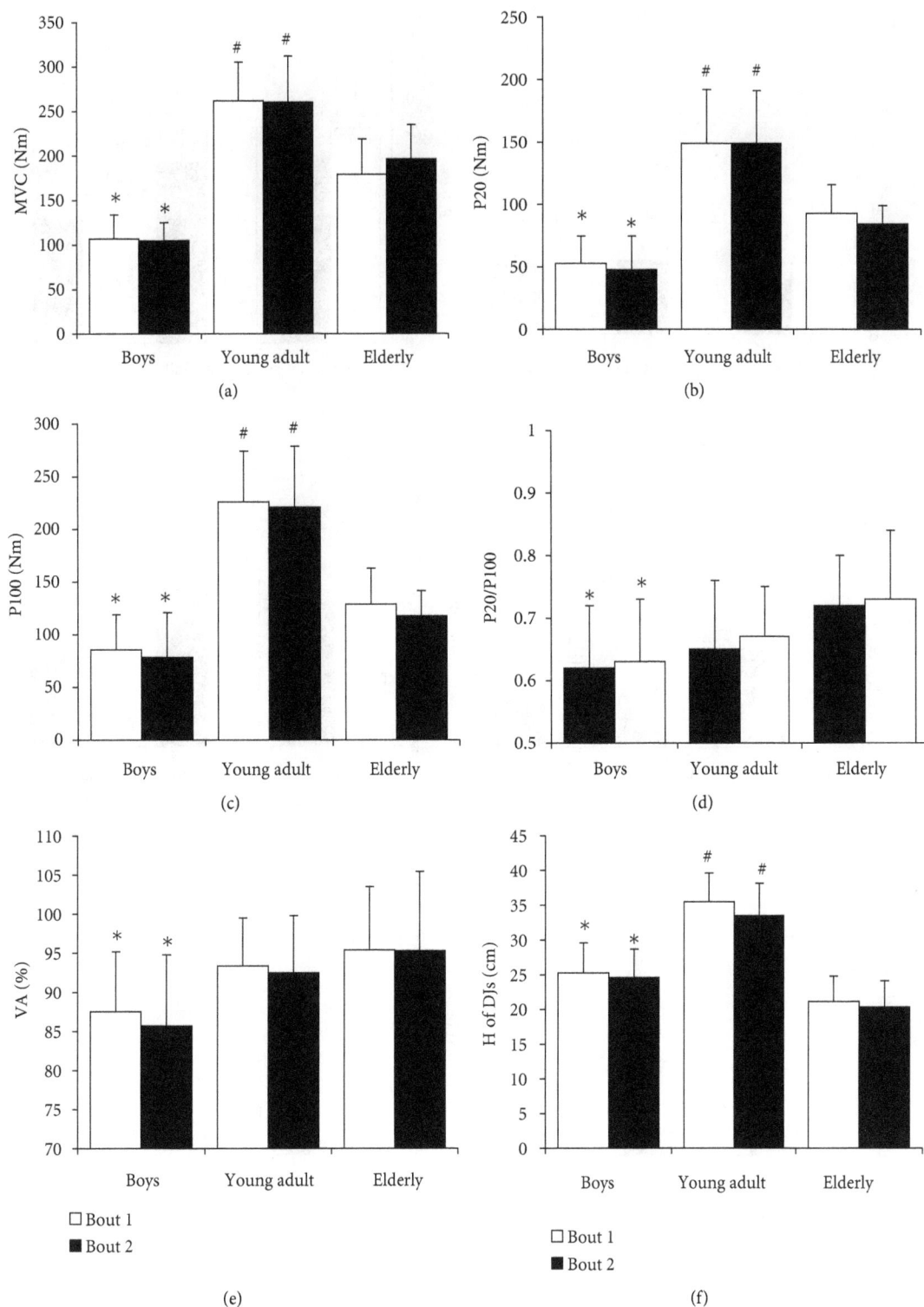

FIGURE 1: Voluntary (MVC (a) maximal voluntary contraction force of quadriceps muscle; $H$ of DJs (f) height of drop jumps; VA (e) quadriceps muscle voluntary activation index) and electrically (P20 (b) and P100 (c) muscle contraction force evoked by stimulating quadriceps muscle at 20 Hz and 100 Hz frequencies, resp.; P20/P100 (d) ratio of P20/P100 indicates the level of low frequency fatigue) induced muscle performance before *bout* 1 and *bout* 2 in boys ($n = 11$), young adult males ($n = 11$), and elderly males ($n = 11$) (mean ± SD). *$P < 0.01$, between elderly males and boys; #$P < 0.001$, compared with young adult males and boys.

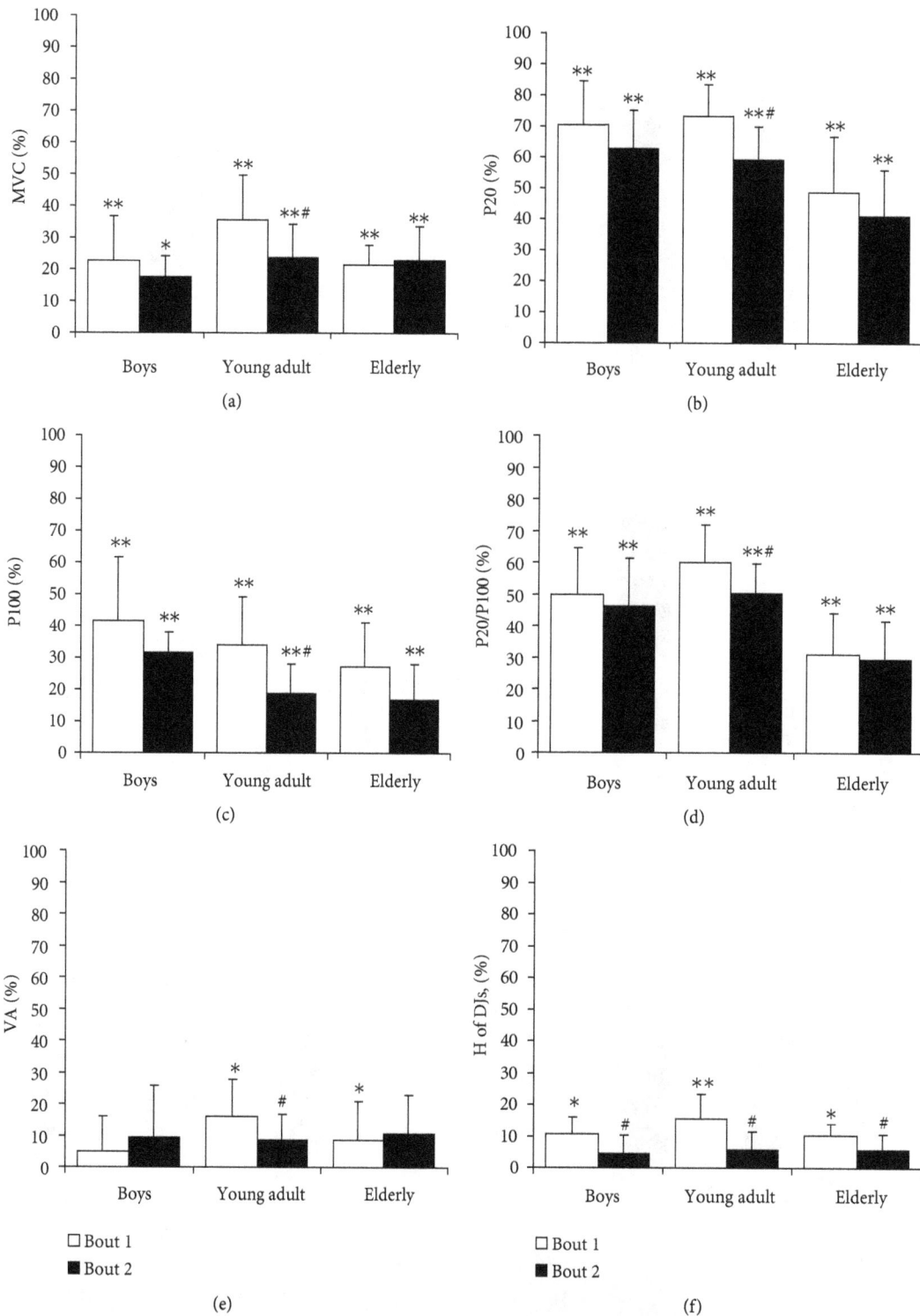

FIGURE 2: Voluntary (MVC (a) maximal voluntary contraction force of quadriceps muscle; $H$ of DJs (f) height of drop jumps; VA (e) quadriceps muscle voluntary activation index) and electrically (P20 (b) and P100 (c) muscle contraction force evoked by stimulating quadriceps muscle at 20 Hz and 100 Hz frequencies, resp.; P20/P100 (d) ratio of P20/P100 indicates the level of low frequency fatigue) induced muscle performance after *bout* 1 and *bout* 2 in boys ($n = 11$), young adult males ($n = 11$), and elderly males ($n = 11$) (mean ± SD). [*,**] $P < 0.05$ and $P < 0.001$, compared with the before-exercise level; [#] $P < 0.05$, compared with *bout* 1.

performance did not recover within 48 h after *bout* 1 in all groups (Figure 3). Voluntarily muscle performance (MVC and *H* of DJs) fully recovered within 48 h after *bout* 2 in all groups. It is of interest that VA did not recover to the initial level within 48 h after *bout* 1 in young adult and elderly males.

*3.5. Changes in CK and Muscle Soreness.* The CK activity in the blood increased significantly in all groups 24 and 48 h after SSE ($P < 0.05$; SP > 80%) (Table 1). However, CK activity in young adult males 24 h and 48 h after SSE was significantly greater than in boys and elderly males ($P < 0.001$; SP > 80%). There was no significant difference in CK 24–48 h after *bout* 1 and *bout* 2 between boys and elderly males. Differences in CK 24 h in boys, young adult males and elderly males between *bout* 1 and *bout* 2 were 240.5 ± 188.8, 587.8 ± 568.5, and 437.0 ± 110.2 IU/L, respectively ($P < 0.01$ between groups; SP > 80%). Muscle soreness in boys, young adult males and elderly males was 5.7 ± 1.8, 6.2 ± 2.1, and 3.7 ± 1.1 points, respectively, 24 h after *bout* 1 and 4.9 ± 2.8, 6.3 ± 2.1, and 3.1 ± 1.9 points 48 h after *bout* 2 ($P < 0.05$, between groups). Muscle soreness in boys, young adult males and elderly males was 3.1 ± 1.2, 3.2 ± 1.4, and 1.7 ± 0.8 points, respectively, 24 h after *bout* 1 ($P < 0.05$, between elderly males, young adult males, and boys), and 1.7 ± 0.8, 2.1 ± 0.1, and 1.3 ± 0.8 points 48 h after *bout* 2 ($P < 0.05$, between elderly males and young adult males). Differences in muscle soreness 24 h after *bout* 1 and *bout* 2 between boys, young adult and elderly males were 2.4 ± 1.6, 3.1 ± 1.7, and 2.0 ± 0.8 points, respectively ($P > 0.05$ between young adult males and elderly males). Thus, RBE in the CK indicator was significantly greater in young adult males than in boys and elderly males, while in boys it was greater than in elderly males. However, there was no significant difference in RBE of muscle soreness between groups.

# 4. Discussion

The main findings of the present study are as follows: (1) indirect symptoms of muscle damage were more significantly seen in young adult males than in elderly males and boys after the first stretch-shortening exercise bout; (2) RBE was significantly greater in young adult males than in elderly males and boys in the majority of variables; (3) RBE in boys and elderly males was more evident from changes in periphery (muscles); changes were evident in young adults in both, that is, voluntary activation and skeletal muscles.

*4.1. The Main Causes of Changes in Neuromuscular Function after Eccentric Exercise.* The main reasons for a decrease in voluntary and electrically induced muscle performance after SSE (Figure 2) are damage to force-bearing structures [2, 22, 23], changes in the excitation-contraction coupling system [4, 6, 24], and voluntary activation of muscle [19, 25]. The following indirect symptoms of muscle damage manifested within 48 h of SEE: muscle soreness, elevated plasma CK activity, decreased P20, P100, MVC, and *H* of DJs, and increased low-frequency fatigue (Figure 2). The decreases in VA after SSE show that, in our study, central fatigue

only occurred in young adult males. Peripheral fatigue was greater than central fatigue, as changes in voluntary and electrically induced torque were greater than changes in VA. SSE decreased P20 to a greater extent than P100, indicating that the muscles were subjected to LFF. LFF is characterized by a relative loss of force at low stimulation frequencies [4, 5, 20, 26].

*4.2. Why Are the Muscles of Boys and Elderly Males More Resistant to Exercise-Induced Muscle Damage Than Those of Young Adult Males?* The results of our research show that indirect symptoms of exercise-induced muscle damage after SSE are more markedly expressed in young adult males than in boys and elderly males (Figures 2 and 3 and Table 1). This coincides with the results of research done by other scientists indicating that the muscles of adults are less resistant to exercise-induced damage than those of children [8–11]. Since previous research is scarce and has been carried out applying different physical loads and measuring different indirect symptoms in subjects of different age, it is rather difficult to draw generalized conclusions because of age effects on exercise-induced muscle damage. Furthermore, the reasons why the muscles of children are more resistant to exercise-induced muscle damage remain unclear. Soares et al. [9] have shown that 12-year-old children experience less muscle damage (they judged muscle damage on the basis of increase in CK, muscle pain, and decrease in MVCF) than adults, when exercise was performed by elbow extension with the maximal intensity. Webber et al. [8] compared muscle soreness and CK levels in children and adults following a single bout of downhill running (30 min). Although CK levels were elevated 24 h after exercise in both groups, the increase was significantly greater in adults than in children. However, muscle soreness ratings 24 h after exercise were similar in both groups. Duarte et al. [10] drew similar conclusions in their study of markers of muscle overuse in boys, although they made no direct comparison with adults.

One of the possible interpretations why the muscles of adults are less resistant to fatigue when performing SSE is the fact that the muscles of adults contain relatively more fast twitch muscle fibers (type II) than do the muscles of children and elderly males [14–16].

One of the possible interpretations of the fact why the muscles of adults are more sensitive to damage than those of children is that, according to Webber et al. [8], adults, because of their greater body weight, generate more force per fiber unit during eccentric contractions compared with children, thus resulting in greater damage and a greater release of CK into the blood serum. In our study, the average *H* of DJs of young adults, elderly males, and boys was not the same. Therefore, the mechanical stimuli of both young adult and elderly males are also probably not the same.

One can find contradictory data as to resistance of the fast-type and slow-type muscle fibers to eccentric-contraction-induced damage. It has been established, for instance, that when performing eccentric exercise, type IIb muscle fibers seem to be preferentially damaged [27, 28]. Data from research done by other authors indicate that LFF

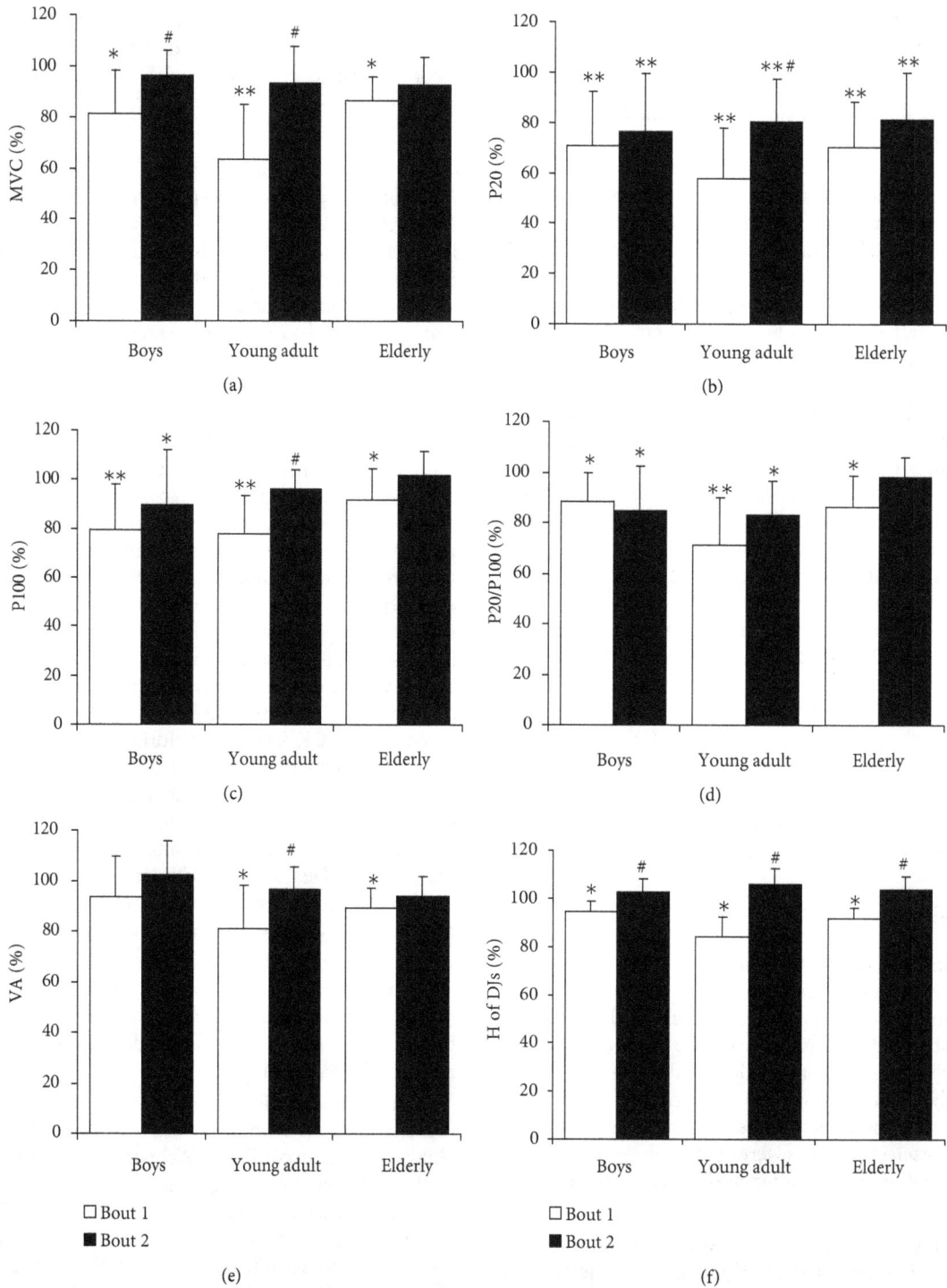

FIGURE 3: Voluntary (MVC (a) maximal voluntary contraction force of quadriceps muscle; $H$ of DJs (f) height of drop jumps; VA (e) quadriceps muscle voluntary activation index) and electrically (P20 (b) and P100 (c) muscle contraction force evoked by stimulating quadriceps muscle at 20 Hz and 100 Hz frequencies, resp.; P20/P100 (d) ratio of P20/P100 indicates the level of low frequency fatigue) induced muscle performance 48 h (percent from before) after *bout* 1 and *bout* 2 in boys ($n = 11$), young adult males ($n = 11$), and elderly males ($n = 11$) (mean ± SD). [*,**]$P < 0.05$ and $P < 0.001$, compared with before; [#]$P < 0.05$, compared with *bout* 1.

was not different between most of the high-fit and fit subjects [29] and that increasing the muscle oxidative capacity of isometric electrical stimulation training did not protect muscles against eccentric-contraction-induced damage [30]. Skurvydas et al. [18] have shown that muscle fatigue of elite track-and-field sprinters is not greater than that of long-distance runners when repeated SSE is being performed. Therefore, the authors of the present paper have some doubts as to the universal character of the conclusion made by Friden et al. [27] that fast-twitch muscle fibers (type IIb) are more sensitive to exercise-induced damage. Thus, irrespective of the attractive hypothesis that the muscles of children and elderly men are damaged less than those of young adults because their muscles contain fewer type IIb muscle fibers, we cannot assert with conviction that this is the only reason why the muscles of children are less sensitive to exercise-induced damage.

There are no grounds whatever to conclude that the boys and elderly males were not able to fully activate their muscles, for example, type IIb muscle fibers, when performing jumps since they were strongly motivated during SSE (i.e., they were informed of HJ and were asked to perform each jump as high as possible). Research done by other scientists has shown that boys of this age are capable of achieving complete activation of their leg muscles during maximal voluntary contractions [31].

*4.3. Why Are the Muscles of Elderly Males and Boys More Resistant to LFF Than the Muscles of Young Adult Males?* If the muscles of adult males contain a greater number of type IIb muscle fibers than do those of boys and elderly males, then it is understandable why the muscles of boys and elderly males are more resistant to LFF (Figure 2), since it has been established that in cat skeletal muscles, fatigue-resistant motor units are less susceptible to LFF compared with fast-fatigable motor units [32]. LFF is characterized by a relative loss of force at low frequencies of stimulation, and it is important to mention that the force is not impaired or there is only a relatively low impairment at high frequencies [4, 7, 20, 26]. In our case, there was a decrease in the force evoked not only by low stimulation frequencies (20 Hz), but by high stimulation frequencies (100 Hz) as well (Figure 2).

It has been shown that recovery of force and $[Ca^{2+}]_i$ after fatigue follows a complex time course. Recently, it has been found that the decrease in $Ca^{2+}$ release from sarcoplasmic reticulum associated with fatigue (particularly with LFF) has at least two components: (1) a metabolic component, which recovers within 1 h and (2) a component dependent on the elevation of the $[Ca^{2+}]_i$-time integral, which recovers more slowly [33]. We think that in our study, immediately after the exercise, this metabolic component had an influence on muscle fatigue.

*4.4. RBE.* After *bout* 2, we observed a much less pronounced decrease in voluntary and electrically stimulated torque, as well as lower muscle soreness and plasma CK activity, than after *bout* 1. Our data are in agreement with those of other researchers who have concluded that the adaptation

response provides protection against further damage [3, 5]. Different mechanisms have been suggested to explain RBE. The proposed rebuild processes include removal of weakened sarcomeres, longitudinal addition of sarcomeres and strengthening of the cell membrane [24], remodeling of sarcomeres [34], increased connective tissue [35], and/or reorganization of the intermediate filaments [36].

Isometric torque recovery was significantly greater after the second SSE bout in all groups, but this improvement was accompanied by a higher level of voluntary activation only in young adult males (Figure 3). Apparently, a single bout of SSE enlarges the neural drive or prevents central fatigue during a subsequent exercise bout in adult male, at least when full muscle function recovery has been allowed to occur.

A postexercise decrease in central activation was observed in the present study. Thus, SSE-induced muscle damage is associated with a reduction in the adult subjects' ability to voluntarily activate the knee extensor muscles (Figure 2). This agrees with the findings of Prasartwuth et al. [25] that reduced voluntary activation contributes to force loss after eccentric exercises with the elbow flexors. In our study, central fatigue in young adults after *bout* 1 was significantly greater than in boys and elderly males; however, there was no significant difference in VA after *bout* 2. This, however, indicates that only in young adult males, is RBE manifested as well by the change in voluntary activation of exercising muscle.

## 5. Conclusions

In conclusion, (1) indirect symptoms of muscle damage (muscle soreness, CK levels, and prolonged impairment of muscle function during both voluntary and electrically stimulated contractions at low (10–20 Hz) and high frequencies (100 Hz)) were more significantly evident in young adult males than in elderly males and boys after the first stretch-shortening exercise bout; (2) RBE was more pronounced in young adult males than in boys and elderly males.

## Conflict of Interests

The authors declare that they have no conflict of interests.

## References

[1] P. M. Clarkson and M. J. Hubal, "Exercise-induced muscle damage in humans," *American Journal of Physical Medicine and Rehabilitation*, vol. 81, no. 11, Supplement, pp. S52–S69, 2002.

[2] C. Byrne, C. Twist, and R. Eston, "Neuromuscular function after exercise-induced muscle damage. Theoretical and Applied Implications," *Sports Medicine*, vol. 34, no. 1, pp. 49–69, 2004.

[3] T. C. Chen, K. Nosaka, and P. Sacco, "Intensity of eccentric exercise, shift of optimum angle, and the magnitude of repeated-bout effect," *Journal of Applied Physiology*, vol. 102, no. 3, pp. 992–999, 2007.

[4] A. Skurvydas, M. Brazaitis, S. Kamandulis, and S. Sipaviciene, "Peripheral and central fatigue after muscle-damaging exercise is muscle length dependent and inversely related," *Journal of*

*Electromyography and Kinesiology*, vol. 20, no. 4, pp. 655–660, 2010.

[5] A. Skurvydas, M. Brazaitis, and S. Kamandulis, "Repeated bout effect is not correlated with intraindividual variability during muscle-damaging exercise," *Journal of Strength and Conditioning Research*, vol. 25, no. 4, pp. 1004–1009, 2011.

[6] A. Skurvydas, M. Brazaitis, T. Venckunas, and S. Kamandulis, "Predictive value of strength loss as an indicator of muscle damage across multiple drop jumps," *Applied Physiology, Nutrition and Metabolism*, vol. 36, no. 3, pp. 353–360, 2011.

[7] S. Kamandulis, A. Skurvydas, M. Brazaitis, L. Škikas, and J. Duchateau, "The repeated bout effect of eccentric exercise is not associated with changes in voluntary activation," *European Journal of Applied Physiology*, vol. 108, no. 6, pp. 1065–1074, 2010.

[8] L. M. Webber, W. C. Byrnes, T. W. Rowland, and V. L. Foster, "Serum creatine kinase activity and delayed onset muscle soreness in prepubescent children: a preliminary study," *Pediatric Exercise Science*, vol. 1, pp. 351–359, 1989.

[9] J. M. C. Soares, P. Mota, J. A. Duarte, and H. J. Appell, "Children are less susceptible to exercise-induced muscle damage than adults: a preliminary investigation," *Pediatric Exercise Science*, vol. 8, no. 4, pp. 361–367, 1996.

[10] J. A. Duarte, J. F. Magalhaes, L. Monteiro, A. Almeida-Dias, J. M. Soares, and H. J. Appeal, "Exercise-induced signs of muscle overuse in children," *International Journal of Sports Medicine*, vol. 20, no. 2, pp. 103–108, 1999.

[11] V. Marginson, A. V. Rowlands, N. P. Gleeson, and R. G. Eston, "Comparison of the symptoms of exercise-induced muscle damage after an initial and repeated bout of plyometric exercise in men and boys," *Journal of Applied Physiology*, vol. 99, no. 3, pp. 1174–1181, 2005.

[12] A. P. Lavender and K. Nosaka, "Responses of old men to repeated bouts of eccentric exercise of the elbow flexors in comparison with young men," *European Journal of Applied Physiology*, vol. 97, no. 5, pp. 619–626, 2006.

[13] J. Fell and A. D. Williams, "The effect of aging on skeletal-muscle recovery from exercise: possible implications for aging athletes," *Journal of Aging and Physical Activity*, vol. 16, no. 1, pp. 97–115, 2008.

[14] J. Lexell, M. Sjostrom, A. S. Nordlund, and C. C. Taylor, "Growth and development of human muscle: a quantitative morphological study of whole vastus lateralis from childhood to adult age," *Muscle and Nerve*, vol. 15, no. 3, pp. 404–409, 1992.

[15] A. D. Kriketos, L. A. Baur, and J. O'Connor, "Muscle fibre type composition in infant and adult populations and relationships with obesity," *International Journal of Obesity and Related Metabolic Disorders*, vol. 21, no. 9, pp. 796–801, 1997.

[16] F. Brunner, A. Schmid, A. Sheikhzadeh, M. Nordin, J. Yoon, and V. Frankel, "Effects of aging on type II muscle fibers: a systematic review of the literature," *Journal of Aging and Physical Activity*, vol. 15, no. 3, pp. 336–348, 2007.

[17] C. Bosco, P. V. Komi, and J. Tihanyi, "Mechanical power test and fiber composition of human leg extensor muscles," *European Journal of Applied Physiology and Occupational Physiology*, vol. 51, no. 1, pp. 129–135, 1983.

[18] A. Skurvydas, V. Dudoniene, A. Kalvénas, and A. Zuoza, "Skeletal muscle fatigue in long-distance runners, sprinters and untrained men after repeated drop jumps performed at maximal intensity," *Scandinavian Journal of Medicine and Science in Sports*, vol. 12, no. 1, pp. 34–39, 2002.

[19] S. Kamandulis, A. Skurvydas, N. Masiulis, G. Mamkus, and H. Westerblad, "The decrease in electrically evoked force production is delayed by a previous bout of stretch-shortening cycle exercise," *Acta Physiologica*, vol. 198, no. 1, pp. 91–98, 2010.

[20] V. Martin, G. Y. Millet, A. Martin, G. Deley, and G. Lattier, "Assessment of low-frequency fatigue with two methods of electrical stimulation," *Journal of Applied Physiology*, vol. 97, no. 5, pp. 1923–1929, 2004.

[21] S. C. Gandevia, "Spinal and supraspinal factors in human muscle fatigue," *Physiological Reviews*, vol. 81, no. 4, pp. 1725–1789, 2001.

[22] G. L. Warren, D. A. Lowe, and R. B. Armstrong, "Measurement tools used in the study of eccentric contraction-induced injury," *Sports Medicine*, vol. 27, no. 1, pp. 43–59, 1999.

[23] D. G. Allen, "Eccentric muscle damage: mechanisms of early reduction of force," *Acta Physiologica Scandanavica*, vol. 171, pp. 311–319, 2001.

[24] U. Proske and D. L. Morgan, "Muscle damage from eccentric exercise: mechanism, mechanical signs, adaptation and clinical applications," *Journal of Physiology*, vol. 537, no. 2, pp. 333–345, 2001.

[25] O. Prasartwuth, T. J. Allen, J. E. Butler, S. C. Gandevia, and J. L. Taylor, "Length-dependent changes in voluntary activation, maximum voluntary torque and twitch responses after eccentric damage in humans," *Journal of Physiology*, vol. 571, no. 1, pp. 243–252, 2006.

[26] H. Westerblad and D. G. Allen, "Recent advances in the understanding of skeletal muscle fatigue," *Current Opinion in Rheumatology*, vol. 14, no. 6, pp. 648–652, 2002.

[27] J. Friden, J. Seger, and B. Ekblom, "Sublethal muscle fibre injuries after high-tension anaerobic exercise," *European Journal of Applied Physiology and Occupational Physiology*, vol. 57, no. 3, pp. 360–368, 1988.

[28] P. C. D. Macpherson, M. A. Schork, and J. A. Faulkner, "Contraction-induced injury to single fiber segments from fast and slow muscles of rats by single stretches," *American Journal of Physiology*, vol. 271, no. 5, pp. C1438–C1446, 1996.

[29] M. A. Babcock, D. F. Pegelow, B. D. Johnson, and J. A. Dempsey, "Aerobic fitness effects on exercise-induced low-frequency diaphragm fatigue," *Journal of Applied Physiology*, vol. 81, no. 5, pp. 2156–2164, 1996.

[30] T. J. Patel, D. Cuizon, O. Mathieu-Costello, J. Fridén, and R. L. Lieber, "Increased oxidative capacity does not protect skeletal muscle fibers from eccentric contraction-induced injury," *American Journal of Physiology*, vol. 274, no. 5, pp. R1300–R1308, 1998.

[31] A. Y. Belanger and A. J. McComas, "Contractile properties of human skeletal muscle in childhood and adolescence," *European Journal of Applied Physiology and Occupational Physiology*, vol. 58, no. 6, pp. 563–567, 1989.

[32] R. K. Powers and M. D. Binder, "Effects of low-frequency stimulation on the tension-frequency relations of fast-twitch motor units in the cat," *Journal of Neurophysiology*, vol. 66, no. 3, pp. 905–918, 1991.

[33] E. R. Chin, C. D. Balnave, and D. G. Allen, "Role of intracellular calcium and metabolites in low-frequency fatigue of mouse skeletal muscle," *American Journal of Physiology*, vol. 272, no. 2, Part 1, pp. C550–C559, 1997.

[34] J. G. Yu, L. Carlsson, and L. E. Thornell, "Evidence for myofibril remodeling as opposed to myofibril damage in human muscles

with DOMS: an ultrastructural and immunoelectron microscopic study," *Histochemistry and Cell Biology*, vol. 121, no. 3, pp. 219–227, 2004.

[35] T. K. Lapier, H. W. Burton, R. Almon, and F. Cerny, "Alterations in intramuscular connective tissue after limb casting affect contraction-induced muscle injury," *Journal of Applied Physiology*, vol. 78, no. 3, pp. 1065–1069, 1995.

[36] T. M. Lehti, R. Kalliokoski, and J. Komulainen, "Repeated bout effect on the cytoskeletal proteins titin, desmin, and dystrophin in rat skeletal muscle," *Journal of Muscle Research and Cell Motility*, vol. 28, no. 1, pp. 39–47, 2007.

# Ascorbic Acid and BSA Protein in Solution and Films: Interaction and Surface Morphological Structure

**Rafael R. G. Maciel, Adriele A. de Almeida, Odin G. C. Godinho, Filipe D. S. Gorza, Graciela C. Pedro, Tarquin F. Trescher, Josmary R. Silva, and Nara C. de Souza**

*Grupo de Materiais Nanoestruturados, Campus Universitário do Araguaia, Universidade Federal de Mato Grosso, 78600-000 Barra do Garças, MT, Brazil*

Correspondence should be addressed to Nara C. de Souza; ncsouza@ufmt.br

Academic Editor: Rita Casadio

This paper reports on the study of the interactions between ascorbic acid (AA) and bovine serum albumin (BSA) in aqueous solution as well as in films (BSA/AA films) prepared by the layer-by-layer technique. Regarding to solution studies, a hyperchromism (in the range of ultraviolet) was found as a function of AA concentration, which suggested the formation of aggregates from AA and BSA. Binding constant, $K$, determined for aggregates from BSA and AA was found to be about $10^2$ $M^{-1}$, which indicated low affinity of AA with BSA. For the BSA/AA films, it was also noted that the AA adsorption process and surface morphological structures depended on AA concentration. By changing the contact time between the AA and BSA, a hypochromism was revealed, which was associated to decrease of accessibility of solvent to tryptophan due to formation of aggregates. Furthermore, different morphological structures of aggregates were observed, which were attributed to the diffusion-limited aggregation. Since most of studies of interactions of drugs and proteins are performed in solution, the analysis of these processes by using films can be very valuable because this kind of system is able to employ several techniques of investigation in solid state.

## 1. Introduction

Interactions between drugs and proteins have important implications for processes related to health [1]. The interactions can result in the formation of stable complexes (aggregates) that can have significant effects on the distribution, free concentration, and biological activity.

In recent years, studies on the mechanisms of interactions between drugs and proteins have been performed with techniques such as spectroscopy, chromatography, and electrochemical and atomic force microscopy. Several factors can affect these mechanisms, such as concentration, temperature, and pH of solution. Self-assembly layer-by-layer deposition technique (LbL technique) is a convenient method for the formation of films from electrolytes. Through the change of experimental conditions, it is possible to control the film thickness at nanoscale and also manipulate the sequence of layers tuning them according to the wished property. Several experimental investigations have been performed using LbL technique, in which proteins, enzymes, nucleic acids, or carbohydrates are used [2–4]. LbL technique is based on adsorption process which can be understood by analyzing the adsorption kinetics (study of layer growth over time) and adsorption isotherms (study of adsorbed amount versus concentration). In addition, morphological analysis may also reveal information about the interaction between molecules [5].

In this paper, we have investigated the behavior of the interaction of the ascorbic acid (AA) with bovine serum albumin (BSA) protein in the form of aqueous solution and BSA/AA films prepared by the LbL technique. Ascorbic acid can act as an antioxidant catalyst for tissue formation and wound healing, and inhibitor of tumor cell growth. These applications are possible due to interactions of ascorbic acid and human body protein. AA can bind to biological proteins modulating their activities. This bind process is determined by the behavior of interactions between drugs with proteins. In this context, albumin stands out for its ability to bind and transport small molecules [1, 6–9]. To the best of our knowledge there have been no studies on interaction between AA and proteins in LbL films.

## 2. Materials and Methods

Ascorbic acid was purchased from Amresco. Bovine serum albumin, BSA (fraction V, purity 96–100%), was obtained from Acros Organics and used as received. The concentration of the dipping solutions (BSA and AA) for the preparation of BSA/AA films was set at 0.5 g/L. For the solutions of BSA, the pH was adjusted to 7 by adding $NH_4OH$ and to 3 for the solutions of AA by using HCl. The spectra of AA and BSA control solution are shown in the Supplementary Materials available online at http://dx.doi.org/10.1155/2013/461365. Both BSA and AA were used at concentrations less than 0.5 mg/mL, which were obtained by diluting the initial solution with aqueous 1 M HCl or 1 M $NH_4OH$. BSA/AA films were adsorbed on quartz slides (36,0 mm × 14,0 mm × 1,0 mm). The fabrication procedures of layer by layer (LbL) followed essentially those described by Cardoso et al. [10]. The adsorbed amount, which is proportional to absorbance, was monitored by measuring the UV-vis absorption spectra with a double-beam Thermo Scientific spectrophotometer model Genesys 10. Surface morphological structure was investigated by using an LCD digital microscope (model 44340, Celestron, USA) and an atomic force microscope— AFM (model EasyScan II, Nanosurf Instruments, Switzerland) using the taping mode (256 × 256 pixels), under ambient conditions. *Gwyddion* software was used to determine the quantity of the surface forming structures. *ImageJ* software [11] was employed to determine the fractal dimension.

## 3. Results and Discussion

### 3.1. Study of Solutions

*3.1.1. Influence of Ascorbic Acid Concentration.* In order to study the interaction between AA and BSA, we have used UV-vis spectroscopy. This technique is a simple and effective method to investigate the molecular interaction and complex formation [12]. UV-vis analyses were performed for BSA in aqueous solution (pH 7, $c$ = 0.01 g/L) and modified solution after the addition of AA (pH 3) at different concentrations. All of these experiments were carried out using 2.5 mL of BSA aqueous solution contained in a quartz cuvette. The amount of AA added was the same (40 $\mu$L) for all concentrations examined (0.01, 0.03, 0.06, 0.12, 0.25, 0.37, and 0.5 g/L). Figure 1 shows the UV-vis spectra at different concentrations for BSA solution with aliquot of 40 $\mu$L of AA at pH 3. This experiment was repeated for solution without BSA at pH adjusted to 7 (Supplementary Materials) in order to rule out the effect of AA. We have observed a hyperchromism with increasing AA concentration in BSA. This effect can be associated to interaction of BSA with AA [4, 13, 14] and may be indicative of an increase in exposure of tryptophan to the solvent [15] due to a conformational change in the protein [12].

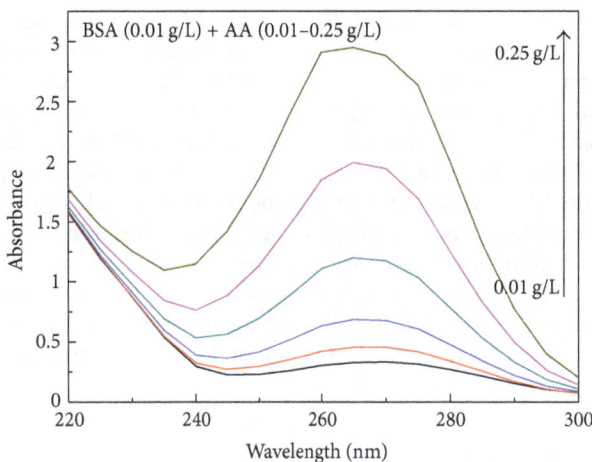

FIGURE 1: Spectra of BSA solution after addition of AA at different concentrations.

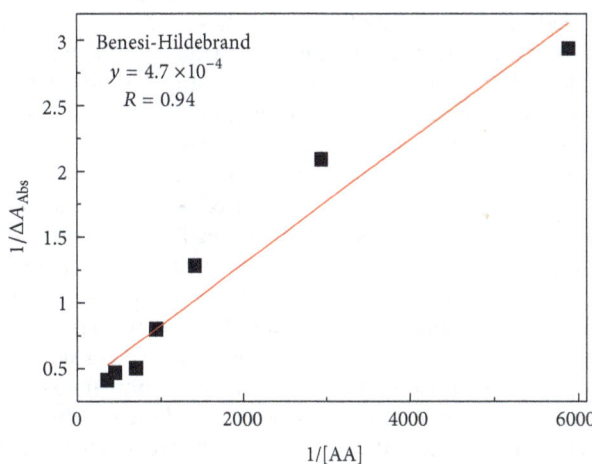

FIGURE 2: Linear regression for the reaction of BSA (0.01 g/L) with AA (0.01 to 0.5 g/L).

*3.1.2. Determination of Binding Constant.* The binding constant, $K$, of AA with BSA was determined from the values of angular and linear coefficients, as shown in Figure 2 [14]. $K$ value was found to be about $7.7 \times 10^2\,M^{-1}$ obtained by using Benesi-Hildebrand equation:

$$\frac{1}{\Delta_{Abs}} = \frac{1}{[BSA]\,[AA]\,\varepsilon K} + \frac{1}{[BSA]\,\varepsilon}, \qquad (1)$$

where [BSA] and [AA] are concentration values in mol/L, $\varepsilon$ is the absorption coefficient, and $\Delta_{Abs}$ is the change in absorbance at 280 nm for BSA bound and free.

The binding constant calculated is similar to that found for antineoplastic cisplatin ($10^2\,M^{-1}$) and far from those found for other drugs, such as azidothymidine (AZT) ($10^6\,M^{-1}$) and aspirin ($10^4\,M^{-1}$). In addition, the bonding constant found here is lower than that determined for AA ($10^4\,M^{-1}$) in lower concentrations examined [6, 16]. This suggests a weak interaction between AA and BSA, which would lead to a high free concentration of AA in blood.

*3.1.3. Influence of Contact Time.* Contact time between solutions is that measured from the moment that they are brought together. The influence of the contact time between AA and BSA on the aggregation process was investigated by using UV-vis spectroscopy, as shown in Figure 3. It is noted that the absorbance decreases with increasing the contact time. This hypochromism may be associated with protein folding and formation of aggregates. During this process, the exposition of tryptophans to solvent would decrease with increasing aggregate sizes leading to a decrease in absorbance. The process of diffusion-limited aggregation can play an important role in the formation of the fractal structures observed in this work [5].

## 3.2. Study of films

*3.2.1. Influence of Concentration.* In solid state, interactions between molecules can exhibit a different behavior from those found in a liquid state. Therefore, the investigation of properties of solid state films, such as surface morphology, can provide insights about these interactions. Here, we have studied the effect of the AA concentration on adsorption process (which is determined by the molecular interactions) of AA onto a single layer of BSA (BSA/AA film).

Figure 4 shows the behavior of an AA layer onto a single layer of BSA. The immersion time used for the solutions of BSA and AA was 10 min. As shown in Figure 4, there are two approximately linear regimes separated by a long plateau in the range of concentrations studied. For the first regime, it is noted that the adsorbed amount of AA increases with increasing concentration. This can be explained considering that as AA concentration increases, there are more molecules near surface of BSA film able to adsorb, and then as the substrate is immersed in the AA solution, the adsorbed amount increases. For the second regime, the linearity indicates that the number of sites for adsorption remains constant in this concentration range [17]. Finally, the third regime, which is an increase again, suggests that a second layer is being formed.

Figure 5 shows an image sequence for a film in which AA was adsorbed onto BSA forming a top layer (BSA/AA film). It is observed that the presence of AA leads to the formation of fractal-shaped aggregates, which have their forms dependent on the AA concentration value. In the case of pure BSA films (0.5 g/L), small aggregates are observed but without fractal structures. It should be noted that although AA films present low coverage ratio of the layer of BSA, the images are reproducible. Fractal structures can be characterized by their fractal dimension. The fractal dimensions were determined using *ImageJ* software, which uses the box-counting method of fractal analysis. The values of fractal dimension, $D_f$, found for each AA concentration employed were 1.69 (0.012 g/L), 1.71 (0.015 g/L), 1.75 (0.187 g/L), 1.87 (0.370 g/L), and 1.84 (0.5 g/L). It was shown that an increase occurs in the fractal dimension with the increase in the AA concentration.

At low concentrations, the aggregates are organized as discrete structures on the surface, whereas for higher concentrations the aggregates are organized as compacted structures. Spontaneous organizations leading to fractal structures are common in natural systems and the understanding of

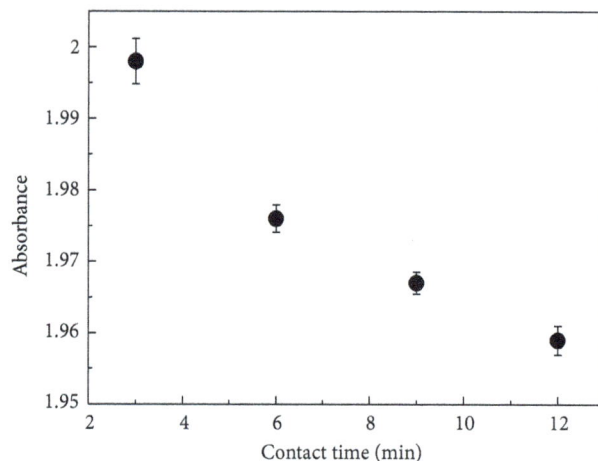

FIGURE 3: Absorbance as a function of contact time for the solution of BSA and AA.

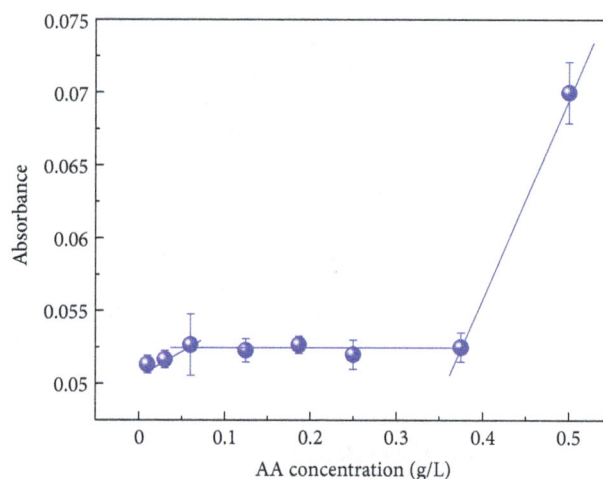

FIGURE 4: Absorbance versus concentration of AA. The immersion time into solutions of BSA and AA was 10 min. Solid lines are a guide to the eyes.

bonding mechanisms of small particles to form large aggregates is interesting in the analysis and control of biomolecular interaction processes [18].

*3.2.2. Influence of Contact Time.* The influence of the contact time between AA and BSA on the surface morphology of the BSA/AA films was investigated. The same volumes of solutions of BSA (0.5 g/L) and AA (0.5 g/L) were brought together each other at different contact times. The LbL films were obtained by immersing the quartz slides by 3 min in the BSA + AA mixture.

When the solution of AA is introduced in the same ratio $(v : v)$ in solution of BSA for different contact times, the hypochromism effect is observed both on film and in solution. This phenomenon may be associated with the formation of aggregates which decreases the accessibility of the tryptophan, thereby reducing the intensity of absorbance.

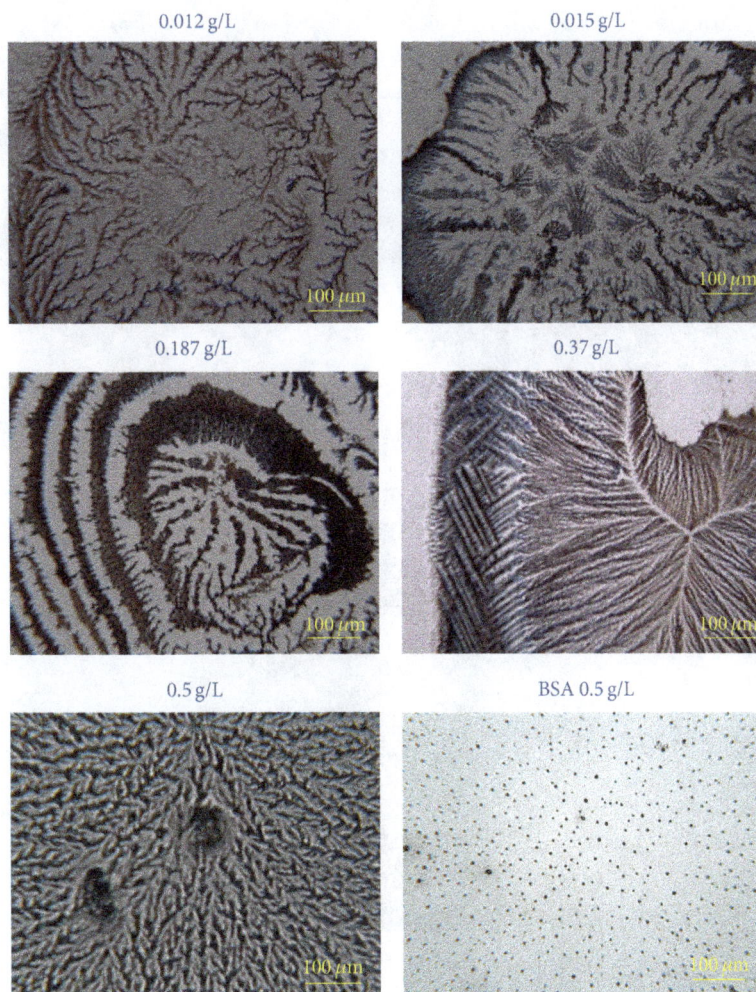

FIGURE 5: One-bilayer BSA/AA film with AA in top layer. AA concentration varied as indicated in the figure. The scale has length of 100 $\mu$m. The last image corresponds to a film with a single layer of BSA.

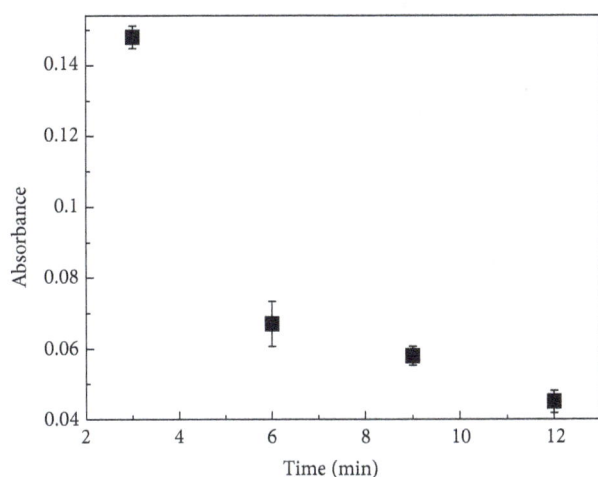

FIGURE 6: Absorbance as a function of contact time for BSA/AA films.

In order to gain an insight about the hypochromism showed in Figure 6, an analysis of surface morphological structure of LbL BSA/AA films was performed. Figure 7 shows images and the corresponding height profiles obtained by atomic force microscopy in scanning window of 25 $\mu$m $\times$ 25 $\mu$m for BSA/AA films.

Fractal-shaped aggregates that corroborate the hypochromism (Figure 6) are noted. These structures are also consistent with those observed in the images obtained by optical microscopy (Figure 5).

Figure 8 displays the number of aggregates as a function of contact time for the morphological structures shown in Figure 7. It was observed that the number of aggregates increases with increasing the contact time. For short contact times, the interaction of BSA may be more intense with the solvent, which could explain the lower number of aggregates. Increasing the contact time, the interaction between BSA and AA should increase favoring an increase in the number of aggregates. The longer the contact time the more stable aggregates structures. Since the number of aggregates formed in solution depends on contact time, it is expected that the films present different morphological structures as this parameter is changed. Furthermore, it is well known that the aggregation in solution depends on experimental factors such as pH, concentration, temperatures or time [9, 19].

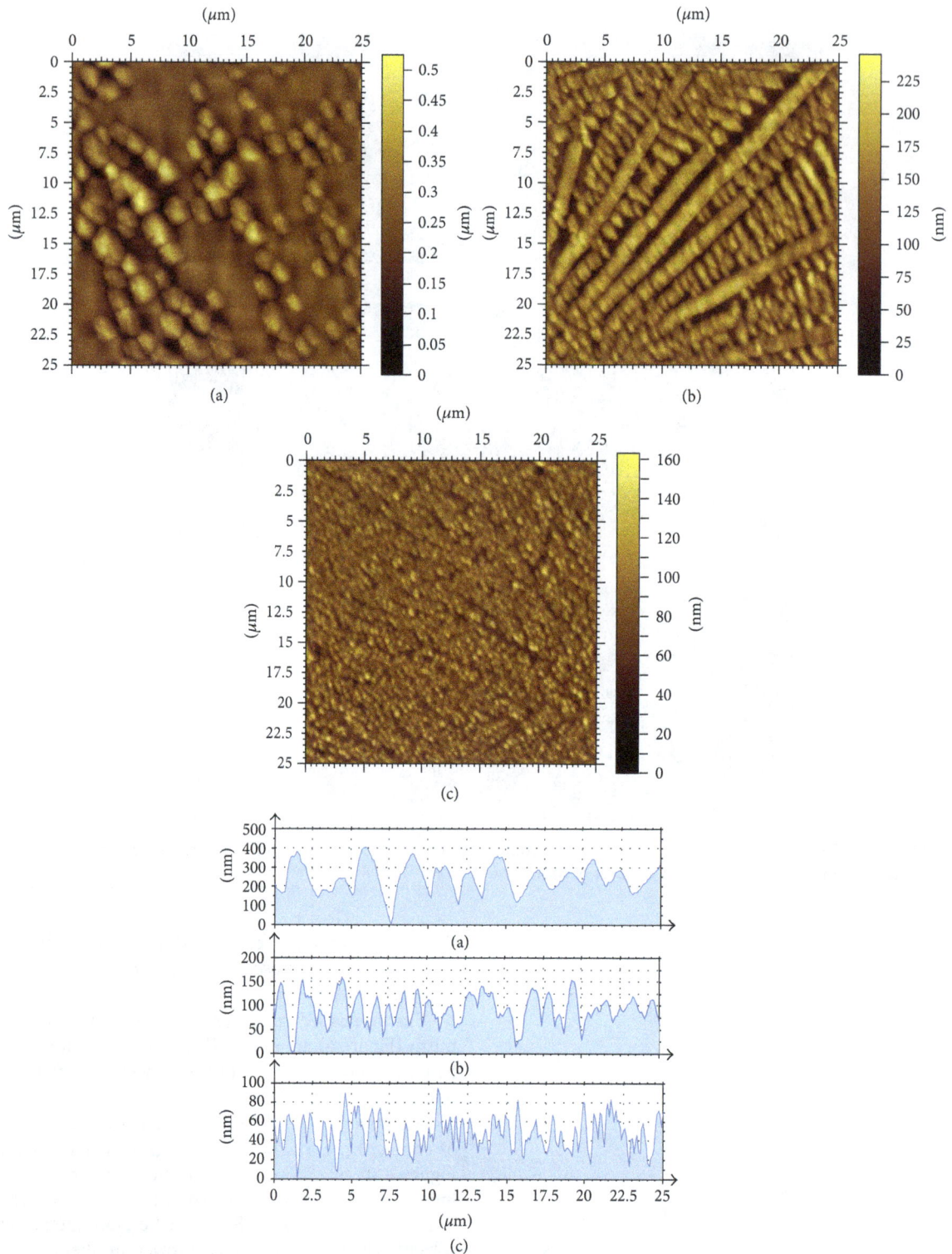

FIGURE 7: Images obtained by atomic force microscopy and corresponding profiles of self-assembled monolayers in time for solutions of BSA + AA obtained after (a) 3 min, (b) 6 min, and (c) 12 min of contact time.

The increase of number of aggregates as a function of the contact time, shown by the results of morphological analysis, corroborates the hypothesis that the aggregates make the tryptophan less accessible and this way the absorbance as a function of contact time decreases.

## 4. Conclusion

We have investigated the interaction of AA with BSA in aqueous solution and also their effects on films by changing the concentration and contact time. In solution, hyperchromism

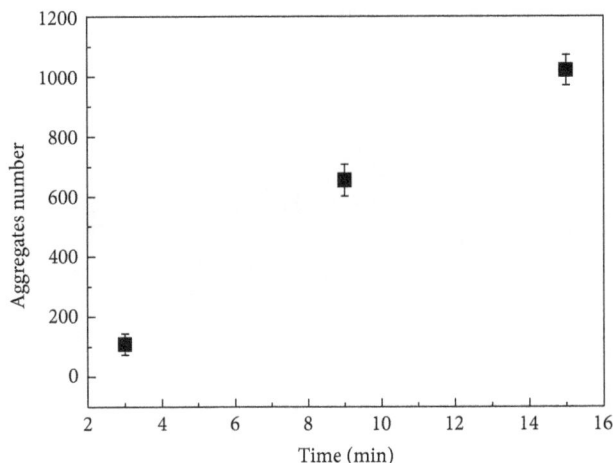

FIGURE 8: Number of aggregates as a function of contact time between BSA and AA for BSA/AA films.

indicated the formation of aggregates of AA and BSA. In addition, the binding constant of AA with BSA was found to be lower than those found for other drugs or even for lower concentrations of AA. This indicates a high free fraction of drugs. From the pharmacological viewpoint, only the free fraction can be transported by blood and other fluids to all tissues of the body. The fraction of drug bound to plasma protein forms a reversible complex, capable of dissociation. As the free part is used by the body, the linked part begins to dissociate. The increase in the free concentration of the drug increases its effect but also accelerates its elimination. For BSA/AA films, different regimes of adsorption (by using UV-vis spectroscopy) and surface morphology structures (by using optical microscopy) as a function of concentration were found. Furthermore, hypochromism as a function of the contact time was found, which was attributed to a decrease of accessibility in tryptophan to solvent due to aggregation. Atomic force microscopy for BSA/AA films revealed that the surface morphological structure of the films also depends on contact time; that is, for different contact times, different number and forms of aggregates were observed. In conclusion, films prepared by the layer-by-layer technique are interesting for drug-protein interaction studies because they exhibit structural organization and keep the active sites in the molecules immobilized. The use of this kind of film could pave the way for new investigations on the interactions of proteins with drugs, which usually employ pharmacokinetics techniques with solution samples.

## Conflict of Interests

Specifically related to AMRESCO, The authors would like to declare that they do not have a direct financial relation with the commercial identities mentioned in the paper that might lead to a conflict of interests.

## References

[1] H. Lin, J. Lan, M. Guan, F. Sheng, and H. Zhang, "Spectroscopic investigation of interaction between mangiferin and bovine serum albumin," *Spectrochimica Acta A*, vol. 73, no. 5, pp. 936–941, 2009.

[2] H. Ai, S. A. Jones, and Y. M. Lvov, "Biomedical applications of electrostatic layer-by-layer nano-assembly of polymers, enzymes, and nanoparticles," *Cell Biochemistry and Biophysics*, vol. 39, no. 1, pp. 23–43, 2003.

[3] P. C. S. F. Tischer and C. A. Tischer, "Nanobiotechnology: platform technology for biomaterials and biological applications the nanostructures," *Biochemistry and Biotechnology Reports*, vol. 1, pp. 32–53, 2012.

[4] P. Lavalle, J.-C. Voegel, D. Vautier, B. Senger, P. Schaaf, and V. Ball, "Dynamic aspects of films prepared by a sequential deposition of species: perspectives for smart and responsive materials," *Advanced Materials*, vol. 23, no. 10, pp. 1191–1221, 2011.

[5] J. R. Silva, J. B. Brito, S. T. Tanimoto, and N. C. de Souza, "Morphological structure characteriztion of PAH/NiTsPc multilayer nanostructured films," *Materials Sciences and Applications*, vol. 2, pp. 1661–1666, 2011.

[6] S. Nafisi, G. Bagheri Sadeghi, and A. Panahyab, "Interaction of aspirin and vitamin C with bovine serum albumin," *Journal of Photochemistry and Photobiology B*, vol. 105, no. 3, pp. 198–202, 2011.

[7] Y. Liu, M.-X. Xie, M. Jiang, and Y.-D. Wang, "Spectroscopic investigation of the interaction between human serum albumin and three organic acids," *Spectrochimica Acta A*, vol. 61, no. 9, pp. 2245–2251, 2005.

[8] H. Wang, L. Kong, H. Zou, J. Ni, and Y. Zhang, "Screening and analysis of biologically active compounds in Angelica sinensis by molecular biochromatography," *Chromatographia*, vol. 50, no. 7-8, pp. 439–445, 1999.

[9] E. Lozinsky, A. Novoselsky, A. I. Shames, O. Saphier, G. I. Likhtenshtein, and D. Meyerstein, "Effect of albumin on the kinetics of ascorbate oxidation," *Biochimica et Biophysica Acta*, vol. 1526, no. 1, pp. 53–60, 2001.

[10] G. Cardoso, R. J. da Silva, R. R. G. Maciel, N. C. de Souza, and J. R. Silva, "Roughness control of layer-by-layer and alternative spray films from Congo red and PAH via Laser light irradiation," *Materials Sciences and Applications*, vol. 3, pp. 552–556, 2012.

[11] C. A. Schneider, W. S. Rasband, and K. W. Eliceiri, "NIH Image to ImageJ: 25 years of image analysis," *Nature Methods*, vol. 9, pp. 671–675, 2012.

[12] R. S. Kumar, P. Paul, A. Riyasdeen et al., "Human serum albumin binding and cytotoxicity studies of surfactant-cobalt(III) complex containing 1,10-phenanthroline ligand," *Colloids and Surfaces B*, vol. 86, no. 1, pp. 35–44, 2011.

[13] K. A. Connors, *Binding Constants—The Measurement of Molecular Complex*, Wiley-Interscience, New York, NY, USA, 1st edition, 1997.

[14] I. D. Kuntz Jr., F. P. Gasparro, M. D. Johnston Jr., and R. P. Taylor, "Molecular interactions and the Benesi-Hildebrand equation," *Journal of the American Chemical Society*, vol. 90, no. 18, pp. 4778–4781, 1968.

[15] P. Daneshegar, A. A. Moosavi-Movahedi, P. Norouzi, M. R. Ganjali, M. Farhadic, and N. Sheibani, "Characterization of paracetamol binding with normal and glycated human serum albumin assayed by a new electrochemical method," *Journal of the Brazilian Chemical Society*, vol. 23, no. 2, pp. 315–321, 2012.

[16] H. A. Tajmir-Riahi, "An overview of drug binding to human serum albumin: protein folding and unfolding," *Scientia Iranica*, vol. 14, no. 2, pp. 87–95, 2007.

[17] C. H. Giles, D. Smith, and A. Huitson, "A general treatment and classification of the solute adsorption isotherm. I. Theoretical," *Journal of Colloid And Interface Science*, vol. 47, no. 3, pp. 755–765, 1974.

[18] A. L. Barabàsi and H. E. Stanley, *Fractal Concepts in Surface Growth*, Brittish Library, 1st edition, 2002.

[19] N. C. De Souza, J. R. Silva, C. A. Rodrigues, L. D. F. Costa, J. A. Giacometti, and O. N. Oliveira Jr., "Adsorption processes in layer-by-layer films of poly(o-methoxyaniline): the role of aggregation," *Thin Solid Films*, vol. 428, no. 1-2, pp. 232–236, 2003.

# 7

# Corticospinal Reorganization after Locomotor Training in a Person with Motor Incomplete Paraplegia

Nupur Hajela,[1,2] Chaithanya K. Mummidisetty,[1] Andrew C. Smith,[1] and Maria Knikou[1,2,3]

[1] Electrophysiological Analysis of Gait and Posture Laboratory, Sensory Motor Performance Program,
  Rehabilitation Institute of Chicago, 345 East Superior Street, Chicago, IL 60611, USA
[2] Department of Physical Medicine and Rehabilitation, Northwestern University Feinberg School of Medicine, Chicago, IL 60611, USA
[3] Department of Physical Therapy and the Graduate Center, The City University of New York, Staten Island, NY 10314, USA

Correspondence should be addressed to Maria Knikou; m-knikou@northwestern.edu

Academic Editor: Francisco Miró

Activity-dependent plasticity as a result of reorganization of neural circuits is a fundamental characteristic of the central nervous system that occurs simultaneously in multiple sites. In this study, we established the effects of subthreshold transcranial magnetic stimulation (TMS) over the primary motor cortex region on the tibialis anterior (TA) long-latency flexion reflex. Neurophysiological tests were conducted before and after robotic gait training in one person with a motor incomplete spinal cord injury (SCI) while at rest and during robotic-assisted stepping. The TA flexion reflex was evoked following nonnociceptive sural nerve stimulation and was conditioned by TMS at 0.9 TA motor evoked potential resting threshold at conditioning-test intervals that ranged from 70 to 130 ms. Subthreshold TMS induced a significant facilitation on the TA flexion reflex before training, which was reversed to depression after training with the subject seated at rest. During stepping, corticospinal facilitation of the flexion reflex at early and midstance phases before training was replaced with depression at early and midswing followed by facilitation at late swing after training. These results constitute the first neurophysiologic evidence that locomotor training reorganizes the cortical control of spinal interneuronal circuits that generate patterned motor activity, modifying spinal reflex function, in the chronic lesioned human spinal cord.

## 1. Introduction

A plethora of studies have shown that the isolated mammalian spinal cord can generate muscle activation patterns suited for locomotion in absence of inputs from the brain [1, 2]. This work led to the notion that neural drive from the brain is needed mostly when environmental constraints increase such as stepping over an obstacle or on an uneven surface [3–5]. However, corticospinal neurons are active during simple locomotion and exhibit a profound step-related modulation in the cat [6–8]. Similarly, corticospinal pathways to leg muscles are activated in a phase-dependent manner during simple treadmill walking in humans, long-latency reflexes of the tibialis anterior (TA) muscle are partly mediated by a transcortical pathway, and impaired transmission in the corticospinal tract is related to gait disability of

individuals with a spinal cord injury (SCI) [9–11]. These findings support the notion of a substantial cortical involvement in human walking.

Because of motor incomplete SCI, the spinal cord is not completely severed and thus some descending fiber tracts and segmental spinal cord circuits remain intact; it is logical to hypothesize that cortical control of spinal neural circuits is reorganized after locomotor training. This hypothesis is supported by the fact that activity-dependent neuroplasticity takes place simultaneously in multiple sites of the central nervous system due to training [12, 13]. Improvements in walking ability have been achieved with locomotor training post-SCI, and changes have been reported in walking speed, step length, and step symmetry [14]. The reported changes are likely the result of task-specific sensorimotor feedback that reorganizes corticospinal and spinal pathways in a functional manner [15, 16]. For example, in 4 people with SCI, functional magnetic resonance imaging showed a greater activation in sensorimotor cortical and cerebellar regions after 36 sessions of body weight supported (BWS) robotic gait training [17]. In individuals with incomplete SCI, 3 to 5 months of daily locomotor training increased the size of the motor evoked potentials (MEPs) in 9 out of 13 muscles tested, increased the maximal MEP, and changed the slope of the MEP input-output curve [18]. The changes in MEP size were significantly correlated to the degree of locomotor recovery, suggesting that corticospinal plasticity was involved, at least in part, in the recovery of walking ability after training [18].

Collectively, we hypothesized that locomotor training reorganizes the cortical control of spinal interneuronal pathways that generate patterned motor activity during locomotion. We tested our hypothesis by establishing the effects of subthreshold transcranial magnetic stimulation (TMS) over the primary motor cortex region on the spinal polysynaptic flexion reflex before and after BWS robotic gait training in one person with motor incomplete paraplegia while at rest and during robotic-assisted stepping. We selected this reflex because the interneuronal circuits that generate the flexion reflex also participate in pattern generation during locomotion, and this reflex is susceptible to descending control [19].

## 2. Materials and Methods

*2.1. Subject.* A 52-year-old woman, 11-year post-SCI, at the level of thoracic 7 due to fall, participated in this study following written consent to the experimental procedures approved by the Northwestern University (Chicago, IL, USA) Institutional Review Board committee and conducted in accordance with the Declaration of Helsinki. Based on neurological examination according to the American Spinal Injury Association guidelines, the subject had an AIS grade D impairment scale at the time of admission to the study. The subject received 35 training sessions (1 hour/day, 5 days/week) with a robotic exoskeleton (Lokomat, Hocoma, Switzerland). Before and after training, electromyographic (EMG) activity was recorded from medial gastrocnemius (MG), peroneus longus (PL), gracilis (GRC) and medial hamstrings (MH) of the right leg, and tibialis anterior (TA) and soleus (SOL) from both legs with bipolar differential electrodes of fixed interelectrode distance (Motion Lab Systems, Baton Rouge, LA, USA). EMG and foot switches data were collected at 2000 Hz with custom-written acquisition software (Labview, National Instruments, Austin, TX, USA). Results of clinical evaluation tests and treadmill parameters before and after training are summarized in Table 1.

*2.2. Neurophysiological Tests Conducted before and after Training.* With the subject seated at rest, the sural nerve of the left leg was stimulated with a pulse train of 30 ms duration once every 10 s with a constant current stimulator (DS7A, Digitimer, Hertfordshire, UK) [20, 21]. Stimulation was delivered by two disposable pregelled Ag-AgCl electrodes

(Conmed Corporation, NY, USA) placed on the lateral malleolus and maintained in place via an athletic wrap. Reflex responses were recorded from the ipsilateral TA muscle. Sural nerve stimulation during testing was delivered at 1.3 times the reflex threshold. No limb movement or pain was present upon stimulation.

Single TMS pulses over the right primary motor cortex (M1) were delivered with a Magstim 200 stimulator (Magstim, Whitland, UK). The double-coned coil was oriented on the skull to produce an induced current in the posterior-to-anterior direction. The optimal position for TMS was determined by varying the position of the coil from the vertex with gradually increasing intensities, until an MEP in the contralateral (left) TA muscle was observed at the lowest stimulation intensities with the subject seated at rest. MEP resting threshold was defined as the stimulus intensity at which three MEPs of at least $100 \, \mu V$ of peak-to-peak amplitude were evoked following five consecutive stimuli with the subject at rest.

After cortical and sural nerve stimulation sites were established, the effects of TMS delivered at 0.9 TA MEP resting threshold on the TA flexion reflex at the conditioning-test (C-T) intervals of 70, 90, 110, and 130 ms were determined with the seated subject. Ten flexion reflexes, each evoked once every 10 s, were recorded under control conditions and following subthreshold TMS. Then, the subject was transferred to standing at 50% BWS, and the TA flexion reflex and MEP thresholds were reestablished. During robotic-assisted stepping, the flexion reflex was conditioned by TMS at $0.9 \times$ TA MEP resting threshold at the C-T intervals of 70 ms and 110 ms before and after training. The subject stepped at 50% BWS and at 1.8 Km/h treadmill speed for both data collection sessions. Stimulation was triggered every 3 steps, based on the signal from the left-foot switch, which was sent randomly across different phases of a step cycle that was divided into 16 equal time windows or bins [21, 22].

*2.3. Data Analysis.* EMG signals during BWS-assisted stepping from the steps before sural nerve and transcranial magnetic stimulation were full-wave rectified, high-pass filtered at 20 Hz, and low-pass filtered at 500 Hz. After full-wave rectification, linear envelopes were obtained at 20 Hz low-pass filter, and the mean EMG amplitude across all steps was determined. Integrated EMG was defined as the area under the linear envelope. This analysis was conducted separately for each muscle during BWS-assisted stepping for both sessions. The overall average of the EMG linear envelope (including all bins) from each muscle was also estimated and compared before and after training with a paired $t$-test.

Flexion reflexes were measured as the area under the full-wave rectified EMG response. The conditioned TA flexion reflex ($n = 10$) recorded at each C-T interval before and after training with the seated subject was expressed as a percentage of the mean size of the associated control flexion reflex. Statistically significant differences between the conditioned flexion reflexes recorded at different C-T intervals before and after training were established with a multiple ANOVA at $2 \times 4$ levels (2: pre-/post-training, 4: C-T intervals) along with Holm-Sidak tests for repeated measures. At each bin of

TABLE 1: Treadmill walking parameters and functional outcomes[1].

| BWS (%) | Speed (Km/h) | R and L foot lifters | Guidance force by the Robot (%) | Clonus | Extensor spasticity (SCATS) | Manual muscle testing | 6 min walk | 30 sec chair-stand test | Time up and go |
|---|---|---|---|---|---|---|---|---|---|
| | | | | Before robotic gait training | | | | | |
| 50 | 1.8 | None | 100 | 1L/0R | 0L/0R | R leg = 24/25 L leg = 16/25 | 269 m using quad cane | 12 reps | 11.3 sec using quad can |
| | | | | After robotic gait training | | | | | |
| 15 | 3.2 | None | 15 | 1L/0R | 0L/0R | R leg = 24/25 L leg = 17/25 | 335 m using quad cane | 21 reps | 9.1 sec using quad cane |

[1] BWS: body weight support; extensor spasticity grade is based on the spinal cord assessment tool for spasticity (SCATS): where subjects are positioned supine, the lower limb is rapidly moved into passive extension, and the severity of quadriceps contraction is scored; R: right, L: left; 0: no reaction to stimulus; 1: mild quadriceps contraction between 1–3 seconds.

the step cycle, the full-wave rectified area of the TA flexion reflex response was calculated and averaged separately for steps with and without sural nerve stimulation and TMS [22]. The average of TA EMGs of non-stimulated steps was subtracted from the average of EMGs of stimulated steps (conditioned reflex) at identical time windows for each bin and was expressed as a percentage of the control flexion reflex recorded with the seated subject. Statistically significant differences between the conditioned flexion reflexes recorded at each bin of the step cycle before and after training were established with a two-way ANOVA at 2 × 16 levels (2: pre-/post- training, 16: bins of the step cycle) along with Holm-Sidak tests for repeated measures. This analysis was conducted separately for flexion reflexes at the C-T intervals of 70 and 110 ms. Alpha was set at 95% for all statistical tests.

## 3. Results

The latency of the TA flexion reflex following sural nerve stimulation measured from the onset of the pulse train was 160 ms, while the latency of the TA MEP was 40 ms before and after training. The EMG activation patterns as a function of the step cycle changed significantly after robotic gait training. Specifically, the SOL EMG burst duration was prolonged during the stance phase (Figure 1(a)), MG displayed an EMG burst during the stance and late swing phases (Figure 1(b)); while the PL EMG burst was enhanced throughout the stance phase (Figure 1(f)). The EMG activation profiles of SOL, MG, PL, and MH muscles are similar to those observed in control subjects during robotic-assisted stepping, but an absent TA activity is noted at early stance and late swing phases when compared to the TA EMG profile observed commonly in control subjects (see Figure 1(b) in [23]). The most pronounced change noted is in the TA muscle in which a burst of activity was present at late stance phase (Figure 1(c)), while before training a clear TA EMG activity was absent. An increase in the overall EMGs amplitude computed across all bins of the step cycle was noted in all leg muscles ($P < 0.05$; Figure 1(g)).

In Figure 2(a), full-wave rectified waveform averages of the TA flexion reflex recorded under control conditions (grey line) and following TMS at 0.9 × MEP resting threshold (black lines) are indicated for recordings taken before and after training. In Figure 2(b), the amplitude of the conditioned TA flexion reflex as a percentage of the control flexion reflex before and after training is indicated. A MANOVA showed that the conditioned long-latency TA flexion reflex was statistically significantly different before and after training ($F_{1,8} = 81.7$, $P < 0.05$), and that the amplitude of the conditioned flexion reflex did not vary across C-T intervals tested for recordings taken before and after training ($F_{3,24} = 1.4$, $P > 0.05$).

The changes observed after training during robotic-assisted stepping were more complex compared to the uniform flexion reflex depression observed with the seated subject. In Figure 3, the mean amplitude of the long-latency TA flexion reflex following TMS at 0.9 × MEP resting threshold at the C-T intervals of 70 ms and 110 ms as a function of the step cycle is indicated. A two-way ANOVA at 2 × 16 levels (2: pre/post training, 16: bins of the step cycle) showed that the TA flexion reflex at the C-T interval of 70 ms was statistically significantly different across bins ($P < 0.001$). Pairwise multiple comparisons (Holm-Sidak tests) showed that the conditioned flexion reflex at bins 1, 2, 5, 6, 7, 9, 11, 12, 13, 15, and 16 was statistically significantly different before and after training ($P < 0.05$). These results suggest that after training, the conditioned TA flexion reflex at the C-T interval of 70 ms was significantly enhanced during the stance phase, followed by a depression from early swing until midswing (bins 9–13) when compared to the conditioned flexion reflex recorded before training (Figure 3(a)). A two-way ANOVA at 2 × 16 levels (2: pre-/post- training, 16: bins of the step cycle) showed that the TA flexion reflex at the C-T interval of 110 ms was statistically significantly different across bins ($P < 0.001$). Pairwise Holm-Sidak tests for multiple comparisons showed that the conditioned flexion reflex throughout the stance phase (bins 2–8) was facilitated, followed by a significant depression at early swing

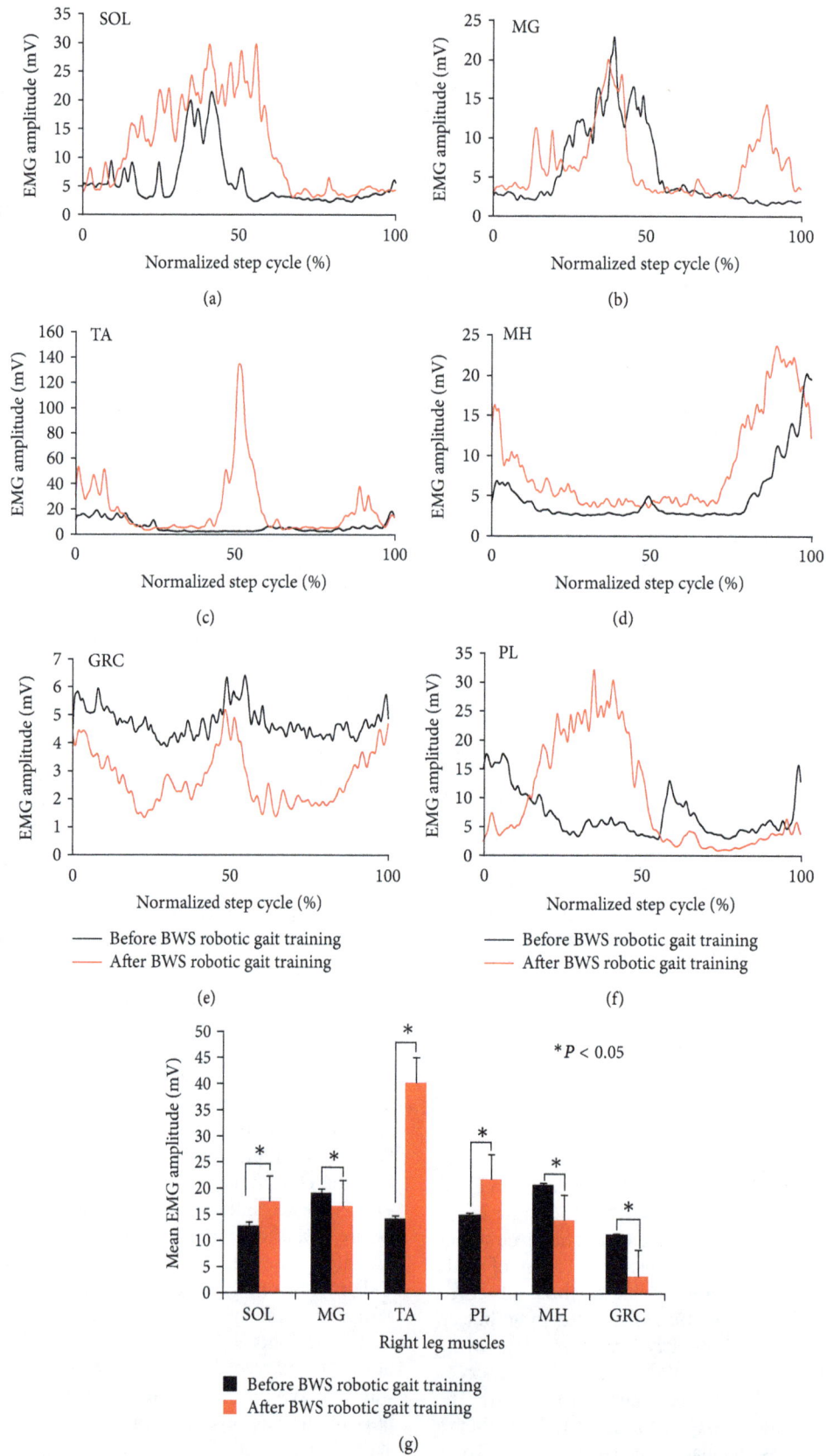

FIGURE 1: EMG activity during robotic-assisted stepping before and after training. (a)–(f) EMG activity of the right side muscles during robotic-assisted stepping at 50% BWS and at 1.8 Km/h before and after training as a function of the step cycle. (g) Mean EMG amplitude for stepping before (black squares) and after (red squares) 35 sessions of robotic gait training. EMG: electromyography; SOL: soleus; MG: medial gastrocnemius; TA: tibialis anterior; PL: peroneus longus; MH: medial hamstrings; GRC: gracilis.

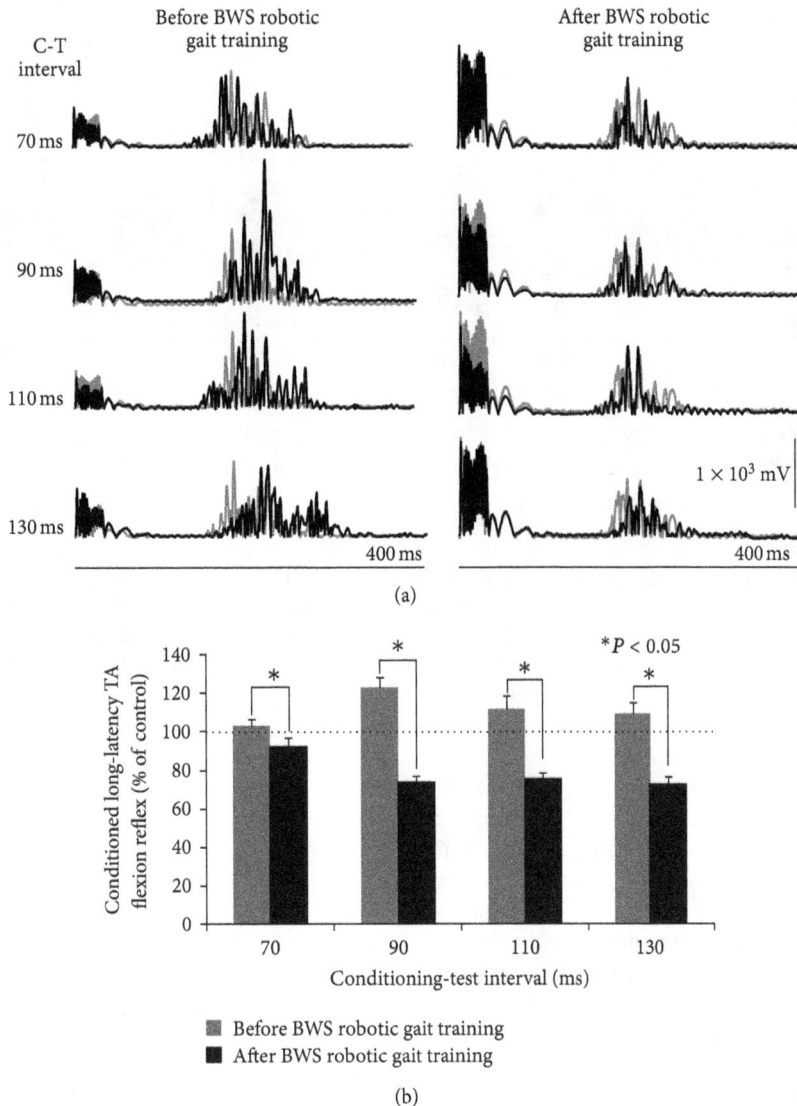

FIGURE 2: Effects of subthreshold TMS on the TA flexion reflex while seated before and after BWS robotic gait training. (a) Full-wave rectified waveform averages ($n = 10$) of the control tibialis anterior (TA) flexion reflex (grey line) and the conditioned flexion reflex following single pulse transcranial magnetic stimulation (TMS) of the right primary motor cortex at 0.9 TA motor evoked potentials (MEPs) resting threshold. (b) Mean amplitude of the conditioned TA flexion reflexes recorded before and after BWS robotic gait training with the seated subject. The conditioning-test interval is denoted on the abscissa. Asterisks indicate statistically significant differences between the conditioned TA flexion reflexes recorded before and after training. Error bars denote the SEM.

phase (bins 11, 12) and a significant facilitation at swing-to-stance transition phase (bins 15, 16) ($P < 0.05$) (Figure 3(b)).

## 4. Discussion

Locomotor training with a robotic exoskeleton reorganized the cortical control of spinal interneuronal circuits and modified the flexion reflex function at rest and during assisted stepping in a person with a chronic motor incomplete SCI. Before training and with the seated subject, subthreshold TMS resulted in facilitation of the long-latency TA flexion reflex, but after training a pronounced reflex depression was evident. Corticospinal actions on the flexion reflex changed

in a more complex pattern during robotic-assisted stepping. After training, corticospinal facilitation of the flexion reflex at early and midstance was replaced with depression at early and midswing followed by facilitation at late swing. Two possible explanations for these changes are that the residual intact supraspinal connections were reorganized or that new supraspinal connections with spinal networks were formed with locomotor training as a result of activity-dependent mechanisms driven by task-specific sensory cues [12, 13, 24]. These sensory cues included load alternation and leg positioning with kinematics of the hips, knees, and ankles timed to the step cycle in a physiologic pattern predetermined by the robotic exoskeleton system.

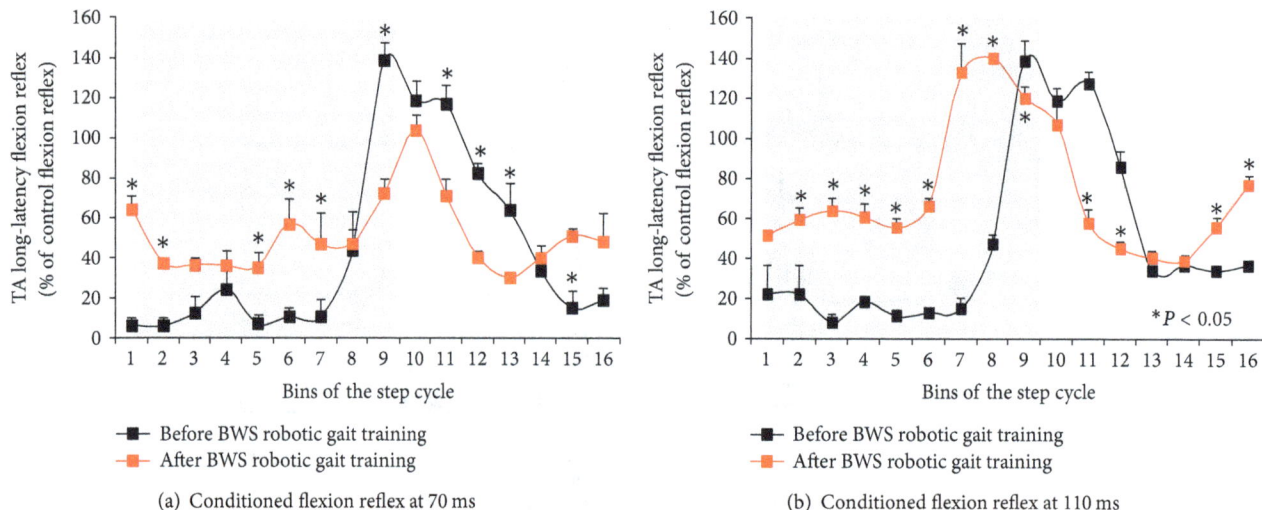

(a) Conditioned flexion reflex at 70 ms

(b) Conditioned flexion reflex at 110 ms

FIGURE 3: Changes in cortical control of the flexion reflex after 30 sessions of BWS robotic gait training during robotic-assisted stepping. The mean normalized long-latency tibialis anterior (TA) flexion reflex following single pulse transcranial magnetic stimulation (TMS) of the right primary motor cortex at $0.9 \times$ TA motor evoked potentials (MEPs) at the conditioning-test interval of 70 (a) and 110 (b) ms is indicated as a function of the step cycle. Asterisks indicate suppressive and/or facilitatory conditioned flexion reflexes after locomotor training compared to those observed before training based on the $P$ value computed from pairwise multiple comparisons (two-way ANOVA along with Holm-Sidak tests). Grey squares denote the stance phase. Error bars denote the SEM.

Activity-dependent plasticity involves both physiological and structural changes that alter the anatomical connectivity of neurons [24–26]. We are not able to effectively assess which anatomical connections exist after the injury and which change with training. Nonetheless, the neuronal pathways and circuits that may have changed due to training are intracortical and interhemispheric inhibitory circuits, corticospinal monosynaptic connections with TA alpha motoneurons, and oligo- or polysynaptic cortical connections with flexion reflex afferent (FRA) interneurons. The rationale for proposing these neuronal pathways is based on the demonstrated effects of subthreshold TMS on the spinal motoneurons through intracortical and interhemispheric inhibitory circuits [27–30], and on the fact that TMS delivered $0.9 \times$ MEP resting threshold, it may have produced corticospinal motor volleys that affected the excitability state of FRA interneurons and TA alpha motoneurons. Because of the long latency of the flexion reflex as well as that the conditioning reflex effects were observed at long C-T intervals, it is likely that monosynaptic excitation of TA alpha motoneurons by corticospinal volleys was absent and that corticospinal descending volleys affected FRA interneurons after a polysynaptic relay [30].

Sural nerve stimulation largely excited $A\beta$ (or group II) sensory afferents mediating tactile information. The conduction velocity of these afferents ranges from 30 to $70 \, \text{m} \, \text{s}^{-1}$ while during contraction is $45 \, \text{m} \, \text{s}^{-1}$ [31]. Further, the conduction velocity of the early D (or direct) wave after scalp stimulation recorded with epidural electrodes at the thoracic 5 ranges from 62 to $70 \, \text{m} \, \text{s}^{-1}$ [32]. This means that impulses from $A\beta$ fibers reached the spinal cord about 14–30 ms after the first pulse of the reflex stimulus pulse train, while corticospinal motor volleys reached the spinal

cord approximately 10 ms following TMS. Because changes in motoneuronal excitation following M1 excitation can last as long as 80 to 100 ms, it is apparent that at the C-T intervals used in this study, there was ample time for TMS to affect the excitability state of FRA interneurons that produce polysynaptic reflex actions on $\alpha$-motoneurons.

Our finding—that corticospinal control of spinal cord neural circuits was reorganized after locomotor training—is important and constitutes the first proof of principle for this therapeutic strategy based on neurophysiological evidence. The changes in the corticospinal pathways we observed here may be linked to improvements of walking ability and balance. After locomotor training, the person was able to walk 335 m within 6 min compared to 269 m before training, while significant improvements were noted on balance-related motor tasks and speed of walking (Table 1). Clinical studies have demonstrated that locomotor training improves walking ability and cardiovascular function in people with motor incomplete SCI [33, 34]. Taken together, we propose that recovery of walking ability is mediated through reorganization of corticospinal actions on spinal interneuronal circuits modifying reflex function during walking.

At this point, it should be noted that a key limitation of this study is that data was collected from one patient, and thus generalization to a specific SCI population should be cautioned. Further, the subject received only 35 sessions of robotic gait training. Rehabilitation of these patients to achieve restoration of movement and walking is a long-term process, while reorganization of corticospinal control of spinal reflex circuits may differ after 60 or 90 training sessions. Thus, the corticospinal reorganization we observed here, evident by the modulation pattern of the flexion reflex following TMS with the seated subject and during robotic-assisted stepping, may reflect a specific stage of the task-dependent plasticity of corticospinal neural circuits [35]. It is apparent that further research is needed to outline the neurophysiological changes associated with corticospinal reorganization due to locomotor training and the role of corticospinal neural plasticity in restoration of walking ability after SCI.

## 5. Conclusion

We demonstrate in this study, for the first time, that cortical actions on spinal interneuronal circuits are reorganized after locomotor training in one person with chronic motor incomplete SCI. This neural reorganization may be the result of newly formed supraspinal connections with spinal networks or potentiation of inactive residual intact supraspinal connections due to training. Further research is needed to link reorganization of corticospinal neural pathways to locomotor training-mediated restoration of walking ability as well as phases of neuroplasticity over time.

## Abbreviations

BWS:  Body weight support
C-T:  Conditioning-test
EMG:  Electromyographic
MEP:  Motor evoked potentials
SCI:  Spinal cord injury
TA:  Tibialis anterior
TMS:  Transcranial magnetic stimulation.

## Conflict of Interests

The author(s) declare that they have no financial interests or potential conflict of interests with respect to the research, authorship, and/or publication of this paper to report.

## Authors' Contribution

N. Hajela conducted the experiments and contributed to data analysis. C. K. Mummidisetty conducted the experiments and contributed to data analysis and figure development. A. C. Smith administered the locomotor training sessions, clinically evaluated the patient, and contributed to data analysis. M. Knikou developed the idea and the experimental protocol, contributed to data analysis, and wrote the paper. All of the authors approved the final version of the paper.

## Acknowledgments

The authors wish to thank the research subject for her dedication and motivation. This study was supported by the New York State Department of Health (NYSDOH), Spinal Cord Injury Research Trust Fund, Wadsworth Center (Contract C023690), and the Craig H. Neilsen Foundation (83607) and was conducted at the Rehabilitation Institute of Chicago. Funding sources had no involvement in study design, data collection, data analysis, or data interpretation.

## References

[1] S. Rossignol, G. Barrière, A. Frigon et al., "Plasticity of locomotor sensorimotor interactions after peripheral and/or spinal lesions," *Brain Research Reviews*, vol. 57, no. 1, pp. 228–240, 2008.

[2] V. R. Edgerton, N. J. K. Tillakaratne, A. J. Bigbee, R. D. de Leon, and R. R. Roy, "Plasticity of the spinal neural circuitry after injury," *Annual Review of Neuroscience*, vol. 27, pp. 145–167, 2004.

[3] I. N. Beloozerova and M. G. Sirota, "The role of the motor cortex in the control of accuracy of locomotor movements in the cat," *Journal of Physiology*, vol. 461, pp. 1–25, 1993.

[4] I. N. Beloozerova, B. J. Farrell, M. G. Sirota, and B. I. Prilutsky, "Differences in movement mechanics, electromyographic, and motor cortex activity between accurate and nonaccurate stepping," *Journal of Neurophysiology*, vol. 103, no. 4, pp. 2285–2300, 2010.

[5] T. Drew, J. E. Andujar, K. Lajoie, and S. Yakovenko, "Cortical mechanisms involved in visuomotor coordination during precision walking," *Brain Research Reviews*, vol. 57, no. 1, pp. 199–211, 2008.

[6] D. M. Armstrong and T. Drew, "Discharges of pyramidal tract and other motor cortical neurones during locomotion in the cat," *Journal of Physiology*, vol. 346, pp. 471–495, 1984.

[7] D. M. Armstrong and T. Drew, "Locomotor-related neuronal discharges in cat motor cortex compared with peripheral receptive fields and evoked movements," *Journal of Physiology*, vol. 346, pp. 497–517, 1984.

[8] T. Drew, "Motor cortical activity during voluntary gait modifications in the cat. I. Cells related to the forelimbs," *Journal of Neurophysiology*, vol. 70, no. 1, pp. 179–199, 1993.

[9] D. Barthélemy, M. J. Grey, J. B. Nielsen, and L. Bouyer, "Involvement of the corticospinal tract in the control of human gait," *Progress in Brain Research*, vol. 192, pp. 181–197, 2011.

[10] J. Nielsen, N. Petersen, and B. Fedirchuk, "Evidence suggesting a transcortical pathway from cutaneous foot afferents to tibialis anterior motoneurones in man," *Journal of Physiology*, vol. 501, no. 2, pp. 473–484, 1997.

[11] D. Barthélemy, M. Willerslev-Olsen, H. Lundell et al., "Impaired transmission in the corticospinal tract and gait disability in spinal cord injured persons," *Journal of Neurophysiology*, vol. 104, no. 2, pp. 1167–1176, 2010.

[12] B. H. Dobkin, "Functional rewiring of brain and spinal cord after injury: the three Rs of neural repair and neurological rehabilitation," *Current Opinion in Neurology*, vol. 13, no. 6, pp. 655–659, 2000.

[13] J. R. Wolpaw and A. M. Tennissen, "Activity-dependent spinal cord plasticity in health and disease," *Annual Review of Neuroscience*, vol. 24, pp. 807–843, 2001.

[14] B. Dobkin, D. Apple, H. Barbeau et al., "Weight-supported treadmill vs over-ground training for walking after acute incomplete SCI," *Neurology*, vol. 66, no. 4, pp. 484–493, 2006.

[15] M. Knikou, "Neural control of locomotion and training-induced plasticity after spinal and cerebral lesions," *Clinical Neurophysiology*, vol. 121, no. 10, pp. 1655–1668, 2010.

[16] M. Knikou, "Plasticity of corticospinal neural control after locomotor training in human spinal cord injury," *Neural Plasticity*, vol. 2012, Article ID 254948, 2012.

[17] P. Winchester, R. McColl, R. Querry et al., "Changes in supraspinal activation patterns following robotic locomotor

therapy in motor-incomplete spinal cord injury," *Neurorehabilitation and Neural Repair*, vol. 19, no. 4, pp. 313–324, 2005.

[18] S. L. Thomas and M. A. Gorassini, "Increases in corticospinal tract function by treadmill training after incomplete spinal cord injury," *Journal of Neurophysiology*, vol. 94, no. 4, pp. 2844–2855, 2005.

[19] C. S. Sherrington, "Flexion-reflex of the limb, crossed extension-reflex and reflex stepping and standing," *The Journal of Physiology*, vol. 40, pp. 28–121, 1910.

[20] M. Knikou, "Plantar cutaneous input modulates differently spinal reflexes in subjects with intact and injured spinal cord," *Spinal Cord*, vol. 45, no. 1, pp. 69–77, 2007.

[21] M. Knikou, "Plantar cutaneous afferents normalize the reflex modulation patterns during stepping in chronic human spinal cord injury," *Journal of Neurophysiology*, vol. 103, no. 3, pp. 1304–1314, 2010.

[22] M. Knikou, C. A. Angeli, C. K. Ferreira, and S. J. Harkema, "Flexion reflex modulation during stepping in human spinal cord injury," *Experimental Brain Research*, vol. 196, no. 3, pp. 341–351, 2009.

[23] M. Knikou, N. Hajela, C. K. Mummidisetty, M. Xiao, and A. C. Smith, "Soleus H-reflex phase-dependent modulation is preserved during stepping within a robotic exoskeleton," *Clinical Neurophysiology*, vol. 122, no. 7, pp. 1396–1404, 2011.

[24] J. N. Sanes and J. P. Donoghue, "Plasticity and primary motor cortex," *Annual Review of Neuroscience*, vol. 23, pp. 393–415, 2000.

[25] D. E. Feldman, "Synaptic mechanisms for plasticity in neocortex," *Annual Review of Neuroscience*, vol. 32, pp. 33–55, 2009.

[26] M. Butz, F. Wörgötter, and A. van Ooyen, "Activity-dependent structural plasticity," *Brain Research Reviews*, vol. 60, no. 2, pp. 287–305, 2009.

[27] J. Valls-Solé, A. Pascual-Leone, E. M. Wassermann, and M. Hallett, "Human motor evoked responses to paired transcranial magnetic stimuli," *Electroencephalography and Clinical Neurophysiology*, vol. 85, no. 6, pp. 355–364, 1992.

[28] V. Di Lazzaro, D. Restuccia, A. Oliviero et al., "Magnetic transcranial stimulation at intensities below active motor threshold activates intracortical inhibitory circuits," *Experimental Brain Research*, vol. 119, no. 2, pp. 265–268, 1998.

[29] T. Kujirai, M. D. Caramia, J. C. Rothwell et al., "Corticocortical inhibition in human motor cortex," *Journal of Physiology*, vol. 471, pp. 501–519, 1993.

[30] J. M. A. Cowan, B. L. Day, C. Marsden, and J. C. Rothwell, "The effect of percutaneous motor cortex stimulation on H reflexes in muscles of the arm and leg in intact man," *Journal of Physiology*, vol. 377, pp. 333–347, 1986.

[31] A. Rossi, A. Zalaffi, and B. Decchi, "Interaction of nociceptive and non-nociceptive cutaneous afferents from foot sole in common reflex pathways to tibialis anterior motoneurones in humans," *Brain Research*, vol. 714, no. 1-2, pp. 76–86, 1996.

[32] M. Inghilleri, A. Berardelli, G. Cruccu, A. Priori, and M. Manfredi, "Corticospinal potentials after transcranial stimulation in humans," *Journal of Neurology Neurosurgery and Psychiatry*, vol. 52, no. 8, pp. 970–974, 1989.

[33] B. Dobkin, H. Barbeau, D. Deforge et al., "The evolution of walking-related outcomes over the first 12 weeks of rehabilitation for incomplete traumatic spinal cord injury: the multicenter randomized Spinal Cord Injury Locomotor trial," *Neurorehabilitation and Neural Repair*, vol. 21, no. 1, pp. 25–35, 2007.

[34] M. Turiel, S. Sitia, S. Cicala et al., "Robotic treadmill training improves cardiovascular function in spinal cord injury patients," *International Journal of Cardiology*, vol. 149, no. 3, pp. 323–329, 2011.

[35] J. R. Wolpaw and J. A. O'Keefe, "Adaptive plasticity in the primate spinal stretch reflex: evidence for a two-phase process," *Journal of Neuroscience*, vol. 4, no. 11, pp. 2718–2724, 1984.

# 8

# Cell Mechanosensitivity: Mechanical Properties and Interaction with Gravitational Field

## I. V. Ogneva

*State Research Center of Russian Federation Institute of Biomedical Problems, Russian Academy of Sciences, 76-a, Khoroshevskoyoe shosse, Moscow 123007, Russia*

Correspondence should be addressed to I. V. Ogneva; iogneva@yandex.ru

Academic Editor: Masamitsu Yamaguchi

This paper addressed the possible mechanisms of primary reception of a mechanical stimulus by different cells. Data concerning the stiffness of muscle and nonmuscle cells as measured by atomic force microscopy are provided. The changes in the mechanical properties of cells that occur under changed external mechanical tension are presented, and the initial stages of mechanical signal transduction are considered. The possible mechanism of perception of different external mechanical signals by cells is suggested.

## 1. Introduction

The appearance of life on Earth and the evolution of all living organisms occurred under the influence of external physical fields, gravity, and electromagnetic fields. The formation of a cell—the basic building unit of life, which is capable of an independent existence under these physical conditions meant that the cell's physical properties had to have been such that they enabled it to exist under the influence of these physical fields.

The most constant external physical field is, certainly, the gravitational field. Since the cell is being formed under the influence of an external mechanical field, its mechanical properties, on the one hand, should be such that they enable it to function under the conditions of this field. On the other, hand however, the cell should also be capable of responding to changes in the external mechanical conditions and adapt to them, while not forgoing its ability to reproduce and maintain itself.

Any mechanical system, for example the cell, in the external field is in tension (from a mechanical point of view), and as such the cell forms its structure and internal mechanical tension in accordance with the vector and amplitude of this external force. A change in the external force (either its vector or amplitude) will naturally cause a change in the mechanical tension of the cell and lead to its deformation. The level of significance, and consequences, of these deformations on the essential activities of the cell will depend on the cell's mechanical properties and the sensitivity of its mechanosensors.

Nevertheless, all cells can be divided into two types: cells that form internal tension only in response to an external force, and cells that are also able to generate their own mechanical force, for example, muscle cells. Muscle cells have a specific structure, including a well-developed cytoskeleton that takes up the larger part of the cell and which forms the contractile apparatus. This submembrane (cortical) cytoskeleton of muscle cells is generally similar to the cortical cytoskeleton of nonmuscle cells, except for in several special points, that is in the projection of Z-disc and M-line on the membrane.

Therefore, the problem of cellular mechanosensitivity can be posed as a set of questions: what are different cells' mechanical properties; what is the magnitude of force capable of causing a cellular response; what are the changes in the cellular mechanical and biochemical properties under the changed external mechanical conditions, and finally, what is the cell mechanosensor and how the cellular response is achieved?

## 2. Mechanical Properties of Cells

*2.1. Cells Capable of Generating a Mechanical Force-Cardiomyocytes and Skeletal Muscle Fibres.* With the advent of atomic force microscopy, experimental studies on the mechanical properties of different cells were intensified [1]. Primarily, the focus was on muscle cells, in one of the first studies the stiffness of the muscle fibres was compared with the stiffness of human umbilical vein endothelial cells.

Evidently, the initial interest in muscle cells was connected with the fact that these cells specialize in generating mechanical force and that their mechanical properties determine the force they can produce.

Consequently, Mathur et al. [2] generated one of the first datasets concerning the mechanical properties of intact muscle fibres of the skeletal muscle and myocardium, in comparison with endothelial cells, by using liquid-based atomic force microscopy. The main working hypothesis was that the Young's modulus and viscosity would differ in these three types of cells because of their different structures and functional roles. The experimental samples used were fibres of the rabbit myocardium, C2C12 myoblasts from C3H adult mice and human umbilical vein endothelial cells (HUVEC). The authors showed that the Young's modulus of the endothelial cells was $E = 6.8 \pm 0.4$ kPa in the area of nucleus, $E = 3.3 \pm 0.2$ kPa on the cell body, and $E = 1.4 \pm 0.1$ kPa at the cell edge. As opposed to the endothelium, systematic changes in the Young's modulus of the skeletal muscle fibres and cardiomyocytes, based on the location of the cantilever contact point on the surface, could not be found. For the cells of the myocardium, the Young's modulus was $= 100.3 \pm 10.7$ kPa, and for the cells of skeletal muscles $E = 24.7 \pm 3.5$ kPa. Thus, the fibres of the myocardium are the stiffest, and consequently more stable to deformation, which in author's opinion, could be explained by their constant rhythmic activity.

A group of researchers headed by Defranchi et al. [3] carried out experiments to study the structure and transversal stiffness of the sarcolemmas of fully differentiated fibres under different conditions. The experiments were carried out on muscle fibres of the skeletal muscles of CD1 mice. Measurements of the mechanical properties were made in contact mode, the maximum applied force to the fibre membrane was 1 nN, and indentation depth varied from several hundreds to thousands of nm. The value of the Young's modulus presented in the study was $E = 61 \pm 5$ kPa.

Thus the values of the Young's modulus found by Defranchi et al. exceeded, by approximately 2.5 times, the results provided by Mathur et al. This is, probably, due to Defranchi et al. using fully differentiated cells in their studies, while Mathur et al. used myoblasts. Such differences in the mechanical properties of muscle fibres at different stages of differentiation can show that the ontogenetic dynamics of protein expression, which forms the structural base of the transversal stiffness, is uneven. The first study addressing this question was conducted by Collinsworth et al. [4]. They studied the mechanical properties of muscle cells at different stages, from myocytes to muscle fibres, and found a significant increase in the Young's modulus on the eighth day after the start of differentiation. Thus, for undifferentiated myoblasts the Young's modulus was $E = 11.5 \pm 1.3$ kPa, while on the 8–10th day of differentiation, it was $E = 45.3 \pm 4.0$ kPa. At that viscosity that was assessed by hysteresis

formed at direct and reversal route of cantilever during registration of force curves was not changed during differentiation. The author's hypothesis concerning the connection between the changes in Young's modulus and the formation of tubulin microtubules was not confirmed experimentally, as the Young modulus and viscosity of muscle cells were not changed after processing with colchicines (in concentrations of $0.4\,\mu g/mL$ for 2 hours) or taxol ($10\,\mu M$ for 2 hours). However, processing with cytochalasin D (in concentrations of $3\,\mu M$ for 5–30 minutes) or blebbistatin (in concentrations of 50 mM for 5–30 minutes) did cause a significant decrease in Young's modulus without an accompanying change in viscosity properties. Because of this, the authors connected changes in stiffness properties of the muscle cells during differentiation with the development of an actin-myosin system. It is important to note that Collinsworth et al. did not analyse the contribution of the nonsarcomere cytoskeletal proteins in the transversal stiffness of muscle fibres, even though processing by cytochalasin D can cause damage of actin filaments of the cortical layer.

We developed a method that enabled us to define the transversal stiffness of different parts of the muscle fibre [5]. Consequently, for different parts of rat fibre's membrane in the relaxed state, the following changes in transversal stiffness were seen: membrane at the projection of the Z-disc—$3.08 \pm 0.14$ pN/nm (m. soleus), $2.24 \pm 0.18$ pN/nm (m. medial gastrocnemius), $2.98 \pm 0.29$ pN/nm (m. tibialis anterior), $16.0 \pm 1.3$ pN/nm (heart); membrane at projection of the M-line—$1.98 \pm 0.07$ pN/nm (m. soleus), $1.53 \pm 0.07$ pN/nm (m. medial gastrocnemius), $1.43 \pm 0.26$ pN/nm (m. tibialis anterior), $9.9 \pm 0.6$ pN/nm (heart); membrane in the area of the semisarcomere—$3.05 \pm 0.03$ pN/nm (m. soleus), $2.87 \pm 0.12$ pN/nm (m. medial gastrocnemius), $2.98 \pm 0.29$ pN/nm (m. tibialis anterior), $7.1 \pm 0.4$ pN/nm (heart) [6–8].

Conducted measurements of the fibres of different skeletal muscles and cardiomyocytes show that the transversal stiffness of myocardial fibre membranes, in all parts, are definitively higher than the transversal stiffness of the membrane of skeletal muscle fibres. The increases in transversal stiffness of the muscle cell membranes are connected to the development of the cytoskeleton, in particular at cell maturation. Evidently, a more developed cortical cytoskeleton can explain the increase in cardiomyocyte stiffness, while with the fibres of skeletal muscles this increase is likely explained by high mechanical load.

2.2. Nonmuscle Cells. Costa et al. [9] studied the mechanical properties of human aorta endothelial cells (HAEC). The measurements were conducted in contact mode in liquid, indentation depth was 200 nm. The authors found that there were two types of cells which differed in Young's modulus: one type had a Young's modulus of $5.6 \pm 3.5$ kPa, while others had one of $1.5 \pm 0.76$ kPa. However, after processing with cytochalasin B (in concentrations of $4\,\mu M$), there were no differences in their mechanical properties while their Young's modulus corresponded to the data of Mathur et al. [2], who also studied endothelial cells but only from the umbilical vein and not from the human aorta. Moreover authors showed that in this case the cells' mechanical properties were defined by an actin cytoskeleton.

A similar approach was used by Martens and Radmacher [10] with the aim of demonstrating the main contributions of the actin cytoskeleton in the stiffness of human fibroblasts, which was assessed before and after addition of blebbistatin and the Rho-kinase inhibitor Y27632. The authors showed that, at addition of blebbistatin, the Young's modulus decreases from 20 kPa to 8 kPa over the course of 30–60 minutes, while processing with Y27632 did not cause any significant mechanical effects. In the authors' opinions therefore, these results prove that the tension caused by myosin determines the cell stiffness.

Cai et al. [11] showed that measurements of a cell's mechanical properties can be used as a diagnostic parameter, for example in the analysis of lymphocyte degeneration. Normal human lymphocytes and human T-lymphoblastic Jurkat cells were examined. Atomic force microscopic images showed that the cell profiles, that is, surface striations, are similar in both types. However, although the stiffness of normal lymphocytes is 2.28 ± 0.49 mN/m, for Jurkat lymphocytes it is 4.32 ± 0.3 mN/m.

In researching pathological forms of erythrocytes in patients with hereditary spherocytosis deficit of glucose-6-dehydrogenase, thalassemia and with anisocytosis, Dulińska et al. [12] showed that their stiffness was increased in comparison to normal cells. Lekka et al. [13] meanwhile assessed erythrocytes in patients with confirmed diagnoses of coronary disease, hypertension, and diabetes mellitus, and compared these with erythrocytes of healthy volunteers. The authors showed that the mean Young's modulus values and the width distribution of its values were markedly higher in patients with diabetes mellitus, and in smokers, in comparison to healthy people. Moreover, the Young's modulus of erythrocytes increased with the age of patients.

According to the patient's age, stiffness of chondrocytes also changes, but in this case it was a decrease that was seen. Hsieh et al. [14] showed that the stiffness of chondrocytes in young men is 0.096 ± 0.009 N/m, and in patients with osteoarthritis (old age), it is 0.035 ± 0.005 N/m.

Except those mentioned above, the most widely studied cells are those of the human bone marrow, which are represented in particular by mesenchymal stem cells (hMSCs) and osteoblasts (hOBs). Docheva et al. [15] studied the topography and mechanical properties of two types of hMSCs: rapidly renewing cells (RS) and flat cells (FC), also in addition to hOBs and cells of the osteosarcoma line MG63. Atomic force microscopy images showed that FC and hOB surfaces are strongly striated, while in RS and MG63 they are comparatively smooth. Moreover RS and MG63 cells were flatter on the fibrous substrates than on smoothed surfaces on the slide. In contrast, in cells with greater surface area, that is, FC and hOBs, flatness was not dependent on the substrate and was more evident in comparison with RS and MG63 cells. The authors tried to explain this result as being due to different degrees of cell adhesion, taking into account the higher content of focal adhesion complexes in FC and hOBs than in RS and MG63 cells. When examining the mechanical properties of these types of cell, Docheva et al. [15] showed

that the Young modulus in normal cells, that is, RS, FC, and hOBs, have comparable values across the three different substrates while in MG63 cells, it significantly increases depending on the collagen of the I type.

Takai et al. [16] also studied the dependence of mechanical properties on the substrate. The objects of their study were MC3T3-E1 osteoblast-like cells, tested on the following different substrates: fibronectin (FN), vitronectin (VN), type-I collagen (COL I), fetal bovine serum (FBS), poly-L-lysine (PLL), and no substrate, that is, tested directly on the slide. The results obtained by the authors showed that the Young's modulus of osteoblasts that were bound to extracellular matrix proteins (FN, VN, COL I, FBS) through integrins, was higher than in similar cells on PLL, or on the slide alone where interactions are nonspecific. Also, formation of F-actin stress-fibrils was observed on FN, VN, COL I, and FBS, while there were only very small quantity fibrils of F-actin found for PLL or on the slide alone. Damage to the actin cytoskeleton decreases the Young's modulus of osteoblasts on FN so that it reaches the level of that for cells on the slide alone. At the time damage of microtubules did not cause any significant effects. Takai et al. [16] suggest that increases in the Young's modulus of osteoblasts on FN occurred because of rebuilding of the actin cytoskeleton in response to interactions with extracellular matrix proteins.

A more detailed study of the influence of the extracellular matrix on the morphology and mechanical properties of cells was conducted by Yim et al. [17]. Here, the authors suggested that nanotopography of the extracellular matrix could influence cell behaviour by changing the interaction with integrins and/or focal adhesion complexes. hMSCs cultured on 350 nm layers formed by stiff TCPS (tissue-culture polystyrene) or soft PDMS (polydimethylsiloxane) were selected as the object of this study. Study results showed that in cells cultured on both the TCPS and PDMS, there was a decrease in expression of some integrin subunits ($\alpha2$, $\alpha6$, $\alpha$V, $\beta2$, $\beta3$, $\beta4$) and a flattening of the actin cytoskeleton occurred when compared with a control, where a hard actin reticulum with a random sequence distribution was observed. Moreover, on the stiff TCPS, the Young's modulus of hMSCs was lower than on the soft PDMS. Furthermore, nanotopography effects did not affect the mechanical properties, although in cells cultured on the PDMS, the value of the Young's modulus was notably lower than in cells cultured on TCPS. From results obtained by Yim et al. [17], it can be concluded that both the nanotopography and substrate stiffness have an influence on the cell mechanical properties, but that the role of nanotopography is dominant in the regulation of the cytoskeleton state.

Thus, the data of different authors enables us to suggest that the stiffness of nonmuscle cells depends significantly on the type of substrate on which they are cultivated. In addition, the nanotopography of the substrate has a substantial significance that can be connected to the formation of adhesion sites and development of a submembranous cytoskeleton. Data concerning the increased stiffness of proliferated cells, which are also characterised by an increase in protein content forming cortical cytoskeleton, also demonstrates this.

## 3. The Magnitude of Force Capable of Causing Cellular Responses

The process of transformation of physical signals into biochemical ones, and the formation of appropriate cellular responses, is called mechanotransduction or mechanosensitivity [18].

Understanding of the molecular basis of mechanotransduction is impossible without knowledge of the amplitude and distribution of forces influencing cells. However data presented in the literature and conducted in this direction are very limited. First Davies et al. [19] suggested that the cellular response to an applied external mechanical force could be mediated by its transfer to the cell nucleus and also through intercellular interactions and adhesive contacts. The central idea of these authors is that the force affects more structures, although it becomes lessened with increasing distance from the point of force application. Nevertheless, this is not always the case as these authors showed in additional work using intermediate filaments as a marker for changing forces [20, 21]. Similar experimental approaches were used by several other groups whose purpose was the mapping of stresses that appeared in the cell in response to the applied external force [22–24]. All these results prove that forces have a complex, heterogeneous nature that is mediated by several proteins and their complexes.

Thus, a key question is what is the magnitude force that is able to cause a cellular response?

Force can be exerted on the cell by different experimental methods, and if the applied force is sufficient, the cellular response can be analysed. In this way, Huang et al. [18] showed that a shear stress equal to 1 Pa is critical to endothelial cells. This force is approximately 1 nN when recalculated for an area of $1 \mu m^2$. If this force is balanced by response of focal-adhesive complex only taken over 1% of the whole square but the load on it increases by 100 times. It is relatively difficult to compare these values with the force measured in experiments with alternative deformation because there are no measurements of the force of cell interaction and elasticity of the substrate. However, by taking into account that the Young's modulus is 1 kPa, a force of 100 Pa will cause a relative deformation equal to 10%.

The analysis of the force of cell interaction and the substrate without overlapping of information can be the alternative for such measurements. The point is that by knowing the Young's modulus of the substrate, it is possible to measure its deformation while putting the cell able to bind on it. While studying fibroblasts, Balaban et al. [25] showed that the force of interaction is $5.5 \, nN/\mu m^2$. Assuming the close packaging of integrins in the focal adhesion complex, the authors assessed the force necessary to cause changes in the conformation of every integrin to be at the level of several pN.

The level of force necessary for a change of conformation in other proteins also can be determined. The force necessary to break the link between two proteins can be taken as the upper assessment. So for the dissociation of fibronectin from integrin, a force of 30–100 pN is required [26]. Meanwhile, a force of 3–5 pN is necessary for the unfolding of fibronectin domains [27]. Ferrer et al. [28] showed that the force required for alpha-actinin to dissociate from actin, and also for filamin, is 40–80 pN.

However, these conformational changes can be significant if they exceed the level of heat noise; $kT$ is approximately $4 \, pN \cdot nm$, so that when comparing with the characteristic deformations on the level of 1–10 nm, the force should be not less than 4 pN. This is comparable to the magnitude of force generated by one myosin head at muscle contraction [29], which increases the assurance of the accuracy of the assessments made by Huang et al. [18].

## 4. Potential Mechanosensors

The extracellular matrix and membrane proteins, mechanosensitive and/or other ion channels, structures of the submembrane (cortical) cytoskeleton, and intracellular structures, in particular, could all act as mechanosensors.

*4.1. Extracellular Matrix and Membrane Proteins.* It has been shown that the applying of stretching force to a culture of neurons or smooth muscle cells, through the extracellular matrix leads to an increase in the polymerisation of microtubules [30, 31]. Integrins, which form connections with various proteins in the extracellular matrix, such as fibronectin and vitronectin, form a primary site of transduction and consequently, can be considered as mechanosensors. From the intracellular side, a number of proteins are concentrated near the focal adhesion complex that directly interact with $\alpha$- or $\beta$-subunits of the integrin heterodimer. These proteins include paxillin, focal adhesion kinase, and caveolin, where paxillin and focal adhesion kinase can connect to a large number of other proteins, thus forming signalling cascades. Besides these, tensin, alpha-actinin, and filamin can also connect with integrins and the cortical cytoskeleton, as they have the appropriate domains, and work in conjunction with integrins and actin [32]. Furthermore, alpha-actinin has more than one domain with which it interacts with actin and form an actin network. The structure of the focal adhesion complex is characterised by large number of proteins, all located in the immediate vicinity of each other, and so this feature complicates the analysis of the contribution of each in mechanotransduction and does not allow for determinations of which have the more dominant roles. However, it is obvious that an external mechanical force can lead to conformational changes in one or several of the proteins of the focal adhesion complex, further triggering the cascade of underlying signalling pathways.

*4.2. Mechanosensitive Ion Channels.* Mechanical stretching of cellular membranes, for example using the patch-clamp technique, changes the transportation activity of mechanosensitive ion channels as a result of conformational changes or changes in the lipid bilayer [33, 34], or in the gate domains of the channel itself [35, 36]. In addition, the majority of channels studied respond to cellular stretching, but not to compression.

Prokaryotic mechanosensitive ion channels have been described in many experimental studies and reviews. One of the most well-characterised mechanosensitive channels is the bacterial MscL, which has a pore with a large diameter and low ion selectivity. This channel possesses the highest conductivity known, about $10^3$ pCm [37], and can be regulated by membrane tension, as shown in experiments using the patch-clamp technique. Increasing the membrane tension, supervised by variation of depth of absorption to a pipette, caused an increase in the conductivity of channels when the forces operating on the channels exceed a certain size [37]. The authors showed that the tension in this case was equal to $10^{-2}$ Pa·m, that is slightly lower, than tension, leading to gap ($6 \times 10^{-2}$ Pa·m) that can have a big physiological value, for example, in the swelling of a bacterial cell that occurs as a consequence of osmotic shock. The results of molecular dynamic modeling [38], based on the data on MscL crystal structure, show that such changes in membrane tension lead to the formation of pores with a diameter of 0.5 nm [39]. At the same time, experiments *in vitro* showed that the diameter of an open pore was 3-4 nm [40]. However, the adequacy of the results of the experimental situations *in vivo* is under question.

Few eukaryotic channels have been identified as mechanosensitive channels: TRP channels, K(2P) channels, MscS-like proteins, and DEG/ENaC channels [41]. The transient receptor potential (TRP) protein superfamily consists of a diverse group of cation channels that have important roles in cells of the nervous system and in nonexcitable cells [42]. It has been shown that the underlying cytoskeleton and scaffolding proteins can influence the regulation of gating in TRP channels [43].

The two-pore-domain K(+) channels, or K(2P) channels, have four transmembrane regions, act as dimers, and are widespread in different tissues (in both excitable and nonexcitable cells). They are thought to play a major role in setting the resting membrane potential of many cell types. K(2P) channels are quasi-instantaneous and noninactivating, and they are active at all membrane potentials and insensitive to the classic K(+) channel blockers. The TWIK-related (TREK)-1 and TWIK-related arachidonic acid-stimulated K(+) (TRAAK) channels were the first cloned polyunsaturated fatty acid-activated and mechanogated K(+) channels [44].

Epithelial sodium channels (ENaCs) are a subfamily of ion channels within the degenerin/ENaC (DEG/ENaC) superfamily. These ion channels are found in different sodium-absorbing epithelia, including the epithelium of the colon, lung, and distal nephron; their activity represents the rate-limiting step for sodium uptake, and thus transepithelial water movement [45]. There is growing evidence concerning the activation of ENaC by mechanical forces, and at least laminar shear stress seems to be an adequate stimulus of physiological significance [46, 47]. Highly selective epithelial $Na^+$ channels are expressed in various vertebrate epithelia where they are exposed to shear forces such as the distal nephron [46, 48], airway epithelia [49], vascular tissue [50–52], and sensory nerve endings, indicating participation in mechanosensitive processes [53].

*4.3. Cortical Cytoskeleton.* Today the role of cortical cytoskeleton in the regulation of ion channels is quite well

established. It has been shown that condensation of cortical actin under the plasma membrane occurs as a result of the phosphatase inhibitor, calyculin A, suppressing a depot-dependent input of calcium in smooth muscle cells in culture [54], and also cytochalasin D [55]. With the use of patch-clamp techniques, it was shown that actin microfilaments take part in the regulation of chloride channels [33, 56], Na+-K+-ATPase [57], electroexcitable sodium channels in brain cells [58], and sodium channels in cells of reabsorbable epithelium [59]. Dismantling of actin filaments by cytochalasin D leads to the activation of sodium channels in K562 cell lines, while actin polymerisation on the cytoplasmic side of an external membrane of a cell causes channel inactivation [60]. Thus, the fragmentation of actin filaments, as associated with a plasma membrane, caused by cytosolic actin-connecting Ca-sensitive proteins that are similar to endogenous gelsolin, can be a major factor in inducing the activity of sodium channels in response to increasing intracellular concentrations of calcium ions in K562 cell lines [61, 62].

In addition to the examples mentioned above, there are data which indicate that an association with the lipid microdomains of cholesterol-rich plasma membranes (rafts) could be the essential factor in defining the activity of integrated membrane proteins, including ionic channels [63–68]. The disruption of the membrane structure and raft integrity, caused by a decrease in the level of membrane cholesterol, interferes with execution of cellular functions, including reorganisation of the actin network [66, 69]. Therefore, it was shown that a partial extraction of membrane cholesterol using methyl-beta cyclodextrin, at concentrations of 2.5 or 5 mM, inhibited mechanosensitive activation of channels in K562 cell lines [70, 71]. In cells with a lowered cholesterol content, there was an observed increase in the threshold of activation, and a decrease in the probability of channels being in an open state. Thus, measurements of mechanosensitive flows in various conditions, and complementary data from fluorescent microscopy, indicate that suppression of activity of mechanosensitive channels is mediated by the reorganisation of actin, which is initiated, according to the authors, by disruption of raft integrity due to decreases in the levels of membrane cholesterol [70, 71].

Considering that the initiation of many intracellular signalling pathways is a result of membrane-bound proteins that amplify, at increased speed, the lateral diffusion of a signal, Jalali et al. [72] offered another theory of mechanotransduction. Using fluorescent labels, they estimated the speed of their migration in phospholipids bilayer, upon applying a shearing stress, and the dependence on how this stress was increased, that is, smoothly or in steps. In the case of a step-wise increase, the diffusion factor increased, while in case of gradual increase of stress this decreased. In their review, Huang et al. [18] assume, being guided by data from Butler et al. [73], that the effect of increases in the lateral diffusion is connected with ERK and JNK activation. Nevertheless, it still remains unclear how shearing stress leads to changes in membrane fluidity.

*4.4. Intracellular Structures.* It is well known that the action of external forces can lead to changes in the levels of gene expression. In combination of facts that the forces enclosed through the membrane-connected receptors, in certain cases, can lead to nucleus deformations [74], it is possible to assume a direct influence of external forces on chromatin and, as a result, on expression level [18]. Forces in this case could be transduced through the cytoskeletal network to the nuclear envelope, and then, through the laminin network to chromatin.

In addition, the action of external forces could be transduced to microtubules, leading to their disruption, depolymerisation, and the initiation of a signalling pathway [75].

It should be noted that conformational changes in various proteins could potentially be applied to a mechanosensory role, but there is no direct, practical proof of this. However, there is at least one example that a biochemical response can be caused by conformational changes of proteins. As discussed above, the folded domains of fibronectin can be exposed through the application of a stretching force to the molecule, which leads to fibril formation. This process was investigated both experimentally and through molecular-dynamic modelling methods [76, 77] and, as result, it was shown that a force of 3–5 pN is sufficient for the unfolding of domains, and that a subsequent force of 5 pN can lead to a molecule lengthening fivefold in comparison to its initial length [77, 78]. These levels of force are comparable to that, which according to estimates, can initiate mechanotransduction.

Much less is known, however, concerning the various intracellular proteins (e.g., Src-family kinases, vinculin, mDia, and ROCK) which also can be specific "molecular switches," undergoing conformational changes in response to an external force [79]. In fact, any protein participating in a mechanotransduction from extracellular contacts of a cell can act as a mechanosensor and stimulate the unfolding of both integrin isoforms [80] in addition to the proteins associated with them [81]. Proteins of the focal adhesion complex are also primary candidates for roles as mechanosensors. This becomes particularly obvious in the case of experimental data which shows that the stretching of cells, through the use of detergent (for removal of the cellular membrane), on a pliable substratum can lead to communication strengthening between focal adhesion kinase and paxillin, in the region of focal adhesion [82]. As the cellular membrane was removed in these experiments, the ion channels could not participate in this response.

According to theories proposed by Ingber [83], the cytoskeleton as a whole reacts to changes in mechanical tension through the extracellular matrix, and the associated integrins, leading to the reorganisation of microfilaments and microtubules. In tandem with this, the cortical cytoskeleton, which supports the plasma membrane through the formation of a rigid 3D-framework, is in tension in an external mechanical field [84].

Therefore, by summarising the studies above (Figure 1), it is possible to see that practically all possible mechanisms of primary mechanotransduction depend on the condition of the submembrane cortical cytoskeleton structure which determines the mechanical properties of various types of cells, and this is ultimately reflected in stiffness of cells.

Together, a number of experiments that focused on the influence of external mechanical conditions indicated that changes in the vector and modulation of external forces lead to changes in the structural/functional properties of both muscle (specialised for the generation of mechanical tension in cells) and nonmuscle cells.

## 5. Nonmuscle Cell Responses to Changes in External Mechanical Conditions

The changes in cell orientation within a gravitational field can be performed with the help of a horizontal clinostat and RPM—a random position machine [85–90].

Different types of cells have been cultivated under changing gravitational vector conditions. Most of the cells grown under these conditions showed changes in their cell profiles, as well as disorganisation of microtubules and microfilaments, and an increase in the number of apoptotic cells in the culture [91–96], in addition to alterations in mitochondrial localisation and clustering behaviour [56].

The data in the literature show that the influences of gravity vector changes in embryonic stem cells, decreased their ability to form embryoid bodies as a result of clinostation [89, 97]. Conversely, according to data from the same authors, the clinostation of embryoid bodies significantly increased the number of beta-III tubulin-positive cells (early neuroblasts). However, it also gave rise to an insignificant decrease in the number of MAP2-positive cells (late neuroblasts) over the course of spontaneous neuronal differentiation [89, 97].

Mesenchymal stem cells are also sensitive to microgravity conditions; however, there is no common hypothesis in the interpretation of the results obtained from these studies [96, 98–101]. Modelling of microgravity effects, meanwhile, suggested an inhibition of osteogenic differentiation and activation of adipogenic differentiation of mesenchymal stem cells [88, 96, 98–100, 102–104].

According to data cited in the literature, the changes in various cultivated cells can be connected to reorganisation of the actin cytoskeleton [90, 94], particularly, with F-actin destruction, which leads to Rho-dependent signalling pathway activation [98–100], and which can also regulate MAP-kinase cascades as a result of an increase of phosphorylated $ERK1/2^{MAPK}$ levels [96, 98, 103]. The change in differentiation potential of stem cells, after real or modelled microgravity changes, can also be connected to cytoskeleton reorganisation, as some of cytoskeletal structures may take part in the determination of the differentiation pathway [105, 106]. In addition, 24 hours of microgravity modelling caused transient changes in gene expression in mesenchymal stem cells. Some of these genes encoded actin cytoskeleton proteins and the elements associated with this. The modelling also decreased the capability for adhesion in these cells [90].

Thus, the change in the external mechanical (gravitational) vector leads to structural/functional changes in nonmuscle cells. These changes are seen in alterations in division speed and development potential, and are probably connected with cytoskeleton reorganisation. Nevertheless, the primary levels of nonmuscle cellular responses to a change in

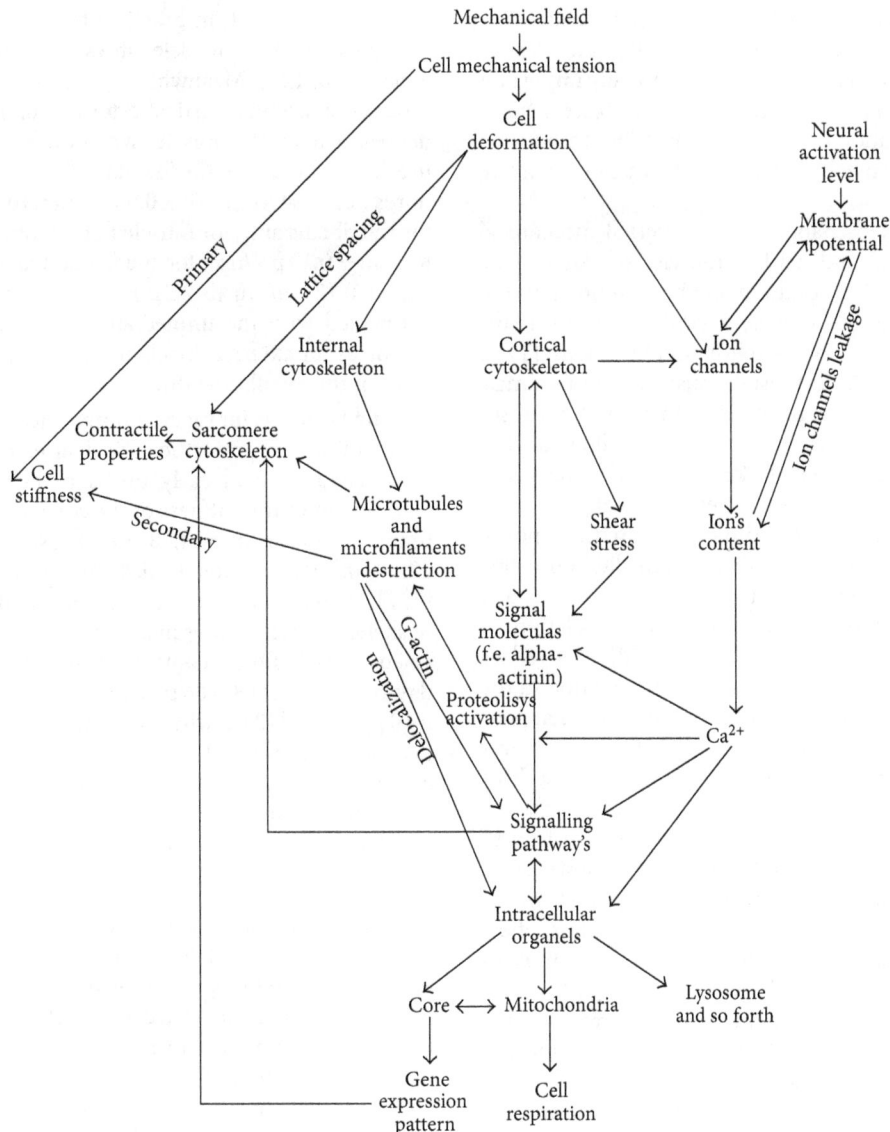

FIGURE 1: General mechanotransduction schemas. Changes in the external mechanical load cause a change in the internal mechanical tension of the cell and its deformation. Deformations can occur in the ion channels causing changes in their permeability for different ions, for example for calcium, which is a secondary messenger and can activate some signalling pathways. Moreover, deformations can also occur in the cytoskeleton, both in the sarcomere (for muscle cells) and cortical cytoskeleton, causing the release of different signalling molecules and activation of downstream signalling pathways. The final result will be a change in the cell's mechanical properties, its functional activity, and formation of adaptive patterns in gene expression.

external mechanical conditions are little studied and so the search for the mechanosensor will not end soon.

## 6. Changes in Original Mechanical Properties of Muscle Cells and Their Cellular Respiration in Response to Changes in External Mechanical Conditions

Results of numerous studies in conditions of real and simulated gravitational unloading testify that negative changes in various bodies and tissues are formed as a result of the action of microgravity.

Skeletal muscles are especially vulnerable to a gravity-free state as the specialised organ that executes position and motor functions. A subject of many studies is the m. soleus for which it has been shown that long exposures to microgravity conditions leads to essential decreases in weight, and changes in fibre atrophy [107–109]. In addition, a decrease in the functionality of both the whole muscle [110, 111] and single fibres [112] also takes place.

Exposure to microgravity also causes various changes in the human cardiovascular system, particularly a cephalic fluid-shift in the cranial direction [113, 114] and a change in the heart's systolic volume [115–117].

Nevertheless, a substantive problem which is obstructing space development by humans, in particular flight to Mars, is the early readaptation period to gravity. Especially important is the reloading of muscle and cardiovascular system functionality, the acceleration of which is impossible without an understanding of the mechanism of development of adaptive changes.

Differently directed changes in the external mechanical conditions of skeletal and cardiac muscles of rats can be achieved by means of a common method; antiorthostatic hindlimb suspension of animals, according to the Ilyin-Novikov method with Morey-Holton modifications [118]. On the one hand, the antiorthostatic suspension of animals leads to hindlimb disuse, while on the other hand, it causes an increase in the mechanical load on cardiomyocytes. In addition, the orientation of skeletal muscle fibres and cardiomyocytes in the gravitational field will change.

According to the scarce data connected with the response of skeletal muscles cells to gravity disuse in the literature, one of the first events to occur, after two days of antiorthostatic disuse of the hindlimbs of mice, is the accumulation of calcium ions in the soleus muscle [119, 120]. Above, we described that the maximal calcium ion accumulation in the soleus muscle fibres of rats and a Mongolian gerbils happens after one day of gravity disuse. For the medial gastrocnemius and tibialis anterior muscles, this maximum was seen a short time later—after seven days of antiorthostatic suspension [121]. The increase in the resting calcium level can lead to calpain activation [122, 123] and the following destruction of the muscle fibre structure. However, the means by which it is accumulated remain unknown. Probably, this effect happens via L-type calcium channels, but the mechanosensitive channels of the TRP family may also take part. In any case, the functioning of the channels incorporated in the membrane depends on the condition of the sarcolemma and the cytoskeleton connected with it. Moreover, we can suppose that the mechanical properties of cardiomyocytes will change, in other ways to that of skeletal muscle fibres, as the load on the cardiac muscle increases in microgravity conditions, while the load on the skeletal muscles of hindlimbs decreases.

### 6.1. Stiffness of Muscle Cells.

It is rather difficult to perform a direct evaluation of the native state of the cortical cytoskeleton of muscle fibres. However, the definition of its mechanical properties (transversal stiffness, to be exact) can help in the analysis of its structural changes. In addition, the complicated sarcomere organisation of a muscle fibre enables us to suppose that the transversal stiffness of different parts of the sarcolemma (Z-disk, M-line, and the part between them) will differ from each other. The differentiation of stiffness factors is of great interest, for example, in regarding the signalling role of different proteins of the costamere, which probably depends on its structure.

The methodology based on this above concept enabled us to define the transversal stiffness both of a contractile apparatus and of a membrane with a cortical cytoskeleton [5].

It was shown that the transversal stiffness of different parts of the soleus muscle contractile apparatus in the relaxed, calcium activated, or rigor states decreases, over the course of gravity disuse. It, in practical terms, does not change for the gastrocnemius muscle fibres and increases for tibialis anterior [8, 121]. Meanwhile, the transversal stiffness of the sarcolemma, with a cortical cytoskeleton in the relaxed state, decreases in all the muscles when under antiorthostatic disuse. This happens on the first day of disuse (the soleus muscle fibres decrease from $3.05 \pm 0.03$ pN/nm to $1.25 \pm 0.07$ pN/nm; for the tibialis anterior muscle fibres from $3.56 \pm 0.28$ pN/nm to $2.46 \pm 0.13$ pN/nm; for medial gastrocnemius fibres from $2.87 \pm 0.12$ pN/nm to $2.21 \pm 0.08$ pN/nm), which could be connected with the immediate change in the gravitational vector as an external mechanical factor, thus changing the load on the membrane [8].

Furthermore, for rat cardiomyocytes, it seems, that under the conditions of antiorthostatic disuse, there was an increase in mechanical load early on as a result of hypovolemia. Consequently, the stiffness of the cortical cytoskeleton membrane increased: from $4.03 \pm 0.11$ pN/nm in the control to $12.3 \pm 0.4$ pN/nm after 14 days under antiorthostatic disuse [6, 7]. At reloading after antiorthostatic disuse, the stiffness of skeletal muscles fibres increased (transversal stiffness of rat soleus muscle fibres: control—$2.94 \pm 0.14$ pN/nm, after 14-day-display—$1.11 \pm 0.06$ pN/nm, after 3-day reloading $2.92 \pm 0.10$ pN/nm) [124], while the stiffness of cardiomyocytes decreased (transversal stiffness of rat left ventricle cardiomyocytes: control—$4.03 \pm 0.11$ pN/nm, after 14-day display—$12.3 \pm 0.4$ pN/nm, after 7-day reloading—$4.3 \pm 0.4$ pN/nm) [6, 7], returning to the control level.

### 6.2. Protein Content.

Changes in the transversal stiffness factors, both of skeletal muscles fibres and cardiomyocytes, under the conditions of antiorthostatic disuse and the following reloading, correlated with the content of nonmuscle isoforms of actin (that form a cortical cytoskeleton) in the membrane fraction. They were also connected with differently directed changes in nonmuscle isoforms of the alpha-actinin content 1 and 4 of the membrane and cytoplasmic fractions [6, 124].

For soleus muscle fibres, it was shown that the content of beta-actin in the cytoplasmic fraction did not change as a result of 14-day suspension, nor in the following 3-day reloading. However the content of beta-actin in the membrane protein fraction decreased after disuse more than threefold when compared with the control level, but after three days of recovery, it did not differ from the control level, which correlated with the changes in sarcolemma transversal stiffness [124]. Meanwhile, for cardiomyocytes the content of gamma-actin in the cytoplasmic protein fraction also remained at the level of the control during disuse and reloading. However, the content of gamma-actin in the membrane protein fraction significantly increased during the first day of antiorthostatic suspension and continued to rise until the fourteenth day, showing similar dynamics to that of the transversal stiffness change. At the same time, the content of beta-actin in the cytoplasmic and membrane fractions did not change under the conditions of antiorthostatic disuse and the following reloading [6]. It should be noted that the increase of nonmuscle F-actin (beta-actin) content was noticed in cat cardiomyocytes during hypertrophy stimulation [8], although the content of gamma-actin was not defined during this experiment.

The change in the content of nonmuscle actin isoforms, particularly in the membrane fraction, will lead to a structural change in the cortical cytoskeleton as well as changes in actin-binding protein content, particularly alpha-actinin-1 and alpha-actinin-4. These alpha-actinin nonmuscle isoforms can bind calcium ions in micromolar concentrations, whereas calcium concentrations of more than $10^{-7}$ M fully inhibit alpha-actinin and actin binding [125], as well as lead to the phosphorylation of tyrosine by focal adhesion kinase inside the actin-biding domain [110]. Considering data on the increase in the resting calcium levels of muscle fibres in gravity disuse, interest is drawn towards the calcium-sensitive actin-binding proteins-alpha-actinin-1 and alpha-actinin-4.

For rat cardiomyocytes under conditions of antiorthostatic disuse, the content of alpha-actinin-1 in the cytoplasmic fraction of proteins decreased after seven days of disuse, while in the membrane fraction it increased. In the period of 3-day reloading after 14 days of antiorthostatic disuse, the content of alpha-actinin-1 in the membrane fraction decreased, while it increased in the cytoplasmic fraction. After seven days of recovery, both fractions were the same as the control level [6].

At the same time, in cardiomyocytes, the content of alpha-acitnin-4 in the membrane protein fraction increased during the first day, and starting from the third day its content increased from control levels in the cytoplasmic fraction as well. During the reloading period the content of alpha-acinin-4 in the membrane fraction decreased to the control level. As for the cytoplasmic fraction, it also decreased, but did not reach the control level [6]. In addition, in the rat soleus muscle fibres under conditions of antiorthostatic disuse, the content of alpha-actinin-4 did not change in the cytoplasmic protein fractions. In the membrane fraction, the content of alpha-acinin-4 decreased after 14 days of disuse, and increased after 3-day reloading, although it did not reach the same levels as the control [124].

Very little is known about the role of nonmuscle isoforms of alpha-actinin in skeletal muscles cells or cardiomyocytes. It is known that alpha-acinin-1 is expressed in cardiomyocytes [126] and in skeletal muscles cells, as well as alpha-actinin-4 in different stages of differentiation [127]. Alpha-actinin-1 and alpha-actinin-4 are nonmuscle isoforms of alpha-actinin—a protein that belongs to the spectrin family [128]. They function via antiparallel homodimer binding of actin thread ends [129]. In addition, alpha-actinin-4 connects the actin cytoskeleton to the membrane and facilitates interactions of the cortical cytoskeleton with cytoplasmic signalling proteins [130].

Nevertheless, there is some data, showing that the increase in alpha-actinin-4 content in the cytoplasmic fraction is associated with a decrease of alpha-actinin-1, and with the formation of a cancerous pattern in fibroblasts [131]. There are also results showing that the malignancy of cells, particularly lymphocytes, is followed by an increase of their stiffness, which was measured with the help of atomic force microscopy [11].

Therefore, the increase of cell stiffness is connected with the development of a cortical cytoskeleton, the decrease of alpha-actinin-1 and the increase of alpha-actinin-4 in the protein membrane fraction. Therefore, it can be supposed that the development of a cortical cytoskeleton will lead to the increase in alpha-actinin-1 and alpha-actinin-4 contents in membrane protein fractions.

*6.3. Cell Respiration.* Data obtained by Goffart et al. [127] showed that alpha-actinin-4 can bind to the promoter region of the cytochrome *c* gene, causing an increase in its expression that can influence the efficiency of cellular respiration.

For muscle cells, it was shown that the functions of mitochondria can influence their shape, due to extension or compression of the membrane, that can be mediated by the cytoskeleton [109, 132]. In addition, Milner et al. [133] found that abnormal accumulations of clusters of subsarcolemmal mitochondria, and also swelling of mitochondria with degeneration mitochondrial matrices, in the fibres of m. soleus of null-desmin mice. Moreover, these authors determined the intensity of cellular respiration and showed that both the level of oxygen uptake and the ADP dissociation constant markedly decreased in comparison with values in the control mice [133].

Since, in the early stages of gravity disuse, the relative desmin content in m. soleus rat fibres decreases [8, 122], this suggests that the speed of cellular respiration will decrease. These findings showed that cellular respiration decreased after three days of gravity disuse, achieved its minimum after seven days, and increased to the control level on the fourteenth day of simulated disuse [134]. A three-day reloading period leads to some decrease in the cellular respiration, but recovery of the respiration parameters up to the control level was observed after seven days of readaptation [135]. After two days of immobilisation of rat hindlimbs, a decrease in the respiration speed (by 37%) was determined in the subsarcolemmal mitochondria extracted from the m. soleus [136]. The ADP-stimulating speed of cellular respiration of the vastus lateralis muscle of two monkeys, under gravity-free conditions for 15 days, decreased by 28% and 32% [137]. Meanwhile, according to Bigard et al. [138], a disuse, for three weeks, did not cause definitive changes in the oxygen uptake of skinned soleus fibres.

During work on cardiomyocytes, Saks et al. [139] showed that the intensity of oxygen uptake depends on the state of the cytoskeleton. Our findings showed that the basal speed of cellular respiration of rat cardiomyocytes did not, in practical terms, change during antiorthostatic disuse; it did increase, but insignificantly, during the first day. Meanwhile on adding glutamate and malate to the medium, the respiration speed and maximum respiration speed significantly increased after just one day, and remained high for the whole period of disuse up to the fourteenth day. Thus, after a 3-day-reloading after 14 days of antiorthostatic disuse, all indicated parameters significantly decreased in comparison to the control level [6]. However, Bigard et al. [138] did not find any changes in the intensity of cellular respiration in cardiomyocytes after three weeks of antiorthostatic suspension. The discrepancy between our results and those of Bigard et al. [138] can be

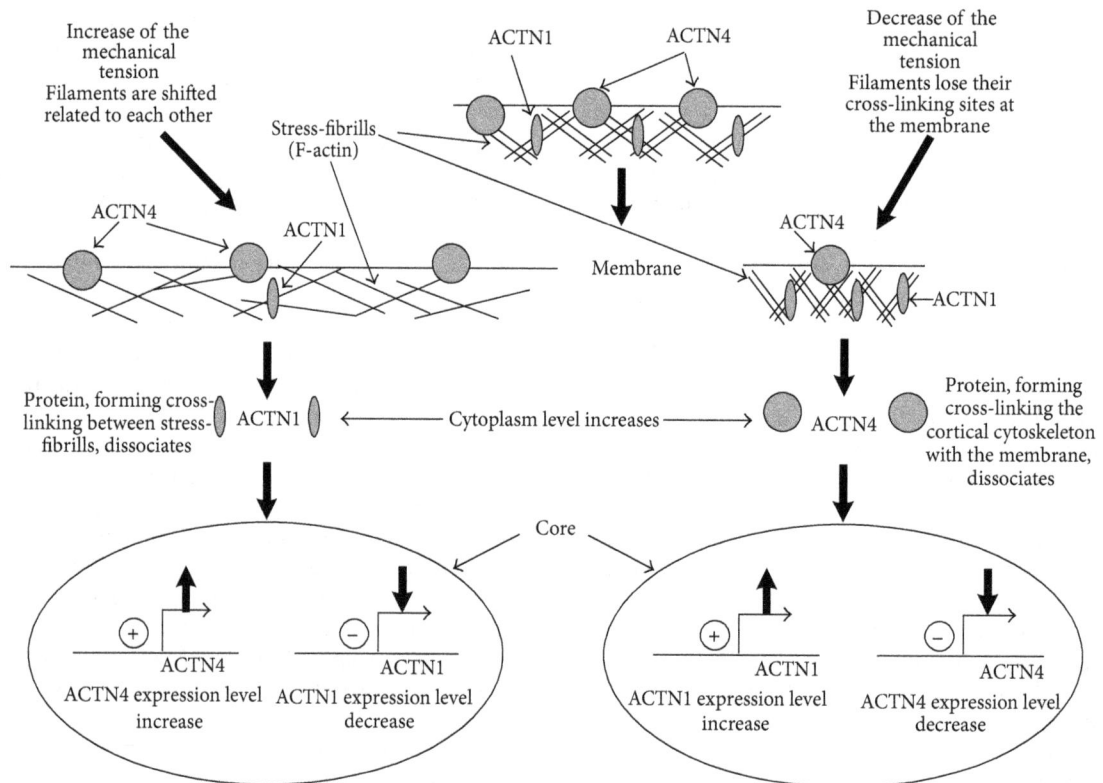

FIGURE 2: Hypothetical mechanism for earlier cellular responses to changes in mechanical conditions. The principal difference between stretch and compression is characterized by the dissociation of different molecules from the cortical cytoskeleton, for example, alpha-actinin-1 at stretching and alpha-actinin-4 at compression. Under cell stretch, there are cortical cytoskeleton deformations and subsequent shifts in actin filaments relative to each other in the stress fibres. This increases the probability of dissociation of the proteins that connect actin filaments, for example, alpha-actinin-1. Under cell compression, this happens predominantly via membrane deformation so that the conformation of the alpha-actinin-4 binding sites (e.g., due to cholesterol raft convergence) can change. This will lead to a release of proteins which connect with the membrane, for example, alpha-actinin-4. The release of alpha-actinin-4 causes the activation of the expression of the alpha-actinin-1 gene and repression of own expression. This occurs similarly for alpha-actinin-1. The release of different proteins causes activation of different pathways and formation of the response to the increase or decrease in mechanical load. The proposed mechanism is only hypothetical and therefore needs to be checked experimentally.

accounted for by different durations of the display. Although the primary increase in maximum respiration speed and respiration speed on the exogenic substrates shows that the numbers of mitochondria, and/or concentrations of the respiration chain complexes in them, can increase, this does not cause the increase in the basal respiration speed in the context of unchanged quantities of endogenic substrates. The increase in the relative content of desmin in rat cardiomyocytes under conditions of antiorthostatic disuse [6], which is necessary for the definition of mitochondria localisation and regulation of the permeability of their membranes, in combination with data on the increase of other oxidative enzyme and mitochondrial creatin kinase contents, prove this hypothesis [140].

## 7. Conclusion and Perspectives

The gravitational field is the most constantly acting factor across the whole evolutionary development of all living organisms on Earth. Therefore, it is logical to suggest that mechanosensory mechanisms explained acts of the primary receptions of the mechanical force could be universal for the different cells. Indeed, they can be connected with the most ancient of cell structures, the membrane with the cortical cytoskeleton.

The data presented in the literature prove that cell mechanical properties, particularly stiffness and the Young's modulus, are defined mostly by the state of the actin cytoskeleton and not by the microtubule system. However, the means of forming a state of tension through the system of thin filaments is not yet clear enough. There are at least two possible ways of realising this function. The first being that actin forms a reticulum which is stiff enough to play the role of a skeleton. The second is through actin-myosin interactions, which are not exclusive to just the muscle cells. The findings of Martens and Radmacher [10] on human fibroblasts showed that cell stiffness can be explained by the tension created by myosin transferring along the actin filaments.

There is no doubt that external surroundings influence cell mechanical properties, most likely by means of reorganisation of the actions of the cytoskeleton. The data on cells cultured on different substrates proves this suggestion. Both

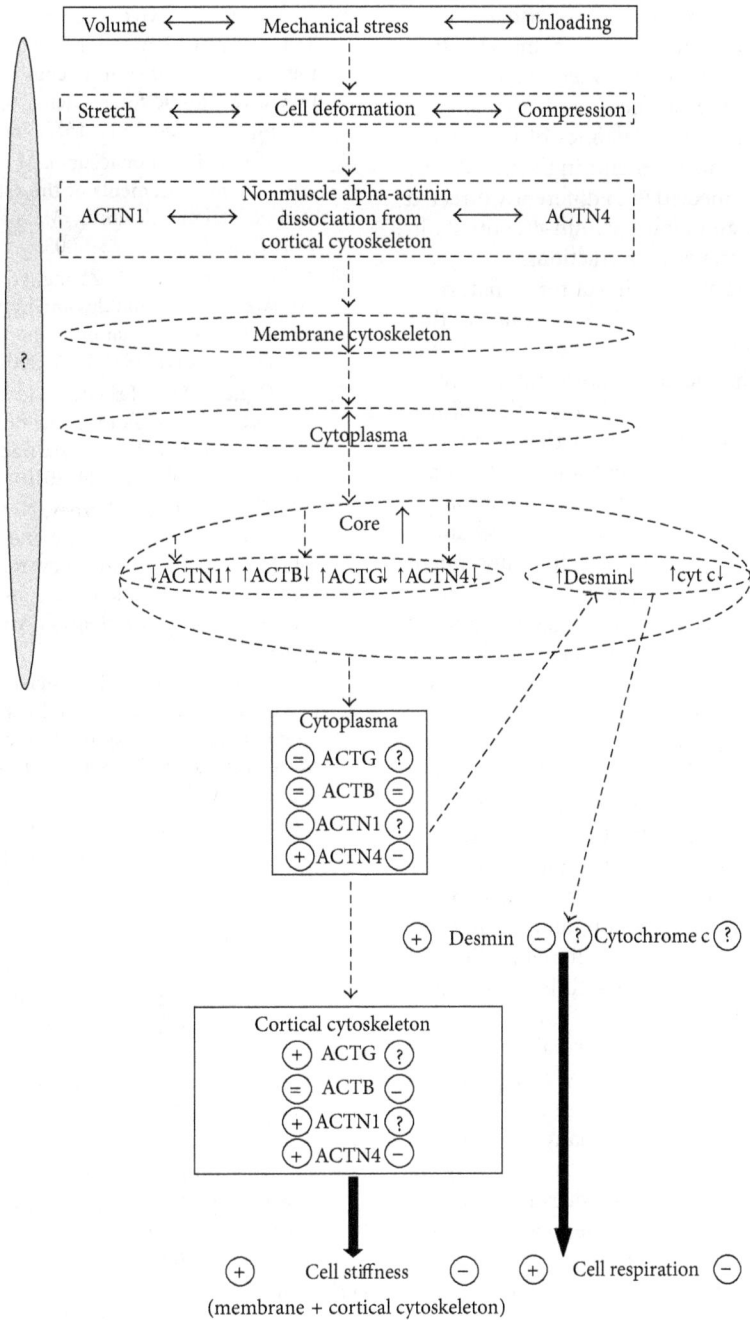

FIGURE 3: Hypothetical mechanism for the development of adaptive responses of skeletal muscle fibres and cardiomyocytes to the change in mechanical conditions. An increase in the external mechanical load on the cardiomyocytes in early stages of antiorthostatic disuse in rats should naturally cause an increase in the cell's ability to resist it, that is, an increase in cell stiffness and development of the cytoskeleton, in addition to an intensification of the cell respiration. Decrease of the load on the rat skeletal muscle fibres during antiorthostatic disuse does not require the development of the cortical cytoskeleton, and as a result the cell's stiffness should decrease. Hypothetical links are shown by dashed arrowheads and contours. −/+: decrease/increase in protein content, from the left/to right—for load/disuse.

the topography of the substrate and its stiffness are of great importance. The data, obtained with the use of agents that damage the actin network meanwhile, show a leading role for the thin filament system in forming the cell structure and its rapid responses to changes in external conditions.

Additionally, it was shown that different pathologic processes, especially malignant transformation of cells, can cause changes in the mechanical properties that tend to increase cell stiffness. Evidently, this is also connected with the impaired regulation of the actin cytoskeleton state that takes place during the increase in cell proliferation speed. On the other hand, aging of the organism causes decreases in the stiffness of some cells that can be also connected with a change in cytoskeleton state.

A change in the mechanical conditions also causes a change in cell stiffness, in particular in skeletal muscle fibres and cardiomyocytes. The decrease in the external mechanical force causes a decrease in the stiffness, and an increase, which in itself is correlated to the dynamics of nonmuscle actin isoform (beta- and gamma-) contents in the membrane fraction of muscle cells, is connected with differently directed changes in alpha-actinin-1 and alpha-actinin-4 contents in the cytoplasmic and membrane protein fractions.

The increase in an external mechanical force, naturally, should cause an increase in a cell's ability to resist it, that is, an increase in cell stiffness and development of the cytoskeleton. A decrease in the force does not require a developed cytoskeleton, and as a result the cell stiffness should decrease. Thus, it can be supposed that regulation of the actin cytoskeleton state plays the leading role in the change of speed of cell division, development of pathologic processes, and responses to the change in cellular mechanical conditions (hypo- and hyper-gravity), causing subsequent alterations in the cell's mechanical properties.

Thus, by summarising the data, we can suggest (Figures 2 and 3) that a decrease in mechanical load causes the decrease in skeletal muscles fibre stiffness, likely due to the decrease in nonmuscle actin isoforms and alpha-actinin content in the membrane protein fraction, and also the decrease in cell respiration intensity that takes place in the skeletal muscles. At the increase of mechanical load on cells, in particular on the cardiomyocytes, during antiorthostatic disuse in rats, causes an intensification of the processes of cellular respiration, an increase of nonmuscle actin and alpha-actinin content in the membrane protein fraction, and an increase in cell stiffness also. Therefore, there are differently directed changes in the alpha-actinin-1 and alpha-actinin-4 content in the cytoplasmic protein fractions.

The results discussed here suggest several approaches to the regulation of cell mechanosensitivity and present a range of possibilities for prevention of the negative effects of microgravity and correction of negative changes connected with hypogravity syndrome. For example, the use of gene constructs or any preparations which provide stabilisation of the cortical cytoskeleton could prevent the negative effects of exposure to a zero-gravity environment. However, the problem of perception of mechanical stimuli by different cells still requires further study.

## Acknowledgments

The financial support of the Russian Foundation of Basic Research (RFBR Grant 10-04-00106-a) and the program of fundamental research SSC RF-IBMP RAS are greatly acknowledged.

## References

[1] Ph. Carl and H. Schillers, "Elasticity measurement of living cells with an atomic force microscope: data acquisition and processing," *Pflugers Archiv European Journal of Physiology*, vol. 457, no. 2, pp. 551–559, 2008.

[2] A. B. Mathur, A. M. Collinsworth, W. M. Reichert, W. E. Kraus, and G. A. Truskey, "Endothelial, cardiac muscle and skeletal muscle exhibit different viscous and elastic properties as determined by atomic force microscopy," *Journal of Biomechanics*, vol. 34, no. 12, pp. 1545–1553, 2001.

[3] E. Defranchi, E. Bonaccurso, M. Tedesco et al., "Imaging and elasticity measurements of the sarcolemma of fully differentiated skeletal muscle fibres," *Microscopy Research and Technique*, vol. 67, no. 1, pp. 27–35, 2005.

[4] A. M. Collinsworth, S. Zhang, W. E. Kraus, and G. A. Truskey, "Apparent elastic modulus and hysteresis of skeletal muscle cells throughout differentiation," *American Journal of Physiology*, vol. 283, no. 4, pp. C1219–C1227, 2002.

[5] I. V. Ogneva, D. V. Lebedev, and B. S. Shenkman, "Transversal stiffness and young's modulus of single fibers from rat soleus muscle probed by atomic force microscopy," *Biophysical Journal*, vol. 98, no. 3, pp. 418–424, 2010.

[6] I. V. Ogneva, T. M. Mirzoev, N. S. Biryukov, O. M. Veselova, and I. M. Larina, "Structure and functional characteristics of rat's left ventricle cardiomyocytes under antiorthostatic suspension of various duration and subsequent reloading," *Journal of Biomedicine and Biotechnology*, vol. 2012, Article ID 659869, 11 pages, 2012.

[7] I. V. Ogneva and I. B. Ushakov, "The transversal stiffness of skeletal muscle fibers and cardiomyocytes in control and after simulated microgravity," in *Book 'Atomic Force Microscopy Investigations into Biology-From Cell to Protein'*, chapter 15, pp. 325–354, 2012.

[8] I. V. Ogneva, "Transversal stiffness of fibers and desmin content in leg muscles of rats under gravitational unloading of various durations," *Journal of Applied Physiology*, vol. 109, no. 6, pp. 1702–1709, 2010.

[9] K. D. Costa, A. J. Sim, and F. C. P. Yin, "Non-Hertzian approach to analyzing mechanical properties of endothelial cells probed by atomic force microscopy," *Journal of Biomechanical Engineering*, vol. 128, no. 2, pp. 176–184, 2006.

[10] J. C. Martens and M. Radmacher, "Softening of the actin cytoskeleton by inhibition of myosin II," *Pflugers Archiv European Journal of Physiology*, vol. 456, no. 1, pp. 95–100, 2008.

[11] X. Cai, J. Cai, S. Dong, H. Deng, and M. Hu, "Morphology and mechanical properties of normal lymphocyte and Jurkat revealed by atomic force microscopy," *Shengwu Gongcheng Xuebao/Chinese Journal of Biotechnology*, vol. 25, no. 7, pp. 1107–1112, 2009.

[12] I. Dulińska, M. Targosz, W. Strojny et al., "Stiffness of normal and pathological erythrocytes studied by means of atomic force microscopy," *Journal of Biochemical and Biophysical Methods*, vol. 66, no. 1–3, pp. 1–11, 2006.

[13] M. Lekka, M. Fornal, G. Pyka-Fościak et al., "Erythrocyte stiffness probed using atomic force microscope," *Biorheology*, vol. 42, no. 4, pp. 307–317, 2005.

[14] C. H. Hsieh, Y. H. Lin, S. Lin, J. J. Tsai-Wu, C. H. Herbert Wu, and C. C. Jiang, "Surface ultrastructure and mechanical property of human chondrocyte revealed by atomic force microscopy," *Osteoarthritis and Cartilage*, vol. 16, no. 4, pp. 480–488, 2008.

[15] D. Docheva, D. Padula, C. Popov, W. Mutschler, H. Clausen-Schaumann, and M. Schieker, "Researching into the cellular shape, volume and elasticity of mesenchymal stem cells, osteoblasts and osteosarcoma cells by atomic force microscopy: stem Cells," *Journal of Cellular and Molecular Medicine*, vol. 12, no. 2, pp. 537–552, 2008.

[16] E. Takai, K. D. Costa, A. Shaheen, C. T. Hung, and X. E. Guo, "Osteoblast elastic modulus measured by atomic force microscopy is substrate dependent," *Annals of Biomedical Engineering*, vol. 33, no. 7, pp. 963–971, 2005.

[17] E. K. F. Yim, E. M. Darling, K. Kulangara, F. Guilak, and K. W. Leong, "Nanotopography-induced changes in focal adhesions, cytoskeletal organization, and mechanical properties of human mesenchymal stem cells," *Biomaterials*, vol. 31, no. 6, pp. 1299–1306, 2010.

[18] H. Huang, R. D. Kamm, and R. T. Lee, "Cell mechanics and mechanotransduction: pathways, probes, and physiology," *American Journal of Physiology*, vol. 287, no. 1, pp. C1–C11, 2004.

[19] P. F. Davies, K. A. Barbee, M. V. Volin et al., "Spatial relationships in early signaling events of flow-mediated endothelial mechanotransduction," *Annual Review of Physiology*, vol. 59, pp. 527–549, 1997.

[20] B. P. Helmke, A. B. Rosen, and P. F. Davies, "Mapping mechanical strain of an endogenous cytoskeletal network in living endothelial cells," *Biophysical Journal*, vol. 84, no. 4, pp. 2691–2699, 2003.

[21] B. P. Helmke, D. B. Thakker, R. D. Goldman, and P. F. Davies, "Spatiotemporal analysis of flow-induced intermediate filament displacement in living endothelial cells," *Biophysical Journal*, vol. 80, no. 1, pp. 184–194, 2001.

[22] A. R. Bausch, F. Ziemann, A. A. Boulbitch, K. Jacobson, and E. Sackmann, "Local measurements of viscoelastic parameters of adherent cell surfaces by magnetic bead microrheometry," *Biophysical Journal*, vol. 75, no. 4, pp. 2038–2049, 1998.

[23] S. Hu, J. Chen, B. Fabry et al., "Intracellular stress tomography reveals stress focusing and structural anisotropy in cytoskeleton of living cells," *American Journal of Physiology*, vol. 285, no. 5, pp. C1082–C1090, 2003.

[24] H. Huang, C. Y. Dong, H. S. Kwon, J. D. Sutin, R. D. Kamm, and P. T. So, "Three-dimensional cellular deformation analysis with two-photon magnetic manipulator workstation," *Biophysical Journal*, vol. 82, no. 4, pp. 2211–2223, 2002.

[25] N. Q. Balaban, U. S. Schwarz, D. Riveline et al., "Force and focal adhesion assembly: a close relationship studied using elastic micropatterned substrates," *Nature Cell Biology*, vol. 3, no. 5, pp. 466–472, 2001.

[26] P. P. Lehenkari and M. A. Horton, "Single integrin molecule adhesion forces in intact cells measured by atomic force microscopy," *Biochemical and Biophysical Research Communications*, vol. 259, no. 3, pp. 645–650, 1999.

[27] H. P. Erickson, "Reversible unfolding of fibronectin type III and immunoglobulin domains provides the structural basis for stretch and elasticity of titin and fibronectin," *Proceedings of the National Academy of Sciences of the United States of America*, vol. 91, no. 21, pp. 10114–20118, 1994.

[28] J. M. Ferrer, H. Lee, J. Chen et al., "Measuring molecular rupture forces between single actin filaments and actin-binding proteins," *Proceedings of the National Academy of Sciences of the United States of America*, vol. 105, no. 27, pp. 9221–9226, 2008.

[29] J. T. Finer, R. M. Simmons, and J. A. Spudich, "Single myosin molecule mechanics: piconewton forces and nanometre steps," *Nature*, vol. 368, no. 6467, pp. 113–119, 1994.

[30] T. J. Dennerll, H. C. Joshi, V. L. Steel, R. E. Buxbaum, and S. R. Heidemann, "Tension and compression in the cytoskeleton of PC-12 neurites II: quantitative measurements," *Journal of Cell Biology*, vol. 107, no. 2, pp. 665–674, 1988.

[31] A. J. Putnam, K. Schultz, and D. J. Mooney, "Control of microtubule assembly by extracellular matrix and externally applied strain," *American Journal of Physiology*, vol. 280, no. 3, pp. C556–C564, 2001.

[32] S. Liu, D. A. Calderwood, and M. H. Ginsberg, "Integrin cytoplasmic domain-binding proteins," *Journal of Cell Science*, vol. 113, no. 20, pp. 3563–3571, 2000.

[33] S. Sukharev and D. P. Corey, "Mechanosensitive channels: multiplicity of families and gating paradigms," *Science's STKE*, vol. 2004, no. 219, Article ID re4, 2004.

[34] R. Maroto, A. Raso, T. G. Wood, A. Kurosky, B. Martinac, and O. P. Hamill, "TRPC1 forms the stretch-activated cation channel in vertebrate cells," *Nature Cell Biology*, vol. 7, no. 2, pp. 179–185, 2005.

[35] S. Sukharev, M. Betanzos, C. S. Chiang, and H. Robert Guy, "The gating mechanism of the large mechanosensitive channel MscL," *Nature*, vol. 409, no. 6821, pp. 720–724, 2001.

[36] J. Howard and S. Bechstedt, "Hypothesis: a helix of ankyrin repeats of the NOMPC-TRP ion channel is the gating spring of mechanoreceptors," *Current Biology*, vol. 14, no. 6, pp. R224–R226, 2004.

[37] O. P. Hamill and B. Martinac, "Molecular basis of mechanotransduction in living cells," *Physiological Reviews*, vol. 81, no. 2, pp. 685–740, 2001.

[38] G. Chang, R. H. Spencer, A. T. Lee, M. T. Barclay, and D. C. Rees, "Structure of the MscL homolog from Mycobacterium tuberculosis: a gated mechanosensitive ion channel," *Science*, vol. 282, no. 5397, pp. 2220–2226, 1998.

[39] J. Gullingsrud, D. Kosztin, and K. Schulten, "Structural determinants of MscL gating studied by molecular dynamics simulations," *Biophysical Journal*, vol. 80, no. 5, pp. 2074–2081, 2001.

[40] E. Perozo, D. M. Cortes, P. Sompornpisut, A. Kloda, and B. Martinac, "Open channel structure of MscL and the gating mechanism of mechanosensitive channels," *Nature*, vol. 418, no. 6901, pp. 942–948, 2002.

[41] J. Arnadottir and M. Chalfie, "Eukaryotic mechanosensitive channels," *Annual Review of Biophysics*, vol. 39, pp. 111–137, 2010.

[42] C. Montell, "The TRP superfamily of cation channels," *Science's STKE*, vol. 2005, no. 272, Article ID re3, 2005.

[43] J. A. Stiber, Y. Tang, T. Li, and P. B. Rosenberg, "Cytoskeletal regulation of TRPC channels in the cardiorenal system," *Current Hypertension Reports*, vol. 14, no. 6, pp. 492–497, 2012.

[44] F. Lesage and M. Lazdunski, "Molecular and functional properties of two-pore-domain potassium channels," *American Journal of Physiology*, vol. 279, no. 5, pp. F793–F801, 2000.

[45] M. Althaus, R. Bogdan, W. G. Clauss, and M. Fronius, "Mechano-sensitivity of epithelial sodium channels (ENaCs): laminar shear stress increases ion channel open probability," *FASEB Journal*, vol. 21, no. 10, pp. 2389–2399, 2007.

[46] L. M. Satlin, S. Sheng, C. B. Woda, and T. R. Kleyman, "Epithelial Na+ channels are regulated by flow," *American Journal of Physiology*, vol. 280, no. 6, pp. F1010–F1018, 2001.

[47] M. D. Carattino, S. Sheng, and T. R. Kleyman, "Epithelial Na+ channels are activated by Laminar shear stress," *Journal of Biological Chemistry*, vol. 279, no. 6, pp. 4120–4126, 2004.

[48] W. Liu, S. Xu, C. Woda, P. Kim, S. Weinbaum, and L. M. Satlin, "Effect of flow and stretch on the [Ca2+]i response of principal and intercalated cells in cortical collecting duct," *American Journal of Physiology*, vol. 285, no. 5, pp. F998–F1012, 2003.

[49] R. Tarran, B. Button, M. Picher et al., "Normal and cystic fibrosis airway surface liquid homeostasis: the effects of phasic shear stress and viral infections," *Journal of Biological Chemistry*, vol. 280, pp. 35751–35759, 2005.

[50] H. A. Drummond, D. Gebremedhin, and D. R. Harder, "Degenerin/epithelial Na+ channel proteins: components of a vascular mechanosensor," *Hypertension*, vol. 44, no. 5, pp. 643–648, 2004.

[51] H. A. Drummond, M. J. Welsh, and F. M. Abboud, "ENaC subunits are molecular components of the arterial baroreceptor complex," *Annals of the New York Academy of Sciences*, vol. 940, pp. 42–47, 2001.

[52] N. L. Jernigan and H. A. Drummond, "Vascular ENaC proteins are required for renal myogenic constriction," *American Journal of Physiology*, vol. 289, no. 4, pp. F891–F901, 2005.

[53] H. A. Drummond, F. M. Abboud, and M. J. Welsh, "Localization of $\beta$ and $\gamma$ subunits of ENaC in sensory nerve endings in the rat foot pad," *Brain Research*, vol. 884, no. 1-2, pp. 1–12, 2000.

[54] H. T. Ma, R. L. Patterson, D. B. Van Rossum, L. Birnbaumer, K. Mikoshiba, and D. L. Gill, "Requirement of the inositol trisphosphate receptor for activation of store-operated Ca2+ channels," *Science*, vol. 287, no. 5458, pp. 1647–1651, 2000.

[55] J. R. Holda and L. A. Blatter, "Capacitative calcium entry is inhibited in vascular endothelial cells by disruption of cytoskeletal microfilaments," *FEBS Letters*, vol. 403, no. 2, pp. 191–196, 1997.

[56] H. Schatten, M. L. Lewis, and A. Chakrabarti, "Spaceflight and clinorotation cause cytoskeleton and mitochondria changes and increases in apoptosis in cultured cells," *Acta Astronautica*, vol. 49, no. 3–10, pp. 399–418, 2001.

[57] P. Devarajan, D. A. Scaramuzzino, and J. S. Morrow, "Ankyrin binds to two distinct cytoplasmic domains of Na,K-ATPase $\alpha$ subunit," *Proceedings of the National Academy of Sciences of the United States of America*, vol. 91, no. 8, pp. 2965–2969, 1994.

[58] Y. Srinivasan, L. Elmer, J. Davis, V. Bennett, and K. Angelides, "Ankyrin and spectrin associate with voltage-dependent sodium channels in brain," *Nature*, vol. 333, no. 6169, pp. 177–180, 1988.

[59] D. J. Benos, M. S. Awayda, I. I. Ismailov, and J. P. Johnson, "Structure and function of amiloride-sensitive Na+ channels," *Journal of Membrane Biology*, vol. 143, no. 1, pp. 1–18, 1995.

[60] Y. A. Negulyaev, E. A. Vedernikova, and A. V. Maximov, "Disruption of actin filaments increases the activity of sodium-conducting channels in human myeloid leukemia cells," *Molecular Biology of the Cell*, vol. 7, no. 12, pp. 1857–1864, 1996.

[61] A. V. Maximov, E. A. Vedernikova, H. Hinssen, S. Y. Khaitlina, and Y. A. Negulyaev, "Ca-dependent regulation of Na+-selective channels via actin cytoskeleton modification in leukemia cells," *FEBS Letters*, vol. 412, no. 1, pp. 94–96, 1997.

[62] A. V. Maximov, E. A. Vedernikova, and N. YuA, "F-actin network regulates the activity of Na+-selective channels in human myeloid leukemia cells. The role of plasma gelsolin and intracellular calcium," *Biophysical Journal*, vol. 72, no. 2, part 2, p. A.226, 1997.

[63] T. Harder and K. Simons, "Clusters of glycolipid and glycosylphosphatidylinositol-anchored proteins in lymphoid cells: accumulation of actin regulated by local tyrosine phosphorylation," *European Journal of Immunology*, vol. 29, no. 2, pp. 556–562, 1999.

[64] D. A. Brown and E. London, "Structure and function of sphingolipid- and cholesterol-rich membrane rafts," *Journal of Biological Chemistry*, vol. 275, no. 23, pp. 17221–17224, 2000.

[65] T. Nebl, K. N. Pestonjamasp, J. D. Leszyk, J. L. Crowley, S. W. Oh, and E. J. Luna, "Proteomic analysis of a detergent-resistant membrane skeleton from neutrophil plasma membranes," *Journal of Biological Chemistry*, vol. 277, no. 45, pp. 43399–43409, 2002.

[66] D. A. Brown, "Lipid rafts, detergent-resistant membranes, and raft targeting signals," *Physiology*, vol. 21, no. 6, pp. 430–439, 2006.

[67] I. Levitan, A. E. Christian, T. N. Tulenko, and G. H. Rothblat, "Membrane cholesterol content modulates activation of volume-regulated anion current in bovine endothelial cells," *Journal of General Physiology*, vol. 115, no. 4, pp. 405–416, 2000.

[68] V. G. Shlyonsky, F. Mies, and S. Sariban-Sohraby, "Epithelial sodium channel activity in detergent-resistant membrane microdomains," *American Journal of Physiology*, vol. 284, no. 1, pp. F182–F188, 2003.

[69] M. Edidin, "The state of lipid rafts: from model membranes to cells," *Annual Review of Biophysics and Biomolecular Structure*, vol. 32, pp. 257–283, 2003.

[70] A. V. Sudarikova, Y. A. Negulyaev, and E. A. Morachevskaya, "Cholesterol depletion affects mechanosensitive channel gating coupled with F-actin rearrangement," *Proceedings of the Physical Society*, pp. 95–96, 2006.

[71] E. A. Morachevskaya, A. V. Sudarikova, and Y. A. Negulyaev, "Mechanosensitive channel activity and F-actin organization in cholesterol-depleted human leukaemia cells," *Cell Biology International*, vol. 31, no. 4, pp. 374–381, 2007.

[72] S. Jalali, M. A. Del Pozo, K. D. Chen et al., "Integrin-mediated mechanotransduction requires its dynamic interaction with specific extracellular matrix (ECM) ligands," *Proceedings of the National Academy of Sciences of the United States of America*, vol. 98, no. 9, pp. 1042–1046, 2001.

[73] P. J. Butler, T. C. Tsou, J. Y. Li, S. Usami, and S. Chien, "Rate sensitivity of shear-induced changes in the lateral diffusion of endothelial cell membrane lipids: a role for membrane perturbation in shear-induced MAPK activation," *The FASEB Journal*, vol. 16, no. 2, pp. 216–218, 2002.

[74] A. J. Maniotis, C. S. Chen, and D. E. Ingber, "Demonstration of mechanical connections between integrins, cytoskeletal filaments, and nucleoplasm that stabilize nuclear structure," *Proceedings of the National Academy of Sciences of the United States of America*, vol. 94, no. 3, pp. 849–854, 1997.

[75] D. J. Odde, L. Ma, A. H. Briggs, A. DeMarco, and M. W. Kirschner, "Microtubule bending and breaking in living fibroblast cells," *Journal of Cell Science*, vol. 112, no. 19, pp. 3283–3288, 1999.

[76] D. Craig, A. Krammer, K. Schulten, and V. Vogel, "Comparison of the early stages of forced unfolding for fibronectin type III modules," *Proceedings of the National Academy of Sciences of the United States of America*, vol. 98, no. 10, pp. 5590–5595, 2001.

[77] M. Gao, D. Craig, V. Vogel, and K. Schulten, "Identifying unfolding intermediates of FN-III10 by steered molecular dynamics," *Journal of Molecular Biology*, vol. 323, no. 5, pp. 939–950, 2002.

[78] V. Vogel, W. E. Thomas, D. W. Craig, A. Krammer, and G. Baneyx, "Structural insights into the mechanical regulation of molecular recognition sites," *Trends in Biotechnology*, vol. 19, no. 10, pp. 416–423, 2001.

[79] B. Geiger, A. Bershadsky, R. Pankov, and K. M. Yamada, "Transmembrane extracellular matrix-cytoskeleton crosstalk," *Nature Reviews Molecular Cell Biology*, vol. 2, no. 11, pp. 793–805, 2001.

[80] M. E. Chicurel, R. H. Singer, C. J. Meyer, and D. E. Ingber, "Integrin binding and mechanical tension induce movement of mRNA and ribosomes to focal adhesions," *Nature*, vol. 392, no. 6677, pp. 730–733, 1998.

[81] C. Zhong, M. Chrzanowska-Wodnicka, J. Brown, A. Shaub, A. M. Belkin, and K. Burridge, "Rho-mediated contractility exposes a cryptic site in fibronectin and induces fibronectin matrix assembly," *Journal of Cell Biology*, vol. 141, no. 2, pp. 539–551, 1998.

[82] Y. Sawada and M. P. Sheetz, "Force transduction by Triton cytoskeletons," *Journal of Cell Biology*, vol. 156, no. 4, pp. 609–615, 2002.

[83] D. E. Ingber, "Cellular mechanotransduction: putting all the pieces together again," *FASEB Journal*, vol. 20, no. 7, pp. 811–827, 2006.

[84] C. Vera, R. Skelton, F. Bossens, and L. A. Sung, "3-D nanomechanics of an erythrocyte junctional complex in equibiaxial and anisotropic deformations," *Annals of Biomedical Engineering*, vol. 33, no. 10, pp. 1387–1404, 2005.

[85] G. Albrecht-Buehler, "The simulation of microgravity conditions on the ground," *ASGSB Bulletin*, vol. 5, no. 2, pp. 3–10, 1992.

[86] A. Cogoli, "Gravitational physiology of human immune cells: a review of in vivo, ex vivo and in vitro studies," *Journal of Gravitational Physiology*, vol. 3, no. 1, pp. 1–9, 1996.

[87] S. J. Pardo, M. J. Patel, M. C. Sykes et al., "Simulated microgravity using the Random Positioning Machine inhibits differentiation and alters gene expression profiles of 2T3 preosteoblasts," *American Journal of Physiology*, vol. 288, no. 6, pp. C1211–C1221, 2005.

[88] J. G. Gershovich, N. A. Konstantinova, P. M. Gershovich, and L. B. Buravkova, "The effects of prolonged gravity vector randomization on differentiation of precursour cells in vitro," *Journal of Gravitational Physiology*, vol. 14, no. 1, pp. P133–134, 2007.

[89] L. B. Buravkova, Y. A. Romanov, N. A. Konstantinova, S. V. Buravkov, Y. G. Gershovich, and I. A. Grivennikov, "Cultured stem cells are sensitive to gravity changes," *Acta Astronautica*, vol. 63, no. 5-6, pp. 603–608, 2008.

[90] P. M. Gershovich, J. G. Gershovich, and L. B. Buravkova, "Simulated microgravity alters actin cytoskeleton and integrin-mediated focal adhesions of cultured human mesenchymal stromal cells," in *Life in Space for Life on Earth*, fra, July 2008.

[91] D. Sarkar, T. Nagaya, K. Koga, and H. Seo, "Culture in vector-averaged gravity environment in a clinostat results in detachment of osteoblastic ROS 17/2.8 cells," *Environmental Medicine*, vol. 43, no. 1, pp. 22–24, 1999.

[92] B. M. Uva, M. A. Masini, M. Sturla et al., "Clinorotation-induced weightlessness influences the cytoskeleton of glial cells in culture," *Brain Research*, vol. 934, no. 2, pp. 132–139, 2002.

[93] S. Gaboyard, M. P. Blanchard, C. Travo, M. Viso, A. Sans, and J. Lehouelleur, "Weightlessness affects cytoskeleton of rat utricular hair cells during maturation in vitro," *NeuroReport*, vol. 13, no. 16, pp. 2139–2142, 2002.

[94] P. A. Plett, R. Abonour, S. M. Frankovitz, and C. M. Orschell, "Impact of modeled microgravity on migration, differentiation, and cell cycle control of primitive human hematopoietic progenitor cells," *Experimental Hematology*, vol. 32, no. 8, pp. 773–781, 2004.

[95] M. A. Kacena, P. Todd, and W. J. Landis, "Osteoblasts subjected to spaceflight and simulated space shuttle launch conditions," *In Vitro Cellular & Developmental Biology*, vol. 39, no. 10, pp. 454–459, 2004.

[96] Z. Q. Dai, R. Wang, S. K. Ling, Y. M. Wan, and Y. H. Li, "Simulated microgravity inhibits the proliferation and osteogenesis of rat bone marrow mesenchymal stem cells," *Cell Proliferation*, vol. 40, no. 5, pp. 671–684, 2007.

[97] N. A. Konstantinova, L. B. Buravkova, E. S. Manuilova, and I. A. Grivennikov, "The effects of gravity vector randomization on mouse embryonic stem cells in vitro," *Journal of Gravitational Physiology*, vol. 13, no. 1, pp. 149–150, 2006.

[98] M. Zayzafoon, W. E. Gathings, and J. M. McDonald, "Modeled microgravity inhibits osteogenic differentiation of human mesenchymal stem cells and increases adipogenesis," *Endocrinology*, vol. 145, no. 5, pp. 2421–2432, 2004.

[99] V. E. Meyers, M. Zayzafoon, J. T. Douglas, and J. M. McDonald, "RhoA and cytoskeletal disruption mediate reduced osteoblastogenesis and enhanced adipogenesis of human mesenchymal stem cells in modeled microgravity," *Journal of Bone and Mineral Research*, vol. 20, no. 10, pp. 1858–1866, 2005.

[100] V. E. Meyers, M. Zayzafoon, S. R. Gonda, W. E. Gathings, and J. M. McDonald, "Modeled microgravity disrupts collagen I/integrin signaling during osteoblastic differentiation of human mesenchymal stem cells," *Journal of Cellular Biochemistry*, vol. 93, no. 4, pp. 697–707, 2004.

[101] L. Yuge, T. Kajiume, H. Tahara et al., "Microgravity potentiates stem cell proliferation while sustaining the capability of differentiation," *Stem Cells and Development*, vol. 15, no. 6, pp. 921–929, 2006.

[102] M. J. Patel, W. Liu, M. C. Sykes et al., "Identification of mechanosensitive genes in osteoblasts by comparative microarray studies using the rotating wall vessel and the Random Positioning Machine," *Journal of Cellular Biochemistry*, vol. 101, no. 3, pp. 587–599, 2007.

[103] Z. Pan, J. Yang, C. Guo et al., "Effects of hindlimb unloading on ex vivo growth and osteogenic/adipogenic potentials of bone marrow-derived mesenchymal stem cells in rats," *Stem Cells and Development*, vol. 17, no. 4, pp. 795–804, 2008.

[104] Y. Huang, Z. Q. Dai, S. K. Ling, H. Y. Zhang, Y. M. Wan, and Y. H. Li, "Gravity, a regulation factor in the differentiation of rat bone marrow mesenchymal stem cells," *Journal of Biomedical Science*, vol. 16, no. 1, p. 87, 2009.

[105] R. Sordella, W. Jiang, G. C. Chen, M. Curto, and J. Settleman, "Modulation of Rho GTPase signaling regulates a switch between adipogenesis and myogenesis," *Cell*, vol. 113, no. 2, pp. 147–158, 2003.

[106] R. McBeath, D. M. Pirone, C. M. Nelson, K. Bhadriraju, and C. S. Chen, "Cell shape, cytoskeletal tension, and RhoA regulate stem cell lineage commitment," *Developmental Cell*, vol. 6, no. 4, pp. 483–495, 2004.

[107] F. W. Booth and J. R. Kelso, "Effect of hind limb immobilization on contractile and histochemical properties of skeletal muscle," *Pflugers Archiv European Journal of Physiology*, vol. 342, no. 3, pp. 231–238, 1973.

[108] D. Desplanches, M. H. Mayet, E. I. Ilyina-Kakueva, B. Sempore, and R. Flandrois, "Skeletal muscle adaptation in rats flown on Cosmos 1667," *Journal of Applied Physiology*, vol. 68, no. 1, pp. 48–52, 1990.

[109] Y. Capetanaki, R. J. Bloch, A. Kouloumenta, M. Mavroidis, and S. Psarras, "Muscle intermediate filaments and their links to membranes and membranous organelles," *Experimental Cell Research*, vol. 313, no. 10, pp. 2063–2076, 2007.

[110] G. T. Waites, I. R. Graham, P. Jackson et al., "Mutually exclusive splicing of calcium-binding domain exons in chick $\alpha$- actinin," *Journal of Biological Chemistry*, vol. 267, no. 9, pp. 6263–6271, 1992.

[111] K. Lee, Y. S. Lee, M. Lee, M. Yamashita, and I. Choi, "Mechanics and fatigability of the rat soleus muscle during early reloading," *Yonsei Medical Journal*, vol. 45, no. 4, pp. 690–702, 2004.

[112] K. S. McDonald and R. H. Fitts, "Effect of hindlimb unloading on rat soleus fiber force, stiffness, and calcium sensitivity," *Journal of Applied Physiology*, vol. 79, no. 5, pp. 1796–1802, 1995.

[113] W. E. Thornton, T. P. Moore, and S. L. Pool, "Fluid shifts in weightlessness," *Aviation, Space, and Environmental Medicine*, vol. 58, no. 9, pp. A86–A90, 1987.

[114] D. E. Watenpaugh and A. R. Hargens, "The cardiovascular system in microgravity," in *Handbook of Physiology. Environmental Physiology*, vol. 1, part 3, chapter 29, pp. 631–674, American Physiological Society, Bethesda, Md, USA, 4th edition, 1996.

[115] J. V. Nixon, R. G. Murray, C. Bryant et al., "Early cardiovascular adaptation to simulated zero gravity," *Journal of Applied Physiology*, vol. 46, no. 3, pp. 541–548, 1979.

[116] M. W. Bungo, D. J. Goldwater, R. L. Popp, and H. Sandler, "Echocardiographic evaluation of space shuttle crewmembers," *Journal of Applied Physiology*, vol. 62, no. 1, pp. 278–283, 1987.

[117] J. B. Charles and C. M. Lathers, "Cardiovascular adaptation to spaceflight," *Journal of Clinical Pharmacology*, vol. 31, pp. 1010–1023, 1991.

[118] E. Morey-Holton, R. K. Globus, A. Kaplansky, and G. Durnova, "The hindlimb unloading rat model: literature overview, technique update and comparison with space flight data," *Advances in Space Biology and Medicine*, vol. 10, pp. 7–40, 2005.

[119] C. P. Ingalls, G. L. Warren, and R. B. Armstrong, "Intracellular Ca2+ transients in mouse soleus muscle after hindlimb unloading and reloading," *Journal of Applied Physiology*, vol. 87, no. 1, pp. 386–390, 1999.

[120] C. P. Ingalls, J. C. Wenke, and R. B. Armstrong, "Time course changes in [Ca2+]i, force, and protein content in hindlimb-suspended mouse soleus muscles," *Aviation Space and Environmental Medicine*, vol. 72, no. 5, pp. 471–476, 2001.

[121] I. V. Ogneva, V. A. Kurushin, E. G. Altaeva, E. V. Ponomareva, and B. S. Shenkman, "Effect of short-term gravitational unloading on rat and mongolian gerbil muscles," *Journal of Muscle Research and Cell Motility*, vol. 30, no. 7-8, pp. 261–265, 2009.

[122] D. L. Enns, T. Raastad, I. Ugelstad, and A. N. Belcastro, "Calpain/calpastatin activities and substrate depletion patterns during hindlimb unweighting and reweighting in skeletal muscle," *European Journal of Applied Physiology*, vol. 100, no. 4, pp. 445–455, 2007.

[123] E. G. Altaeva, L. A. Lysenko, N. P. Kantserova, N. N. Nemova, and B. S. Shenkman, "The basal calcium level in fibers of the rat soleus muscle under gravitational unloading: the mechanisms of its increase and the role in calpain activation," *Doklady Biological Sciences*, vol. 433, no. 1, pp. 241–243, 2010.

[124] I. V. Ogneva, "Transversal stiffness and beta-actin and alpha-actinin-4 content of the m. Soleus fibers in the conditions of a 3-day reloading after 14-day gravitational unloading," *Journal of Biomedicine and Biotechnology*, vol. 2011, Article ID 393405, 7 pages, 2011.

[125] T. Parr, G. T. Waites, B. Patel, D. B. Millake, and D. R. Critchley, "A chick skeletal-muscle $\alpha$-actinin gene gives rise to two alternatively spliced isoforms which differ in the EF-hand Ca2+-binding domain," *European Journal of Biochemistry*, vol. 210, no. 3, pp. 801–809, 1992.

[126] C. Velez, A. E. Aranega, J. C. Prados, C. Melguizo, L. Alvarez, and A. Aranega, "Basic fibroblast and platelet-derived growth factors as modulators of actin and $\alpha$-actinin in chick myocardiocytes during development," *Proceedings of the Society for Experimental Biology and Medicine*, vol. 210, no. 1, pp. 57–63, 1995.

[127] S. Goffart, A. Franko, C. S. Clemen, and R. J. Wiesner, "$\alpha$-Actinin 4 and BAT1 interaction with the Cytochrome c promoter upon skeletal muscle differentiation," *Current Genetics*, vol. 49, no. 2, pp. 125–135, 2006.

[128] M. J. F. Broderick and S. J. Winder, "Towards a complete atomic structure of spectrin family proteins," *Journal of Structural Biology*, vol. 137, no. 1-2, pp. 184–193, 2002.

[129] H. Youssoufian, M. McAfee, and D. J. Kwiatkowski, "Cloning and chromosomal localization of the human cytoskeletal $\alpha$-actinin gene reveals linkage to the $\beta$-spectrin gene," *American Journal of Human Genetics*, vol. 47, no. 1, pp. 62–72, 1990.

[130] M. D. Baron, M. D. Davison, P. Jones, and D. R. Critchley, "The structure and function of alpha-actinin," *Biochemical Society Transactions*, vol. 15, no. 5, pp. 796–798, 1987.

[131] K. Honda, T. Yamada, R. Endo et al., "Actinin-4, a novel actin-bundling protein associated with cell motility and cancer invasion," *Journal of Cell Biology*, vol. 140, no. 6, pp. 1383–1393, 1998.

[132] Y. Capetanaki and D. J. Milner, "Desmin cytoskeleton in muscle integrity and function," *Sub-Cellular Biochemistry*, vol. 31, pp. 463–495, 1998.

[133] D. J. Milner, M. Mavroidis, N. Weisleder, and Y. Capetanaki, "Desmin cytoskeleton linked to muscle mitochondrial distribution and respiratory function," *Journal of Cell Biology*, vol. 150, no. 6, pp. 1283–1297, 2000.

[134] T. M. Mirzoev, N. S. Biryukov, O. M. Veselova, I. M. Larina, B. S. Shenkman, and I. V. Ogneva, "Parameters of fibers cell respiration and desmin content in rat soleus muscle at early stages of gravitational unloading," *Biophysics*, vol. 57, no. 3, pp. 509–514, 2012.

[135] T. M. Mirzoev, N. S. Biryukov, O. M. Veselova, I. M. Larina, B. S. Shenkman, and I. V. Ogneva, "Content of desmin and energy substrates and cell respiration of the rat's m. soleus fibers under 3- and 7-day reloading after 14-day unloading," *Aviakosmicheskaia i Ekologicheskaia Meditsina*, vol. 46, no. 1, pp. 41–45, 2012.

[136] D. A. Krieger, C. A. Tate, J. McMillin-Wood, and F. W. Booth, "Populations of rat skeletal muscle mitochondria after exercise and immobilization," *Journal of Applied Physiology Respiratory Environmental and Exercise Physiology*, vol. 48, no. 1, pp. 23–28, 1980.

[137] I. N. Belozerova, T. L. Nemirovskaya, and B. S. Shenkman, "Structural and metabolic profile of rhesus monkey m. vastus lateralis after spaceflight," *Journal of Gravitational Physiology*, vol. 7, no. 1, pp. S55–58, 2000.

[138] A. X. Bigard, E. Boehm, V. Veksler, P. Mateo, K. Anflous, and R. Ventura-Clapier, "Muscle unloading induces slow to fast transitions in myofibrillar but not mitochondrial properties. Relevance to skeletal muscle abnormalities in heart failure," *Journal of Molecular and Cellular Cardiology*, vol. 30, no. 11, pp. 2391–2401, 1998.

[139] V. A. Saks, A. Kuznetsov, T. Andrienko et al., "Heterogeneity of ADP diffusion and regulation of respiration in cardiac cells," *Biophysical Journal*, vol. 84, no. 5, pp. 3436–3456, 2003.

[140] T. Kunishima, "Ultrastructural and biochemical enzymatic properties of right ventricular muscles during hindlimb suspension in rats," *Nihon Seirigaku Zasshi*, vol. 55, no. 4, pp. 153–164, 1993.

# 9

# Production of Chemoenzymatic Catalyzed Monoepoxide Biolubricant: Optimization and Physicochemical Characteristics

**Jumat Salimon, Nadia Salih, and Bashar Mudhaffar Abdullah**

*School of Chemical Sciences and Food Technology, Faculty of Science and Technology,*
*Universiti Kebangsaan Malaysia, Bangi, 43600 Selangor, Malaysia*

Correspondence should be addressed to Jumat Salimon, jumat@ukm.my

Academic Editor: Rumiana Koynova

Linoleic acid (LA) is converted to per-carboxylic acid catalyzed by an immobilized lipase from *Candida antarctica* (Novozym 435). This per-carboxylic acid is only intermediate and epoxidized itself in good yields and almost without consecutive reactions. Monoepoxide linoleic acid 9(12)-10(13)-monoepoxy 12(9)-octadecanoic acid (MEOA) was optimized using D-optimal design. At optimum conditions, higher yield% (82.14) and medium oxirane oxygen content (OOC) (4.91%) of MEOA were predicted at $15\,\mu L$ of $H_2O_2$, 120 mg of Novozym 435, and 7 h of reaction time. In order to develop better-quality biolubricants, pour point (PP), flash point (FP), viscosity index (VI), and oxidative stability (OT) were determined for LA and MEOA. The results showed that MEOA exhibited good low-temperature behavior with PP of $-41°C$. FP of MEOA increased to $128°C$ comparing with $115°C$ of LA. In a similar fashion, VI for LA was 224 generally several hundred centistokes (cSt) more viscous than MEOA 130.8. The ability of a substance to resist oxidative degradation is another important property for biolubricants. Therefore, LA and MEOA were screened to measure their OT which was observed at 189 and $168°C$, respectively.

## 1. Introduction

Nowadays, people are concerned about green technology that is more environmental friendly and can save our environment. Green technology concept is good for industry which can be realized with sustainable and renewable resources. Epoxidized oils are currently produced by epoxidation of unsaturated plant oils, soybean, linseed oil [1], and *Jatropha curcas* seed oil [2]. Although there are several methods available to epoxidize the double bonds of unsaturated fatty acids, the only method applied on industrial scale is the Prileshajev epoxidation reaction. In this reaction, a peracid from a short-chain fatty acid and hydrogen peroxide under strong acidic conditions is used as the oxidizing agent. The presence of the strong acid in the reaction mixture, however, causes the formation of side products such as vicinal diols, estolides, and other dimmers. Although a carful choice of the peracid and the reaction conditions can help to minimize the epoxides loss, the selectivity of industrial epoxidation of plant oils rarely exceeds 80% [1]. Monoepoxidized oil has

gained widely application including functional fluids, fuel additives, polyol replacements, pharmaceutical molecules, reactive diluents in cationic and UV cure applications, surfactants, adhesives, sealants, coatings, and biolubricants [3].

The current used of epoxy biolubricant is costly and harmful to environment. This study will be one of the solutions to obtain cheaper biolubricant to replace the existing epoxy biolubricant. A milder and more selective epoxidation process has been suggested [4], wherein a lipase is used to catalyze the peracid formation. In the lipase-mediated epoxidation of unsaturation fatty acid, the acid itself reacts with hydrogen peroxide to form the peracid, which then epoxidized the double bond. The reaction is therefore often referred to as "self-epoxidation reaction," in spite of the fact that the second step proceeds predominantly via an intermolecular process [5]. Among the various lipases studied so far, Novozym 435, a commercial preparation of *Candida Antarctica* lipase B, has been shown to be the most effective. Most of the investigations have involved dilution of the substrate in organic solvent. Recently, lipase-mediated

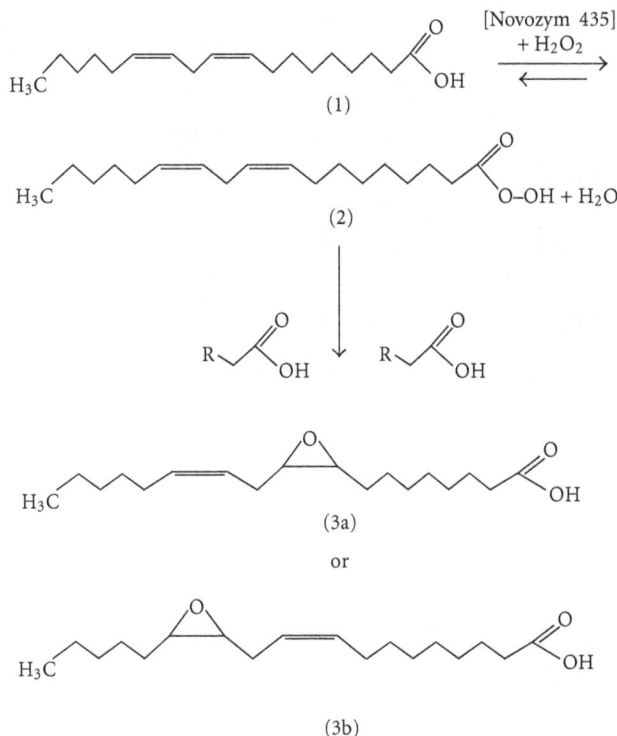

FIGURE 1: Chemoenzymatic MEOA. Notes: linoleic acid (2); perlinoleic acid (3); 9-10-monoepoxy 12-octadecenoic acid (3a); 12-13-monoepoxy 9-octadecenoic acid (3b).

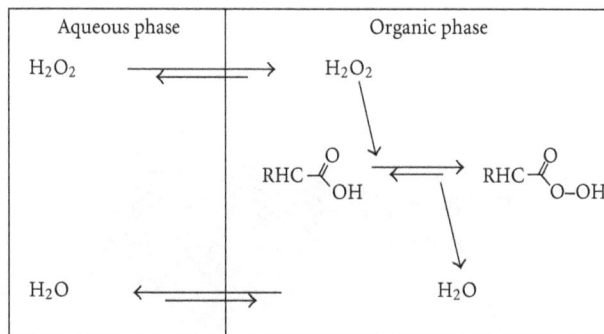

FIGURE 2: Schematic presentation of the mass transport of hydrogen peroxide and water in an organic-water biphasic system.

TABLE 1: Independent variables and their levels for D-optimal design of the MEOA reaction.

| Independent variables | | Variable levels | | |
|---|---|---|---|---|
| | | −1 | 0 | +1 |
| $H_2O_2$ ($\mu$L) | $X_1$ | 15 | 17.5 | 20 |
| Enzyme (mg) | $X_2$ | 80 | 100 | 120 |
| Time (h) | $X_3$ | 6 | 7 | 8 |

epoxidation in a solvent-free medium has also been reported [6].

This study focuses on the physicochemical characteristics such as pour point (PP), flash point (FP), viscosity index (VI), and oxidative stability (OT) of the monoepoxidation process, with the aim to determine optimal reaction conditions with regard to reaction efficiency and enzyme stability using D-optimal design. In this paper, producing of MEOA over diepoxide as biolubricants has been studied for improvement of the physicochemical characteristics (i.e., PP, FP, and VI). Linoleic acid (LA; 18 : 2) was used as the model substrate. Figure 1 demonstrates the scheme for the self-epoxidation reaction of LA. Monoepoxidation of LA results in the mixture of two monoepoxides (cis-9, 10-epoxy 12 c-18 : 1 and cis-12, 13 epoxy 9 c- 18 : 1).

## 2. Methodology

### 2.1. Experimental Procedure.

The enzymatic monoepoxidation was carried out using Novozym 435, a commercial catalyst made up of lipase, from *Candida antartica*, immobilized on a polyacrylate resin. 3 Factors (variables) such as hydrogen peroxide ($\mu$L, $X_1$), enzyme (mg, $X_2$), and ration time (h, $X_3$) were obtained according to [6]. Table 1 shows different ratios of hydrogen peroxide, enzyme, and reaction time using D-optimal design. In a typical chemoenzymatic monoepoxidation of LA 9(12)-10(13)-monoepoxy 12(9)-octadecanoic acid (MEOA), the LA (1.4 g) was dissolved in 10 mL of toluene, and the lipase was added. After stirring

for 15 min, 30% $H_2O_2$ were added, and every 15 min the addition was repeated. Afterwards, the lipase was removed by filtration, the mixture was washed with water to remove excess $H_2O_2$, and the organic phase was dried over anhydrous sodium sulfate, and solvent was evaporated in a vacuum rotary evaporator.

#### 2.1.1. Yield%.

After the laboratory reaction was completed, the actual yield to the theoretical yield was compared to determine the percent yield

$$\frac{\text{actual yield (in grams)}}{\text{theoretical yield (in grams)}} \times 100\% = \text{percent yield.} \quad (1)$$

#### 2.1.2. Oxirane Oxygen Content.

OOC% was calculated according to [7].

#### 2.1.3. Iodine Value.

Iodine value was determined according to [8].

### 2.2. Experimental Design and Statistical Analysis.

To explore the effect of the operation variables on the response in the region of investigation, a D-optimal design at three levels was performed. Hydrogen peroxide ($\mu$L, $X_1$), amount of enzyme (mg, $X_2$), and reaction time (h, $X_3$) were selected as independent variables. The range of values and coded levels of the variables are given in Table 1.

A polynomial equation was used to predict the response as a function of independent variables and their interactions. In this work, the number of independent variables was three,

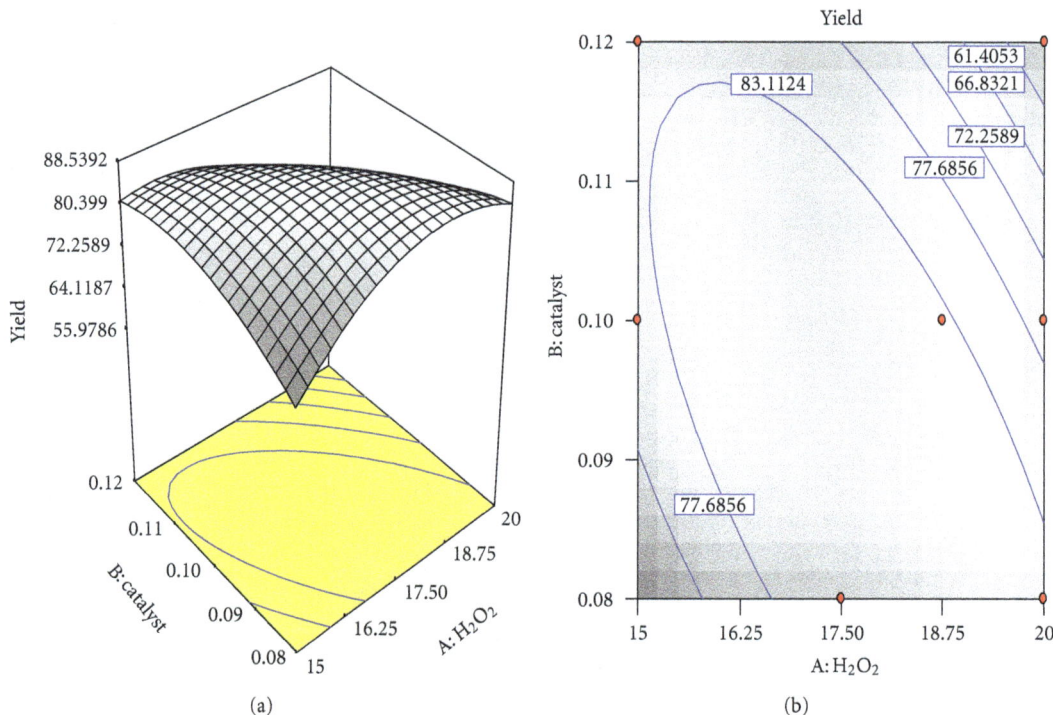

FIGURE 3: Response surface (a) and contour plots (b) for the effect of the $H_2O_2$ ($X_1$, $\mu L$) and catalysts Novozym 435 ($X_2$, mg) on the yield% of MEOA.

and therefore the response for the quadratic polynomials becomes

$$Y = \beta_0 + \sum \beta_i x_i + \sum \beta_{ii} x_i^2 + \sum \sum \beta_{ij} x_i x_j, \qquad (2)$$

where $\beta_0$; $\beta_i$; $\beta_{ii}$; $\beta_{ij}$ are constant, linear, square, and interaction regression coefficient terms, respectively, and $x_i$ and $x_j$ are independent variables. The Minitab software version 14 (Minitab Inc., USA) was used for multiple regression analysis, analysis of variance (ANOVA), and analysis of ridge maximum of data in the response surface regression (RSREG) procedure. The goodness of fit of the model was evaluated by the coefficient of determination $R^2$ and its statistical significance that was checked by the $F$-test.

### 2.3. Physicochemical Characteristics

*2.3.1. Fourier Transforms Infrared Spectroscopy.* Fourier transforms infrared spectroscopy (FTIR) has been carried out according to [8]. FTIR of the products was recorded by using Perkin Elmer Spectrum GX spectrophotometer in the range of 400–4000 cm$^{-1}$ A very thin film of products was covered on NaCl cells (4 mm thickness) and was used for analysis.

*2.3.2. Nuclear Magnetic Resonance Spectroscopy.* Nuclear magnetic resonance spectroscopy for proton $^1$H and $^{13}$C NMR has been carried out according to [8]. $^{13}$C and $^1$H NMR spectra were recorded using a 400 MHz Bruker 300 NMR spectrometer. For these analysis, 20 mg of samples were

dissolved in 1 mL of CDCl$_3$ solvent and introduced into the NMR tube.

*2.3.3. Low Temperature Operability.* The pour point (PP) is defined as the lowest temperature at which a liquid remains pourable (meaning it still behaves as a fluid). This method is routinely used to determine the low-temperature flow properties of fluids. PP values were measured according to [9] using a phase technology analyzer, Model PSA-70S (Hammersmith Gate, Richmond, BC, Canada). Each sample was run in triplicate, and average values rounded to the nearest whole degree are reported. For a greater degree of accuracy, PP measurements were done with a resolution of 1°C instead of the specified 3°C increment. Generally, materials with lower PP exhibit improved fluidity at low temperatures than those with higher PP.

*2.3.4. Flash Point Values.* The flash point of a volatile liquid is the lowest temperature at which it can vaporize to form an ignitable mixture in air. Flash point determination was run according to the American National Standard Method using a Tag Closed Tester [10]. Each sample was run in triplicate, and the average values rounded to the nearest whole degree are reported.

*2.3.5. Viscosity Index Measurements.* Viscosity index (VI) is an arbitrary measure for the change of kinematic viscosity with temperature. It is used to characterize the lubricating oil in the automotive industry. Automated multirange viscometer tubes HV M472 obtained from Walter Herzog

TABLE 2: D-optimal design arrangement and responses for MEOA.

| Run no. | $H_2O_2$ ($X_1$) | Catalyst[a] ($X_2$) | Time[b] ($X_3$) | $Y_1$, yield (%) | $Y_2$, OOC (%) | RCO (%) | $Y_3$, IV (mg/g) | X (%) | SE |
|---|---|---|---|---|---|---|---|---|---|
| 1 | 20 | 80 | 8 | 76.57 | 6.17 | 68.40 | 37.81 | 76.37 | 0.89 |
| 2 | 17.5 | 80 | 7 | 88.57 | 5.48 | 60.75 | 58.95 | 62.53 | 0.97 |
| 3 | 17.5 | 100 | 8 | 72.14 | 7.54 | 83.59 | 32.22 | 79.52 | 1.05 |
| 4 | 20 | 80 | 6 | 72.28 | 6.4 | 70.95 | 40.98 | 73.95 | 0.95 |
| 5 | 20 | 120 | 7 | 54.71 | 5.94 | 65.85 | 64.32 | 59.12 | 1.11 |
| 6 | 20 | 120 | 6 | 60.28 | 7.88 | 87.36 | 30.87 | 80.38 | 1.08 |
| 7 | 15 | 120 | 8 | 73.57 | 5.48 | 60.75 | 53.24 | 66.16 | 0.91 |
| 8 | 20 | 80 | 7 | 81.42 | 5.37 | 59.53 | 56.32 | 64.20 | 0.92 |
| 9 | 15 | 100 | 7 | 75.68 | 5.02 | 55.65 | 74.64 | 52.56 | 1.05 |
| 10 | 18.75 | 100 | 7 | 85.28 | 5.71 | 63.30 | 49.17 | 68.75 | 0.92 |
| 11 | 20 | 100 | 7 | 81.14 | 6.05 | 67.06 | 42.72 | 72.85 | 0.92 |
| 12 | 15 | 120 | 6 | 70.78 | 4.57 | 50.66 | 76.48 | 51.39 | 0.98 |
| 13 | 15 | 100 | 6 | 65.93 | 3.65 | 40.46 | 96.43 | 38.71 | 1.00 |
| 14 | 15 | 120 | 7 | 82.14 | 4.91 | 54.43 | 66.65 | 57.64 | 0.94 |
| 15 | 17.5 | 120 | 8 | 59.28 | 6.74 | 74.72 | 36.37 | 76.63 | 0.97 |
| 16 | 20 | 100 | 6 | 72.85 | 6.51 | 72.17 | 39.76 | 74.73 | 0.90 |
| 17 | 15 | 80 | 8 | 77.14 | 4.34 | 48.11 | 83.85 | 46.71 | 1.03 |
| 18 | 17.5 | 100 | 6 | 80.85 | 3.77 | 41.79 | 87.09 | 44.65 | 0.93 |

Notes: OOC, oxirane oxygen content; RCO, relative percentage conversion to oxirane; IV, iodine value; X, conversion to double bond; SE, oxirane oxygen selectivity.
[a]Catalyst Novozym 435 (mg).
[b]Monoepoxidation time (h).

(Germany) were used to measure viscosity. Measurements were run in a Temp-Trol (Precision Scientific, Chicago, Ill, USA) viscometer bath set at 40.0 and 100.0°C. The viscosity and viscosity index were calculated using ASTM method [11]. Triplicate measurements were made, and the average values were reported.

*2.3.6. Oxidative Stability.* Pressurized DSC (PDSC) experiments were accomplished using a DSC 2910 thermal analyzer from TA Instruments (Newcastle, Del, USA). Typically, a 2 μL sample, resulting in a film thickness of <1 mm, was placed in an aluminum pan hermetically sealed with a pinhole lid and oxidized in the presence of dry air (Gateway Airgas, St. Louis, Mo, USA), which was pressurized in the module at a constant pressure of 1378.95 kPa (200 psi). A 10°C min$^{-1}$ heating rate from 50 to 350°C was used during each experiment. The oxidation onset (OT, °C) was calculated from a plot of heat flow (W/g) versus temperature for each experiment. The sample was run in triplicate, and average values rounded to the nearest whole degree are reported.

## 3. Results and Discussion

In this study, following the reaction experiments, the response surface is approximated by D-optimal design. Hydrogen peroxide is an important reactant for the formation of peracids from fatty acids; hence, the influence of its amount on the monoepoxidation reaction was studied. The addition of $H_2O_2$ solution to the reaction medium (toluene with LA) leads to the formation of two distinct phases: an organic

phase and an aqueous phase. The designs and the responses yield% of MEOA ($Y_1$), OOC% ($Y_2$), and IV mg/g ($Y_3$) are given in Table 2.

The Novozym 435, being adsorbed on a hydrophobic carried, is mainly present in the organic phase, while $H_2O_2$ will be partitioned in both the aqueous and the organic phases, with the concentration being higher in the aqueous phase (Figure 2) [6]. This was determined performing varying amounts of $H_2O_2$ (15, 17.5, and 20 μL) which has been added every 15 min, and the addition was repeated 24 times, Novozym 435 (80, 100, and 120 mg), and different times (6, 7, and 8 h). A stoichiometric excess of the required amount of the peroxide was used to compensate for its possible decomposition by light and temperature.

Table 2 showed that the yield percentage of MEOA, $Y_1$, has increased to 82.14%, while OOC%, $Y_2$, 4.91 and iodine value, $Y_3$, 66.65 which are considerably compared to the theoretical (OOCt) 9.02% and initial iodine value (IV°) 157.35 mg/g. Subsequent experiments were performed using different amounts of $H_2O_2$ 15, 17.5, and 20 μL for every 1.4 gm of LA in one single step. As seen in Table 2, there was a clear increase in the reaction rate (OOC%) and a decrease (IV mg/g) with increasing $H_2O_2$ amount. With 15 μL, monoepoxidation was achieved at 7 h using 120 mg Novozym 435, while total epoxidation was observed within 10 h using 30 μL. Increasing the peroxide amount used for the reaction results in increasing peracid formation. In the state of partial epoxidation, the amount of peracids accumulated is not significant (less than 2% of the total) [4], because the chemical reaction in which they are consumed is

(a)

(b)

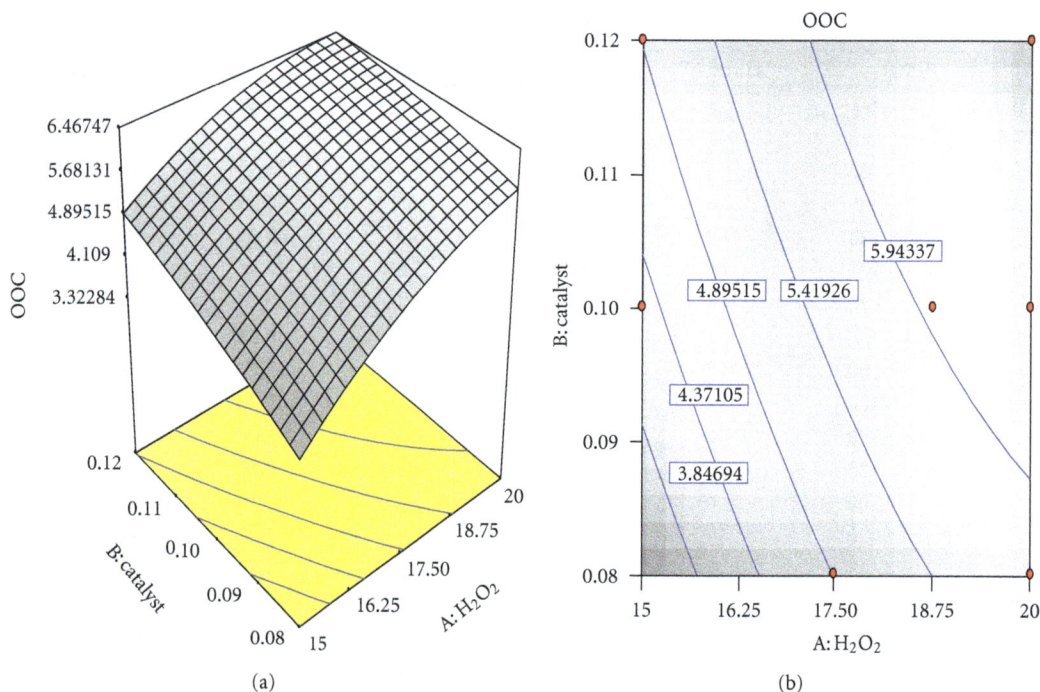

FIGURE 4: Response surface (a) and contour plots (b) for the effect of the $H_2O_2$ ($X_1$, $\mu$L) and catalysts Novozym 435 ($X_2$, mg) on the OOC% of MEOA.

TABLE 3: Regression coefficients of the predicted quadratic polynomial model for response variables of the yield% of MEOA.

| Variables | Coefficients ($\beta$), yield% of MEOA ($Y_1$) | $T$ | $P$ | Notability |
|---|---|---|---|---|
| Intercept | 87.53 | 8.22 | 0.0034 | *** |
| Linear | | | | |
| $X_1$ | −2.82 | 4.91 | 0.0575 | |
| $X_2$ | −4.50 | 10.07 | 0.0131 | ** |
| $X_3$ | −0.14 | 0.010 | 0.9219 | |
| Quadratic | | | | |
| $X_{11}$ | −9.06 | 10.82 | 0.0110 | ** |
| $X_{22}$ | −5.43 | 5.41 | 0.0485 | ** |
| $X_{33}$ | −9.81 | 21.03 | 0.0018 | *** |
| Interaction | | | | |
| $X_{12}$ | −9.74 | 25.22 | 0.0010 | *** |
| $X_{13}$ | −7.14 | 12.99 | 0.0069 | *** |
| $X_{23}$ | −7.80 | 12.83 | 0.0072 | *** |
| $R^2$ | 0.90 | | | |

Notes: $X_1$ = amount of $H_2O_2$; $X_2$ = catalyst Novozym 435; $X_3$ = reaction time; **$P < 0.05$; ***$P < 0.01$. $T$: $F$ test value.
See Table 2 for a description of the abbreviations.

TABLE 4: Regression coefficients of the predicted quadratic polynomial model for response variables of the OOC% for the MEOA.

| Variables | Coefficients ($\beta$), OOC% ($Y_2$) | $T$ | $P$ | Notability |
|---|---|---|---|---|
| Intercept | 5.59 | 2.00 | 0.1717 | |
| Linear | | | | |
| $X_1$ | 1.00 | 12.15 | 0.0082 | *** |
| $X_2$ | 0.58 | 3.25 | 0.1091 | |
| $X_3$ | 0.51 | 2.54 | 0.1494 | |
| Quadratic | | | | |
| $X_{11}$ | −0.37 | 0.36 | 0.5626 | |
| $X_{22}$ | −0.099 | 0.036 | 0.8546 | |
| $X_{33}$ | 0.31 | 0.41 | 0.5404 | |
| Interaction | | | | |
| $X_{12}$ | −0.22 | 0.26 | 0.6267 | |
| $X_{13}$ | −0.55 | 1.50 | 0.2553 | |
| $X_{23}$ | −0.27 | 0.31 | 0.5934 | |
| $R^2$ | 0.69 | | | |

Notes: $X_1$ = amount of $H_2O_2$; $X_2$ = catalyst Novozym 435; $X_3$ = reaction time; **$P < 0.05$; ***$P < 0.01$. $T$: $F$ test value.
See Table 2 for a description of the abbreviations.

very fast. But once all the double bonds are epoxidized, the remaining peracid is not consumed.

The quadratic regression coefficient obtained by employing a least squares method technique to predict quadratic polynomial models for the yield% of MEOA ($Y_1$), OOC% ($Y_2$), and IV mg/g ($Y_3$) is given in Tables 3, 4, and 5. For the yield% of MEOA ($Y_1$), the linear term of Novozym 435 catalyst amount ($X_2$), quadratic terms of $H_2O_2$ ($X_{11}$) and

(a)

(b)

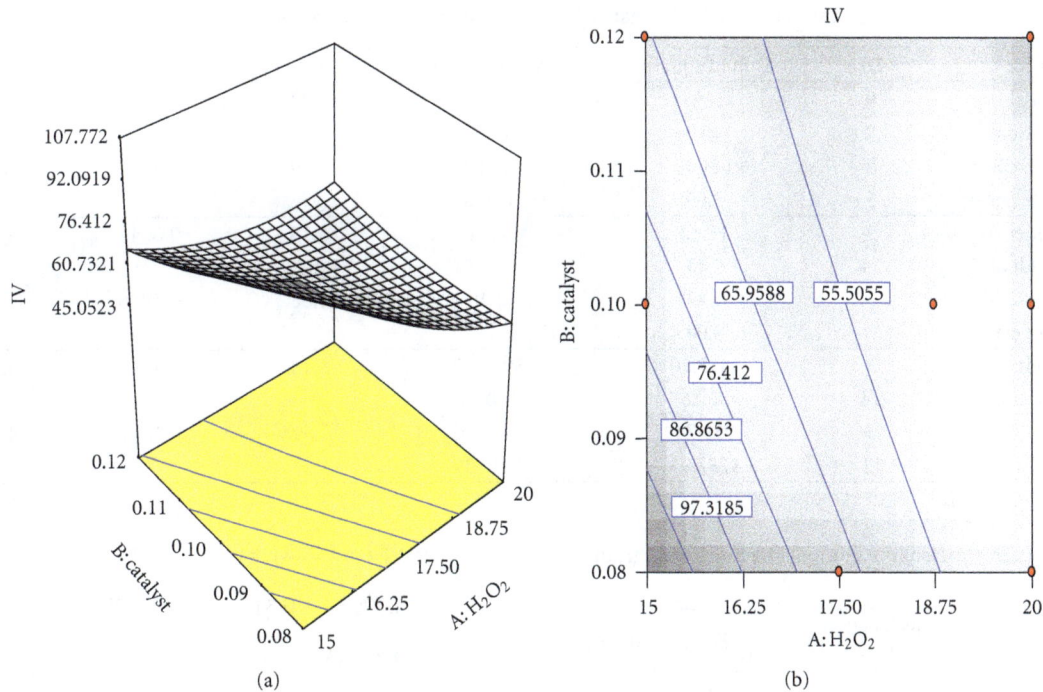

FIGURE 5: Response surface (a) and contour plots (b) for the effect of the $H_2O_2$ ($X_1$, $\mu$L) and catalysts Novozym 435 ($X_2$, mg) on the IV mg/g of MEOA.

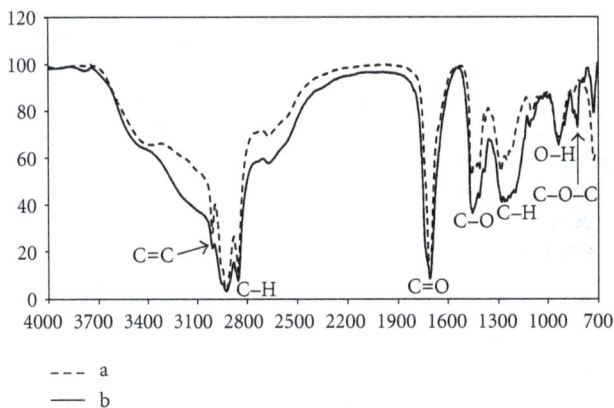

FIGURE 6: FTIR spectrum of the LA (a) and MEOA (b).

TABLE 5: Regression coefficients of the predicted quadratic polynomial model for response variables of the IV mg/g for the MEOA.

| Variables | Coefficients ($\beta$), IV mg/g ($Y_3$) | $T$ | $P$ | Notability |
|---|---|---|---|---|
| Intercept | 56.24 | 3.42 | 0.0489 | ** |
| Linear | | | | |
| $X_1$ | −18.82 | 21.45 | 0.0017 | *** |
| $X_2$ | −8.91 | 3.87 | 0.0849 | |
| $X_3$ | −10.57 | 5.47 | 0.0474 | ** |
| Quadratic | | | | |
| $X_{11}$ | 8.06 | 0.84 | 0.3865 | |
| $X_{22}$ | 4.15 | 0.31 | 0.5935 | |
| $X_{33}$ | −4.41 | 0.42 | 0.5369 | |
| Interaction | | | | |
| $X_{12}$ | 11.59 | 3.50 | 0.0985 | |
| $X_{13}$ | 10.81 | 2.92 | 0.1259 | |
| $X_{23}$ | 5.42 | 0.61 | 0.4580 | |
| $R^2$ | 0.79 | | | |

Notes: $X_1$= amount of $H_2O_2$; $X_2$= catalyst Novozym 435; $X_3$= reaction time; **$P < 0.05$; ***$P < 0.01$. $T$: $F$ test value.
See Table 2 for a description of the abbreviations.

Novozym 435 catalyst amount ($X_{22}$) were significant ($P < 0.05$). The interaction between $H_2O_2$ ($X_{11}$) and Novozym 435 catalyst amount ($X_{12}$) and the interaction between $H_2O_2$ ($X_{11}$), and reaction time ($X_{13}$) were significant ($P < 0.05$), while its quadratic term of reaction time ($X_{33}$) was highly significant ($P < 0.01$).

Highly significant ($P < 0.01$) terms of OOC% ($Y_2$) and IV mg/g ($Y_3$) for the $H_2O_2$ ($X_1$) were linear, while linear term of IV mg/g for the reaction time ($X_3$) was significant ($P < 0.05$). The coefficients of independent variables ($H_2O_2$; $X_1$, catalyst Novozym 435; $X_2$, and reaction time; $X_3$) were determined for the quadratic polynomial models Tables 3, 4 and 5.

The lack of fit $F$-value for all the responses showed that the lack of fit is not significant ($P > 0.05$) relatively to the pure error. This indicates that all the models predicted for the responses were adequate. Regression models for data on responses $Y_1$, $Y_2$, and $Y_3$ were highly significant ($P < 0.01$) with satisfactory $R^2$. However, $R^2$ for $Y_2$ (0.69) was lower although the model was significant. Table 6 summarizes the

TABLE 6: Analysis of variance (ANOVA) for all the responses of MEOA.

|       | Source      | Df | Sum of squares | Mean square | F value | Prob > F |                 |
|-------|-------------|----|----------------|-------------|---------|----------|-----------------|
| $Y_1$ | Model       | 9  | 1298           | 144.22      | 8.22    | 0.0034   | Significant     |
|       | Residual    | 8  | 140.36         | 17.55       |         |          |                 |
|       | Lack-of-fit | 3  | 144.05         | 48.02       | 0.69    | 0.5763   | Not significant |
|       | Pure error  | 5  | 3.69           | 0.738       |         |          |                 |
| $Y_2$ | Model       | 3  | 13.62          | 4.54        | 6.74    | 0.0048   | Significant     |
|       | Residual    | 14 | 9.43           | 0.67        |         |          |                 |
|       | Lack-of-fit | 3  | 1.43           | 0.48        | 0.66    | 0.5956   | Not significant |
|       | Pure error  | 11 | 8.00           | 0.727       |         |          |                 |
| $Y_3$ | Model       | 3  | 4423.70        | 598.002     | 8.18    | 0.002    | Significant     |
|       | Residual    | 14 | 2522.67        | 180.19      |         |          |                 |
|       | Lack-of-fit | 3  | 734.74         | 244.91      | 1.51    | 0.2672   | Not significant |
|       | Pure error  | 11 | 1787.93        | 162.539     |         |          |                 |

TABLE 7: The main wavelengths in the FTIR functional groups of LA and MEOA.

| Wavelength of LA[a] | Wavelength of MEOA[b] | Functional group |
|---------------------|-----------------------|------------------|
| 3009                | 3009                  | C=C bending vibration (aliphatic) |
| 2927, 2855          | 2927, 2856            | C–H stretching vibration (aliphatic) |
| 1719                | 1711                  | C=O stretching vibration (carboxylic acid) |
| 1454                | 1454                  | C–H scissoring and bending for methylene group |
| 1284                | 1284                  | C–O stretching asymmetric (carboxylic acid) |
| 937                 | 934                   | C–H bending vibration (alkene) |
| —                   | 820                   | C–O–C oxirane ring |
| 722                 | 723                   | C–H group vibration (aliphatic) |

Notes: Linoleic acid (a); 9(12)-10(13)-monoepoxy 12(9)-octadecanoic acid (b).

analysis of variance (ANOVA) of all the responses of this study.

These results suggest that linear effect of hydrogen peroxide is the primary determining factors for MEOA. Reference [6] also concluded that this variable had a very large effect on the results of their monoepoxidation study. Final equations in terms of actual factors are:

$$Y_1 = +87.53 - 2.82X_1 - 4.50X_2 - 0.14X_3 - 9.06X_1^2$$
$$- 5.43X_2^2 - 9.81X_3^2 - 9.74X_1X_2 - 7.14X_1X_3 \qquad (3)$$
$$- 7.80X_2X_3,$$

$$Y_2 = +5.59 + 1.00X_1 + 0.58X_2 + 0.51X_3 - 0.37X_1^2$$
$$- 0.099X_2^2 + 0.31X_3^2 - 0.22X_1X_2 - 0.55X_1X_3 \qquad (4)$$
$$- 0.27X_2X_3,$$

$$Y_3 = +56.24 - 18.82X_1 - 8.91X_2 - 10.57X_3 + 8.06X_1^2$$
$$+ 4.15X_2^2 4.41X_3^2 + 11.59X_1X_2 + 10.81X_1X_3 \qquad (5)$$
$$+ 5.42X_2X_3.$$

RSM is one of the best ways of evaluating the relationships between responses, variables, and interactions that exist. Significant interaction variables in the fitted models (Tables 3 to 5) were chosen as the axes (amount of $H_2O_2$; $X_1$, catalyst Novozym 435; $X_2$ and reaction time; $X_3$) for the response surface plots. The relationships between independent and dependent variables are shown in the three-dimensional representation as response surfaces. In a contour plot, curves of equal response values are drawn on a plane whose coordinates represent the levels of the independent factors. Each contour represents a specific value for the height of the surface above the plane defined for a combination of the levels of the factors. Therefore, different surface height values enable one to focus attention on the levels of the factors at which changes in the surface height occur [12].

Figures 3 to 5 are the design-expert plots for all the responses. In the MEOA, performing the technique using low amount of $H_2O_2$ would give the desired OOC% of MEOA as shown in Figure 4, while IV (Figure 5) was higher at this condition. As shown in Figures 4 and 5, the increasing amount of $H_2O_2$ led to an increase in the OOC% and reduction of IV mg/g. The relationships between the parameters and MEOA were linear or almost linear. High OOC% could be obtained by using high amount of H2O2 at high reaction time. Experimental variables should be carefully controlled in order to recover a medium percentage of MEOA of interest with reasonable yield [6].

Optimum conditions of the experiment to obtain higher yield% of MEOA and medium OOC% were predicted at amount of $H_2O_2$ $\mu$L of 15, catalyst Novozym 435 of 120 mg,

TABLE 8: The main signals present in $^{13}$C NMR functional groups of LA, MEOA, and diepoxide LA.

| $\delta$ (ppm) LA[a] | $\delta$ (ppm) MEOA[b] | $\delta$ (ppm) diepoxide LA | Assignment |
|---|---|---|---|
| 24.86–29.79 | 22.69–34.15 | 22.77–29.44 | –CH$_2$–Carbons |
| — | 54.59–57.29 | 54.61–57.32 | (△) epoxide groups |
| 128.27–130.38 | 124.02–132.89 | — | –CH=CH–Olefinic carbons |
| 180.49 | 179.32 | 178.79 | C=O Carboxylic acid |

Notes: Linoleic acid (a); 9(12)-10(13)-monoepoxy 12(9)-octadecanoic acid (b).

TABLE 9: The main signals present in $^{1}$H NMR functional groups of LA, MEOA, and diepoxide LA.

| $\delta$ (ppm) LA[a] | $\delta$ (ppm) MEOA[b] | $\delta$ (ppm) diepoxide LA | Assignment |
|---|---|---|---|
| 0.88–0.91 | 0.86–0.88 | 0.88–0.92 | –CH$_3$ |
| 1.30–2.77 | 1.29–2.33 | 1.34–2.36 | –CH$_2$ |
| — | 2.92–3.12 | 2.99–3.13 | –CH–O–CH– |
| 5.35–5.36 | 5.38–5.49 | — | –CH=CH– |
| 7.27 | 7.27 | 7.27 | –COOH |

Notes: Linoleic acid (a); 9(12)-10(13)-monoepoxy 12(9)-octadecanoic acid (b).

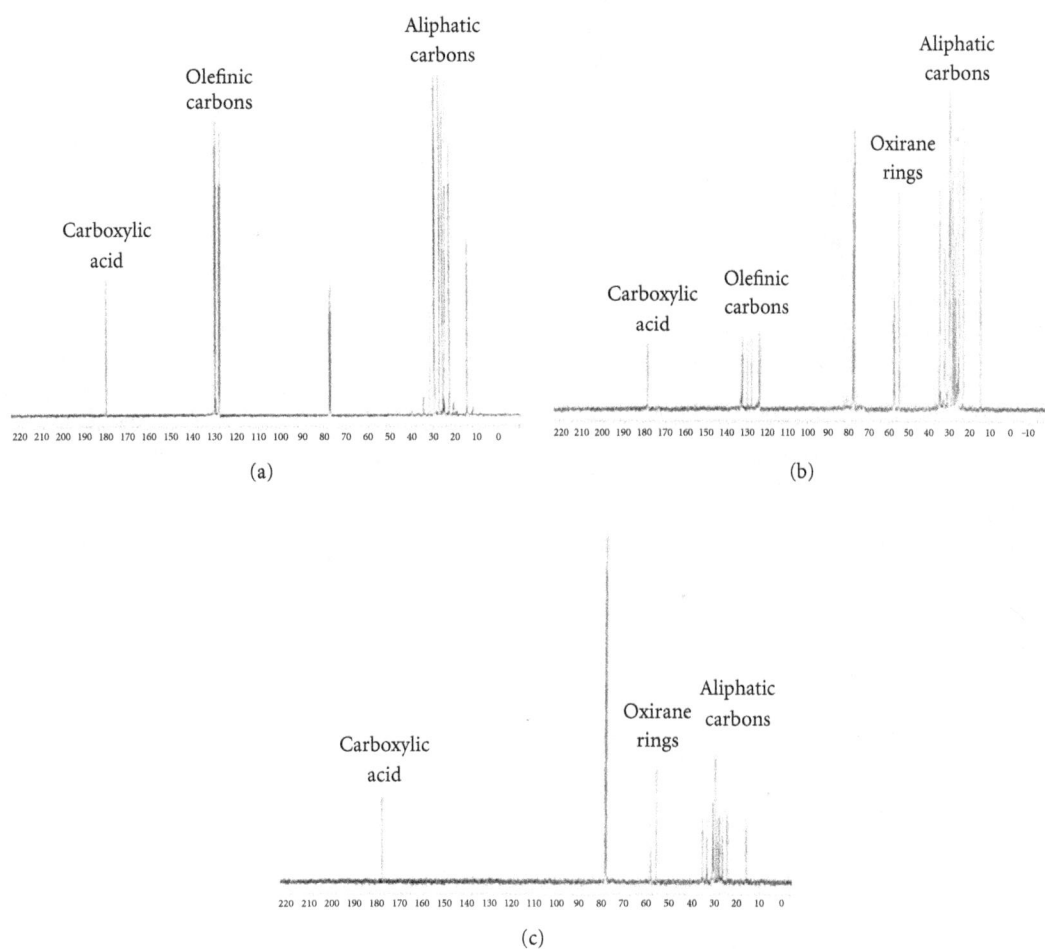

FIGURE 7: $^{13}$C NMR spectrum of LA (a), MEOA, (b) and diepoxide LA (c).

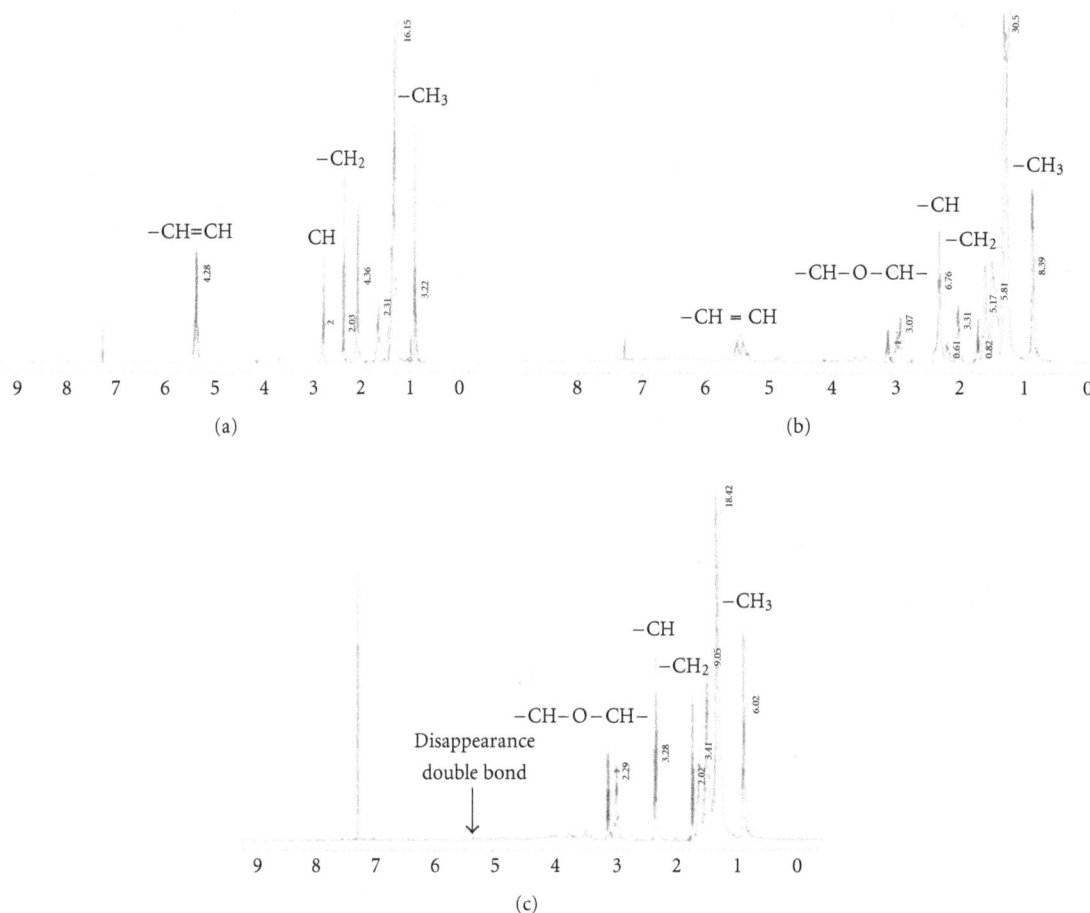

FIGURE 8: $^1$H NMR spectrum of LA (a), MEOA (b) and di-epoxide LA (c).

and 7 h of reaction time. At this condition, the yield% of MEOA was 82.14%, 4.91% of OOC, and 66.65 mg/g of IV.

In order to prove the present oxirane ring of MEOA, final product was tested by FTIR. The comparison between LA (a) and MEOA (b), FTIR spectra is shown in Figure 6. The main peaks and their assignment to functional groups are given in Table 7. Oxirane ring of MEOA can be detected at wavenumber 820 cm$^{-1}$.

For the carboxylic acid carbonyl functional groups (C=O), FTIR spectrum showed absorption bands of LA and MEOA at 1719 and 1711 cm$^{-1}$, respectively, while stretching vibration peak of C=C can be detected at wavenumber 3009 cm$^{-1}$ [13], and peaks at 2927–2856 cm$^{-1}$ indicated the CH$_2$ and CH$_3$ scissoring of LA and MEOA based on Figures 6(a) and 6(b). FTIR spectrum also showed absorption bands at 722, 723 for (C-H) group.

Figure 7(a) indicates the $^{13}$C NMR spectrum of LA. The $^{13}$C spectroscopy shows the main signals assignment of the LA as shown in Table 8(a). The signals at 180.49 ppm refer to the carbon atom of the carbonyl group (carboxylic acid). The signals at 128.27–130.38 ppm refer to the unsaturated carbon atoms (olefinic carbons); 24.86–29.79 ppm due to methylene carbon atoms in fatty acid moieties of LA [8].

Figures 7(b) and 7(c) can be confirmed by the oxirane ring of MEOA 54.59–57.29 ppm and diepoxide LA at about 54.61–57.32 ppm. Indeed, it appeared that the signals were present in the MEOA, as four peaks of roughly equal intensity (132.89, 132.72, 130.15, and 124.02 ppm) were observed in the alkenic carbon region in the $^{13}$C NMR spectrum Figure 7(b), while they disappeared in the diepoxide LA (Figure 7(c)) [14]. The $^{13}$C-NMR spectra indicate the existence of carbonyl group (carboxylic acid) in their structure MEOA 179.32 ppm and diepoxide LA at about 178.79 ppm. The other distinctive signals were aliphatic carbons MEOA at about 22.69–29.38 ppm and di-epoxide LA at about 22.77–29.44 ppm, which are common for these types of compounds [15].

$^1$H NMR spectroscopy shows the main signals assignments in LA, MEOA and di-epoxide LA as shown in Table 9. The $^1$H NMR spectra for the products show some of the key features for a typical (-CH-O-CH-) at about 2.92–3.12 ppm of MEOA, and about 2.99–3.13 ppm of di-epoxide LA (Figure 8(b)). The distinguishable groups are the protons of the terminal methyl of the fatty acid chain. The signals at 0.88–0.86 ppm referred to the methylene group (-CH$_3$) of LA Figure 8(a) which also appear in MEOA 0.86–0.88 ppm and di-epoxide LA 0.88–0.92 ppm (Figures 8(b) and 8(c)) next to

TABLE 10: Physicochemical properties of LA and MEOA.

| Properties | LA[a] | MEOA[b] |
|---|---|---|
| Pour point ($^\circ$C) | −2 | −41 |
| Flash point ($^\circ$C) | 115 | 128 |
| Viscosity index ($^\circ$C) | 224 | 130.8 |
| Oxidative stability OT ($^\circ$C) | 189 | 168 |

Notes: Linoleic acid (a); 9(12)-10(13)-monoepoxy 12(9)-octadecanoic acid (b).

the terminal methyl (-CH$_2$) at 1.30–2.77 ppm of LA, 1.29–2.33 of MEOA, and 1.34–2.36 ppm of di-epoxide LA.

However, the methane proton signals (-CH=CH-) were shifted upfield at about 5.35–5.36 ppm of LA and 5.38–5.49 ppm of MEOA [16], while they disappeared in di-epoxide LA. Another distinctive feature is the hydroxyl proton (-COOH) of the carboxylic acid at about 7.27 ppm.

This approach is used here to improve the low-temperature flow behavior of fatty acids by monoepoxide ring. MEOA improves the PP at −41$^\circ$C significantly comparing with LA at −2$^\circ$C. Monoepoxidation to produce MEOA was the most effective decreasing the PP to −41$^\circ$C (Table 10). It can be assumed that the presence of oxirane ring on the fatty acid creates a steric barrier around the individual molecules and inhibits crystallization, resulting in lower PP [17].

Flash point is another important factor in determining how well oil will behave as a potential biolubricant. The flash point is often used as a descriptive characteristic of oil fuel, and it is also used to describe oils that are not normally used as fuels. Flash point refers to both flammable oils and combustible oils. There are various international standards for defining each, but most agree that oils with a flash point less than 43$^\circ$C are flammable, while those having a flash point above this temperature are combustible [18]. Table 10 has shown the improvement in flash point of MEOA which increased to 128$^\circ$C comparing with 115$^\circ$C of LA that means the result agrees with the various international standards.

The efficiency of the biolubricant in reducing friction and wear is greatly influenced by its viscosity. Generally, the least viscous biolubricant which still forces the two moving surfaces apart is desired. If the biolubricant is too viscous, it will require a large amount of energy to move; if it is too thin, the surfaces will rub and friction will increase. The viscosity index highlights how a biolubricants viscosity changes with variations in temperature. Many biolubricant applications require performing across a wide range of conditions, for example, in an engine. Automotive biolubricants must reduce friction between engine components when it is started from cold (relative to engine operating temperatures) as well as when it is running (up to 200$^\circ$C). The best oils (with the highest VI) will not vary much in viscosity over such a temperature range and therefore will perform well throughout. In MEOA, decreased viscosity index (VI) to 130.8 of MEOA is the result of less double bond (Table 10).

The ability of a substance to resist oxidative degradation is another important characteristic of biolubricants. Therefore, MEOA was screened to measure their oxidation

stability using PDSC through determination of OT. PDSC is an effective method for measuring oxidation stability of oleochemicals in an accelerated mode [19]. The OT is the temperature at which a rapid increase in the rate of oxidation is observed at a constant, high pressure (200 psi). A high OT would suggest high oxidation stability of the material. OT was calculated from a plot of heat flow (W/g) versus temperature that was generated by the sample upon degradation and by definition. In this chapter, oxirane ring of MEOA did not show improvement in the oxidation stability of OT for MEOA at 168$^\circ$C comparing with LA at 189$^\circ$C (Table 10).

## 4. Conclusion

RSM and 3-factor D-optimal design were employed for optimization of MEOA. From the present study, it is evident that hydrogen peroxide is the most critical parameter influencing the chemoenzymatic monoepoxidation reaction. Increase in the hydrogen peroxide amount has a strong effect on the reaction kinetics; however, a large excess of hydrogen peroxide results in accumulation of peracid in the final product. Based on the results obtained, MEOA had a positive influence on the low-temperature properties PP and FP.

## Acknowledgment

The authors thank UKM and the Ministry of Science and Technology for research Grants UKM-GUP-NBT-08-27-113 and UKM-OUP-NBT-29-150/2011.

## References

[1] M. R. G. Klaas and S. Warwel, "Complete and partial epoxidation of plant oils by lipase-catalyzed perhydrolysis," *Industrial Crops and Products*, vol. 9, no. 2, pp. 125–132, 1999.

[2] P.-P. Meyer, N. Techaphattana, S. Manundawee, S. Sangkeaw, W. Junlakan, and C. Tongurai, "Epoxidation of soybean oil and *Jatropha* oil," *Thammasat International Journal of Science Technology*, vol. 13, pp. 1–5, 2008.

[3] A. Adhvaryu and S. Z. Erhan, "Epoxidized soybean oil as a potential source of high-temperature lubricants," *Industrial Crops and Products*, vol. 15, no. 3, pp. 247–254, 2002.

[4] S. Warwel and M. R. G. Klaas, "Chemo-enzymatic epoxidation of unsaturated carboxylic acids," *Journal of Molecular Catalysis B*, vol. 1, no. 1, pp. 29–35, 1995.

[5] M. R. G. Klaas and S. Warwel, "Lipase-catalyzed preparation of peroxy acids and their use for epoxidation," *Journal of Molecular Catalysis A*, vol. 117, no. 1–3, pp. 311–319, 1997.

[6] C. Orellana-Coca, S. Camocho, D. Adlercreutz, B. Mattiasson, and R. Hatti-Kaul, "Chemo-enzymatic epoxidation of linoleic acid: parameters influencing the reaction," *European Journal of Lipid Science and Technology*, vol. 107, no. 12, pp. 864–870, 2005.

[7] S. W. Lin, T. T. Sue, and T. Y. Ai, *PORIM Test Methods*, vol. 1, PORIM, Bandar Baru Bangi, Malaysia, 1995.

[8] J. Salimon and N. Salih, "Modification of epoxidized ricinoleic acid for biolubricant base oil with improved flash and pour points," *Asian Journal of Chemistry*, vol. 22, no. 7, pp. 5468–5476, 2010.

[9] ASTM Standard D5949, *Standard Test Method for Pour Point of Petroleum (Automatic Pressure Pulsing Method)*, ASTM, West Conshohocken, Pa, USA.

[10] ASTM Standard D 56-79, *Standard Test Method for Flash Point of Liquids with a Viscosity Less than, 45 Saybolt Universal Seconds (SUS) at 37.8°C* (that don't contain suspended solids and don't tend to form a surface film under test).

[11] ASTM D 2270-93, *Standard Practice for Calculating Viscosity Index from Kinematic Viscosity at 40 and 100°C*, ASTM, West Conshohocken, Pa, USA.

[12] M. Wu, H. Ding, S. Wang, and S. Xu, "Optimization conditions for the purification of linoleic acid from sunflower oil by urea complex fractionation," *Journal of the American Oil Chemist' Society*, vol. 85, pp. 677–684, 2008.

[13] G. Socrates, *Infrared and Raman Characteristic Group Frequencies: Tables and Charts*, John Wily & Sons, Chichester, UK, 3rd edition, 2001.

[14] G. Du, A. Tekin, E. G. Hammond, and L. K. Woo, "Catalytic epoxidation of methyl linoleate," *Journal of American Oil Chemical Society*, vol. 81, no. 5, pp. 477–480, 2004.

[15] K. M. Doll, B. K. Sharma, and S. Z. Erhan, "Synthesis of branched methyl hydroxy stearates including an ester from bio-based levulinic acid," *Industrial and Engineering Chemistry Research*, vol. 46, no. 11, pp. 3513–3519, 2007.

[16] H. S. Hwang and S. Z. Erhan, "Synthetic lubricant basestocks from epoxidized soybean oil and Guerbet alcohols," *Industrial Crops and Products*, vol. 23, no. 3, pp. 311–317, 2006.

[17] B. K. Sharma, K. M. Doll, and S. Z. Erhan, "Ester hydroxy derivatives of methyl oleate: tribological, oxidation and low temperature properties," *Bioresource Technology*, vol. 99, no. 15, pp. 7333–7340, 2008.

[18] J. Salimon, N. Salih, and E. Yousif, "Chemically modified biolubricantbasestocks from epoxidized oleic acid: Improved low temperature properties and oxidative stability," *Journal of Saudi Chemical Society*, vol. 15, pp. 195–201, 2011.

[19] Y. Y. Zhang, T. H. Ren, H. D. Wang, and M. R. Yi, "A comparative study of phenol-type antioxidants in methyl oleate with quantum calculations and experiments," *Lubrication Science*, vol. 16, no. 4, pp. 385–392, 2004.

# Development and Kinematic Verification of a Finite Element Model for the Lumbar Spine: Application to Disc Degeneration

**Elena Ibarz,**[1] **Antonio Herrera,**[2, 3, 4] **Yolanda Más,**[1] **Javier Rodríguez-Vela,**[2, 3, 4] **José Cegoñino,**[1] **Sergio Puértolas,**[1] **and Luis Gracia**[1, 5]

[1] *Department of Mechanical Engineering, University of Zaragoza, 50018 Zaragoza, Spain*
[2] *Department of Surgery, University of Zaragoza, 50009 Zaragoza, Spain*
[3] *Department of Orthopaedic Surgery and Traumatology, Miguel Servet University Hospital, 50009 Zaragoza, Spain*
[4] *Aragón Health Sciences Institute, 50009 Zaragoza, Spain*
[5] *Engineering and Architecture School, University of Zaragoza, María de Luna 3, 50018 Zaragoza, Spain*

Correspondence should be addressed to Luis Gracia; lugravi@unizar.es

Academic Editor: José M. Vilar

The knowledge of the lumbar spine biomechanics is essential for clinical applications. Due to the difficulties to experiment on living people and the irregular results published, simulation based on finite elements (FE) has been developed, making it possible to adequately reproduce the biomechanics of the lumbar spine. A 3D FE model of the complete lumbar spine (vertebrae, discs, and ligaments) has been developed. To verify the model, radiological images (X-rays) were taken over a group of 25 healthy, male individuals with average age of 27.4 and average weight of 78.6 kg with the corresponding informed consent. A maximum angle of 34.40° is achieved in flexion and of 35.58° in extension with a flexion-extension angle of 69.98°. The radiological measurements were 33.94 ± 4.91°, 38.73 ± 4.29°, and 72.67°, respectively. In lateral bending, the maximum angles were 19.33° and 23.40 ± 2.39, respectively. In rotation a maximum angle of 9.96° was obtained. The model incorporates a precise geometrical characterization of several elements (vertebrae, discs, and ligaments), respecting anatomical features and being capable of reproducing a wide range of physiological movements. Application to disc degeneration (L5-S1) allows predicting the affection in the mobility of the different lumbar segments, by means of parametric studies for different ranges of degeneration.

## 1. Introduction

Lumbar pain currently represents a serious problem due to its socioeconomic impact and repercussions. Degenerative disc disease is the most common cause of lumbar pain [1]. Factors which can have an influence on the degenerative process are, amongst others, the loads supported [2] (which in addition can activate the enzymatic processes which play a part in the degeneration [3]), the movements in flexion [4], and the genetics of each individual [5, 6].

The lumbar spine is a complex structure in biomechanical terms. It has to combine flexibility to allow three-dimensional movements and stability to protect the nervous structures, whilst maintaining a biplanar equilibrium in the erect posture with minimum muscular effort. On the other hand, the spine is a viscoelastic structure which modifies its mechanical properties in relation to the intensity of the load [7]. There are numerous studies to determine the ranges of lumbar spine mobility, in addition to others that analyse the forces and loads that influence the movements and displacements produced. The biomechanics of the lumbar spine have been studied both on corpses [8–11] and "in vivo" using simple or biplanar radiographs [12–14] or other associated methods [15]. Other studies to determine mobility have used a variety of systems, associating studies with TV and computer [16], CT [17], electrogoniometer [18], video fluoroscope [19, 20], NMR [21–23], the inclinometer [24], or measurements with goniometer and the distraction between anatomical structures [25]. Animal spines have also been used [26] for laboratory tests, although there are notable differences between these and human spines [27]. The complexity of the lumbar spine, along with the variability of each individual, conditions the difficulty and reproducibility of "in vivo" or "in vitro" biomechanical studies.

Due to the difficulties of being able to experiment on living people, the irregular results that have been published and the differences between human and animal spines, simulation models have been developed through the use of finite elements (FE). These models make it possible to study the lumbar spine in both physiological and pathological conditions, whenever the model is precise enough to adequately reproduce the biomechanics of the lumbar spine. This method, in existence since 1956 [28], was popularized among the scientific community during the 60s decade [29] and has proved to be adequate for the study of the functionality of a physiological unit as complex as the lumbar spine. There are numerous studies dedicated to simulating the different behavioural aspects of the lumbar spine, from a global biomechanical level to the more specific performance problems of some elements or even different pathologies [30–40]. The majority of the models concerning specific problems consider a unique functional unit or two functional units at the most [30, 38–40]. The availability of a model of the complete lumbar spine would allow a complete nonlinear biomechanical analysis of a healthy lumbar spine, as a step towards studying the consequences of disc degeneration and the effects produced by the implantation of different lumbar fixations or disc prosthesis, not only in a specific functional unit but in any level along the lumbar spine, even multiple degeneration levels and fixations. Concerning disc degeneration, different types of studies have been reported. So, in [41] an in vitro study is presented for 44 corpse specimens, classifying degeneration levels according to Thompson criteria [42]. A complete revision of the main factors affecting disc degeneration from a clinical point of view is presented in [43]. A discussion about reliability of in vitro and in vivo models for the study of disc degeneration is included in [44]. In the field of simulation, a model of poroelastic materials, both for the nucleus pulposus and annulus fibrosus, considering only a functional unit is presented in [45]. Other authors apply different mechanical models for the behaviour of the degenerated disc, but including only a functional unit in most of the cases [35, 40, 46–49].

The objective of this work was to develop and verify a complete three-dimensional FE model of the lumbar spine from L1 to Sacrum with the corresponding intervertebral discs, as well as all the ligaments which intervene in the biomechanical behaviour, reproduced with the greater anatomical detail. By means of this model, lumbar spine standard movements can be simulated, verifying the model with the results obtained in radiographic measurements carried out on healthy individuals and comparing it with published results. After kinematic verification, the model has been applied to the study of disc degeneration obtaining the difference of mobility between healthy and pathologic conditions.

## 2. Materials and Methods

In order to get a model as accurate as possible of the lumbosacral spine, a mixed technique has been used. The starting point for obtaining a precise outer geometric representation of the discs and vertebrae is an anatomical model, trademark Somso QS-15 (Figure 1). The individual parts of this model are scanned using a Roland PICZA laser scanner and processed using the programs Dr. Picza 3 and 3D Editor. Each one of these parts is then positioned to achieve the complete model, in accordance to the spatial placement obtained by means of a 3D CT scan in healthy individuals. Figure 1 shows the geometrical accuracy obtained by means of that procedure, with the modelled geometry reproducing all the anatomical relevant aspects. Then the transition from the zone of exterior cortical bone to the zone of interior cancellous bone was obtained by means of statistical averages from CTs of vertebrae in healthy individuals, with results similar to those mentioned in the bibliography [50]. This method combines high accuracy in the external form with an excellent definition of internal interfaces and a perfect correlation among the different anatomical structures.

The mesh of the vertebrae is made by means of tetrahedral elements with quadratic approximation in the I-deas program [51] with a size thin enough to allow a smooth transition from the zone of exterior cortical bone to the zone of interior cancellous bone. The mesh of the discs is essential for the correct reproduction of the biomechanical behaviour of the lumbar spine; in order to do this, each disc is divided into nucleus pulposus and annulus fibrosus with commonly accepted dimensions [50]. Each part is meshed separately, the nucleus by means of tetrahedra and the annulus by means of hexahedra and prisms with quadratic approximation. The mesh sizes must concord with each other and with the vertebrae. Later, nine layers (outer and four double crossed) of concentric fibres are added to the annulus. These layers are modelled by means of tension-only elements, included in the hexahedra matrix, with variable orientation from the most internal to the most external (Figure 3), ranging from 35° to 80°, respecting at most the anatomical disposition [50, 52].

Finally, the ligaments (anterior longitudinal, posterior longitudinal, interspinous, flavum, supraspinous, intertransverse, and iliolumbar) are modelled by means of tetrahedra and prisms with quadratic approximation; in addition, membrane elements have been used for capsular ligaments. The dimensions of those soft tissues correspond to average anatomical measurements [50, 52] (Figure 3). The number of finite elements for every part is shown in Table 1. The total number of elements of the final mesh, obtained after a sensitivity analysis, is 196553. To this respect a mesh refinement was performed in order to achieve a convergence towards a minimum of the potential energy, both for the whole model and for each of its components, with a tolerance of 1% between consecutive meshes.

The bone and ligament properties were taken from the bibliography. Concerning the bone, in [30] it is demonstrated that the centre of the vertebrae is less rigid than in the exterior zone. For this reason the vertebrae are divided into four areas with variable modulus of elasticity (Figure 4). In addition, the

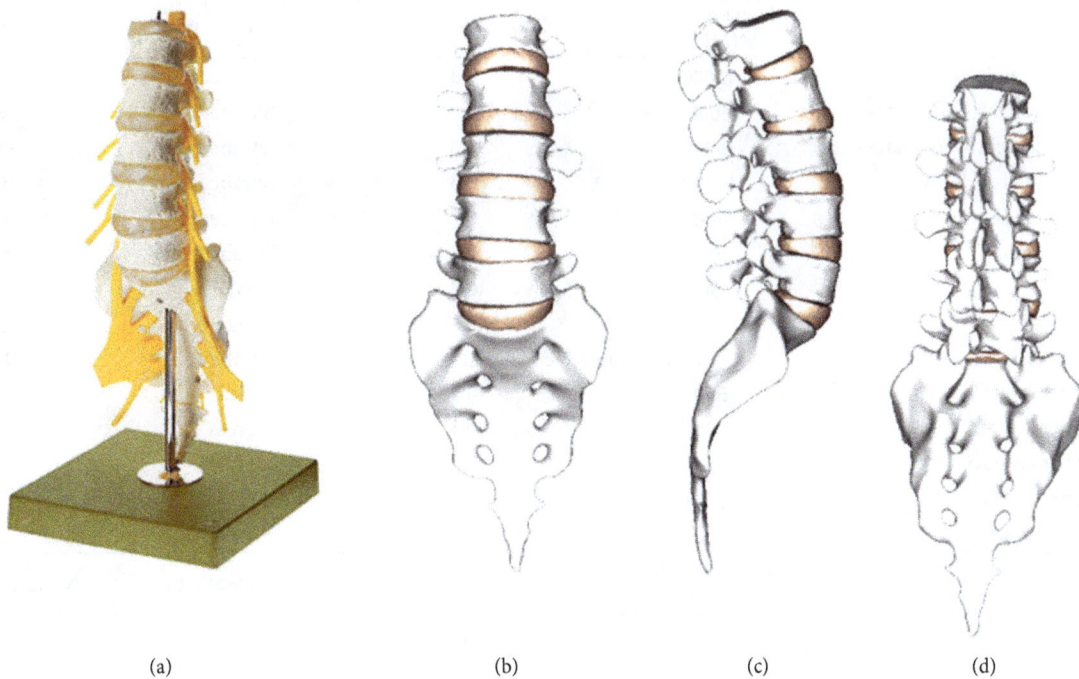

(a)        (b)        (c)        (d)

FIGURE 1: Anatomic model of the lumbar spine and complete geometric model (frontal, lateral, and dorsal views).

corresponding properties are used for the cancellous bone (Table 1).

In the discs, the nucleus pulposus behaves like a non-compressible fluid, which upon being compressed expands towards the exterior tractioning the fibers of the annulus. This behaviour was simulated by means of the hyperelastic Mooney-Rivlin model (incompressible) incorporated in the Abaqus materials library [53]. The fibres of the annulus exhibit a nonlinear only tension behaviour approximated using different linear models for each layer considering their respective range of deformation [30]. The materials of the matrix and cartilage of the apophyses were simulated as elastic materials. Finally, the different ligaments present nonlinear only tension behaviour, included as a bilinear model in the strain range (Table 1) as with most of the reported FEM studies [30, 38].

Four basic movements will be studied: flexion, extension, lateral bending, and rotation (Figure 5), from which any movement of the spine can be obtained. As boundary conditions displacements in the wings of sacrum have been prevented. In all cases the starting point is a compression of $400\,n$, which simulates the precompression due to the body weight. That compression was applied as a follower load from L1 to L5 as is done in [54]. This a standard option in the Abaqus software [53]. Later, by means of an iterative procedure based on optimization techniques, the forces and moments on each vertebra were adjusted until the degrees of rotation in every vertebral segment were achieved according to the specific movement, taking in to account three fundamental muscular groups for flexion-extension

[55]: psoas major as local muscle and rectum and erector spinae as global muscles. For lateral bending and rotation, oblique and multifidus muscles were added. The procedure calculates the force at every considered muscle along the paths of their respective movements (Figure 6, for flexion). Then the associated energy is evaluated as

$$W_{Fi} = \oint_{C_i} \overrightarrow{F_i} \cdot d\overrightarrow{s_i} \tag{1}$$

for the forces (local muscles) and

$$W_{Mj} = \int_0^{\alpha_i} \overrightarrow{M_j} \cdot d\vec{\theta} \tag{2}$$

for the moments (global muscles). The total energy is minimized for every movement:

$$\min W = \sum_{i=1}^{N} W_{Fj} + \sum_{j=1}^{M} W_{Mj} \tag{3}$$

with $N$ and $M$ the number of local and global muscles, respectively, considered in the analyzed movement. As a restriction for the minimization problem, all the forces and moments must be nonnegative.

The correct interaction between the different elements (vertebrae, discs, and ligaments) is essential. For inserting the ligaments, the guidelines set in the anatomy manuals have been followed. Conditions of union between the different vertebral bodies and the corresponding intervertebral discs

TABLE 1: Mechanical properties of materials.

| Material | Young modulus (MPa) | Poisson coefficient | Element type | Number of elements |
|---|---|---|---|---|
| Outer vertebral endplates | 12000 | 0.3 | Tetrahedron | 3578 |
| Intermediate vertebral endplates | 6000 | 0.3 | Tetrahedron | 2244 |
| Centre of the vertebral endplates | 2000 | 0.3 | Tetrahedron | 831 |
| Walls of the vertebral body | 12000 | 0.3 | Tetrahedron | 37205 |
| Cancellous bone (inside vertebrae) | 100 | 0.2 | Tetrahedron | 44954 |
| Posterior vertebra | 3000 | 0.3 | Tetrahedron | 47134 |
| Cartilage | 50 | 0.4 | Wedge | 3086 |
| Annulus fibrosus | 4.2 | 0.45 | Hexahedron | 8288 |
| Nucleus pulposus[*] | Incompressible material | | Tetrahedron | 14410 |
| Annulus fiber layers 1 | 360 | 0.3 | Truss[**] | 592 |
| Annulus fiber layers 2 | 408 | 0.3 | Truss[**] | 592 |
| Annulus fiber layers 3 | 455 | 0.3 | Truss[**] | 592 |
| Annulus fiber layers 4 | 503 | 0.3 | Truss[**] | 592 |
| Annulus fiber layers 5 | 550 | 0.3 | Truss[**] | 296 |

| Ligament | Young modulus (MPa) | Transition strain (%) | Element type | Number of elements |
|---|---|---|---|---|
| Anterior longitudinal ligament | 7.8 20.0 | 12.0 | Wedge[**] | 9046 |
| Posterior longitudinal ligament | 10.0 50.0 | 11.0 | Wedge[**] | 3844 |
| Ligamentum flavum | 15.0 19.0 | 6.2 | Tetrahedron[**] | 3042 |
| Intertransverse ligament | 10.0 59.0 | 18.0 | Tetrahedron[**] | 6678 |
| Capsular ligament | 7.5 33.0 | 25.0 | Membrane[**] | 3220 |
| Interspinous ligament | 8.0 15.0 | 20.0 | Tetrahedron[**] | 2856 |
| Supraspinous ligament | 10.0 12.0 | 14.0 | Tetrahedron[**] | 2657 |
| Iliolumbar ligament | 7.8 20.0 | 12.0 | Wedge[**] | 816 |

[*] $C_{01} = 0.0343$ MPa; $C_{10} = 0.1369$ MPa. An elastic analysis with Young modulus of 1.0 MPa and Poisson ratio of 0.49 was carried out with similar results and a volume change less than 0.6%.
[**] Only tension.

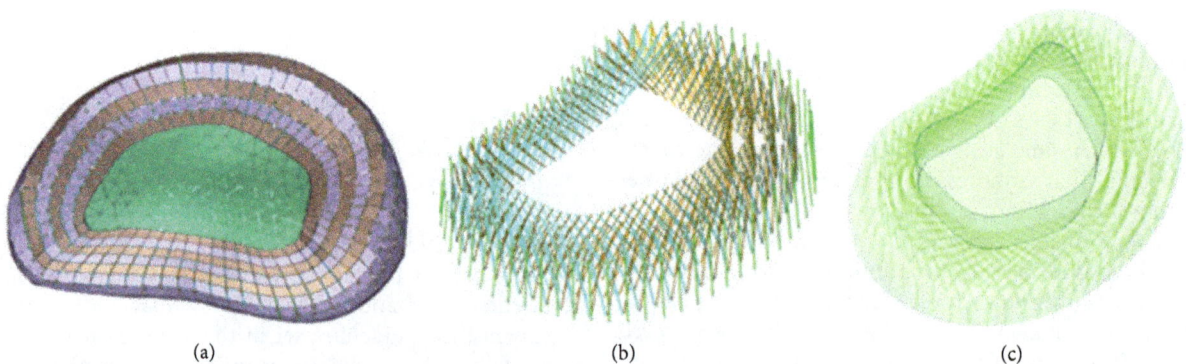

(a)          (b)          (c)

FIGURE 2: Complete FE model including vertebrae, discs, and ligaments (frontal, lateral, and dorsal views).

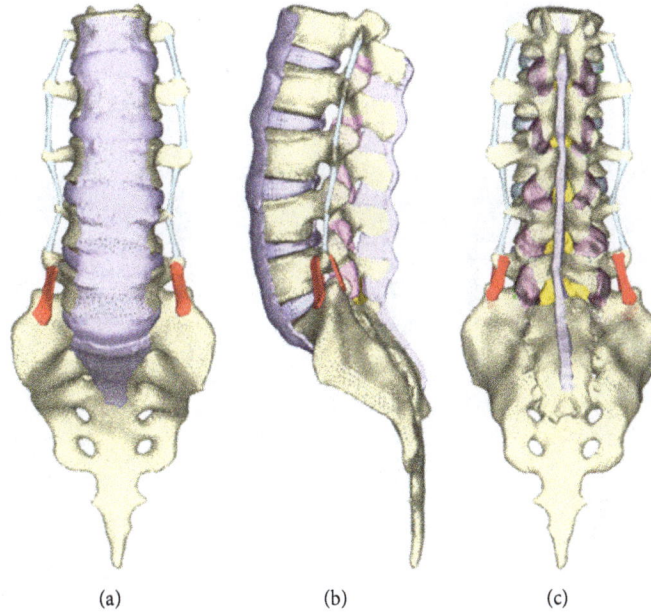

FIGURE 3: Model of the intervertebral disk and its layers of fibres.

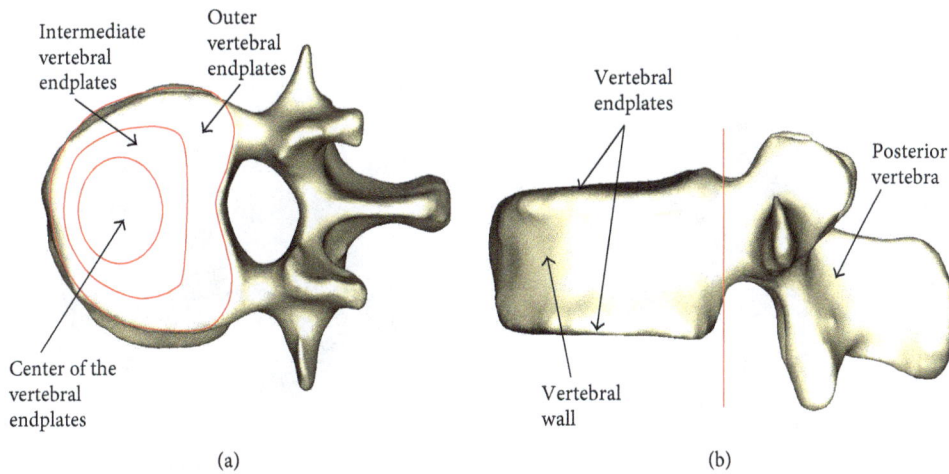

FIGURE 4: Zones of different elastic properties in the vertebral body.

have been established, as it is the most representative of the real anatomy. Because vertebrae and discs were meshed in a separate and independent way in order to get a more accurate definition of the different regions in each of them, the common surfaces between the vertebrae and the discs demand an adequate surface congruency to avoid stress concentrations in isolated points. Then, a joint condition must be established (TIE option, standard in Abaqus software [53]). Finally, contact conditions have been established between the different apophyses which provide a global stability, taking into account that an important part of the loads are transmitted through them. Capsular ligaments were also included in order to a better simulation of physiological conditions. The calculation and postprocessing were carried out using the Abaqus program.

For every movement, the changes in the relative position of the vertebrae in respect to the sacrum are measured by means of perpendicular lines on the upper face of each vertebra, associated with four knots on which the monitoring is carried out (Figure 7). In the same way, another two reference lines are defined on the sacrum. So, for every vertebra, the reference coordinates are [56]:

(1) *frontal plane* (initial node (4); final node (5)):

$$N_{f1}^{Li} \equiv N_{f1}\left(X_{f1} \ Y_{f1} \ Z_{f1}\right), \tag{4}$$

$$N_{f2}^{Li} \equiv N_{f2}\left(X_{f2} \ Y_{f2} \ Z_{f2}\right). \tag{5}$$

(2) *sagittal plane* (initial node (6); final node (7)):

$$N_{s1}^{Li} \equiv N_{s1}\left(X_{s1} \ Y_{s1} \ Z_{s1}\right), \tag{6}$$

$$N_{s2}^{Li} \equiv N_{s2}\left(X_{s2} \ Y_{s2} \ Z_{s2}\right). \tag{7}$$

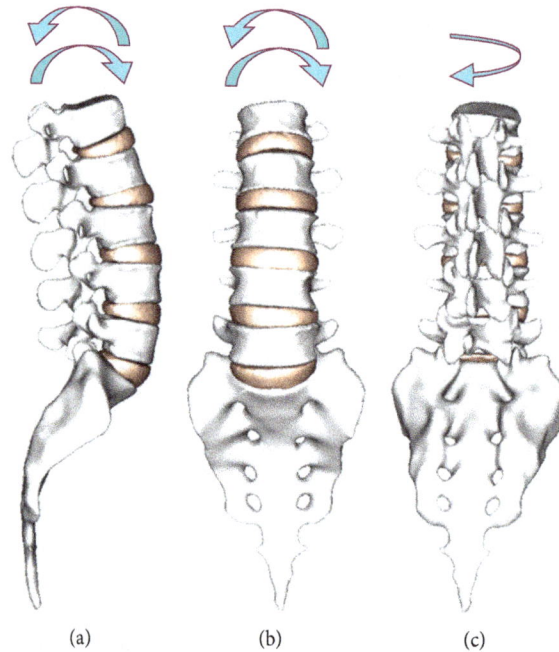

FIGURE 5: Simulated movements: flexion-extension, lateral bending, and axial rotation.

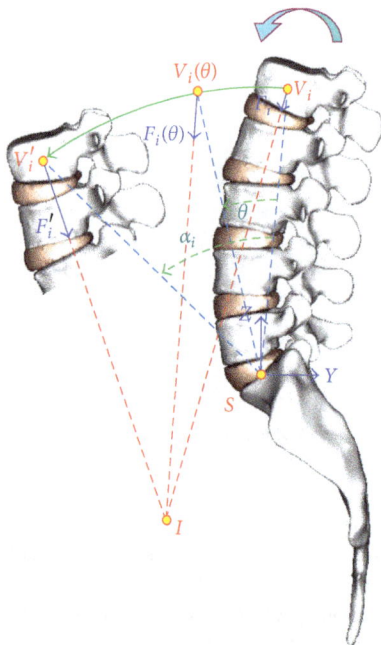

FIGURE 6: Muscle force path in flexion.

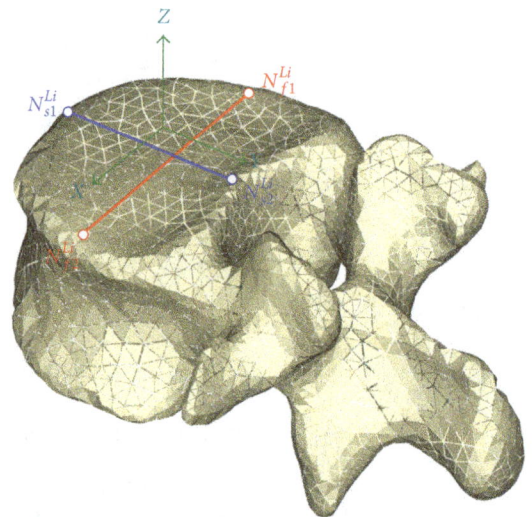

FIGURE 7: Reference points and lines for processing vertebrae mobility.

Using the formulae of analytical geometry, the properties of both lines can be obtained. For the length,

(3) *frontal plane:*

$$L_f = \sqrt{\left(X_{f2} - X_{f1}\right)^2 + \left(Y_{f2} - Y_{f1}\right)^2 + \left(Z_{f2} - Z_{f1}\right)^2}, \quad (8)$$

(4) *sagittal plane:*

$$L_s = \sqrt{\left(X_{s2} - X_{s1}\right)^2 + \left(Y_{s2} - Y_{s1}\right)^2 + \left(Z_{s2} - Z_{s1}\right)^2}, \quad (9)$$

and for the directional cosines, in the general case,

(5) *frontal plane:*

$$l_f = \frac{X_{f2} - X_{f1}}{L_f}, \qquad m_f = \frac{Y_{f2} - Y_{f1}}{L_f},$$

$$n_f = \frac{Z_{f2} - Z_{f1}}{L_f}, \qquad (10)$$

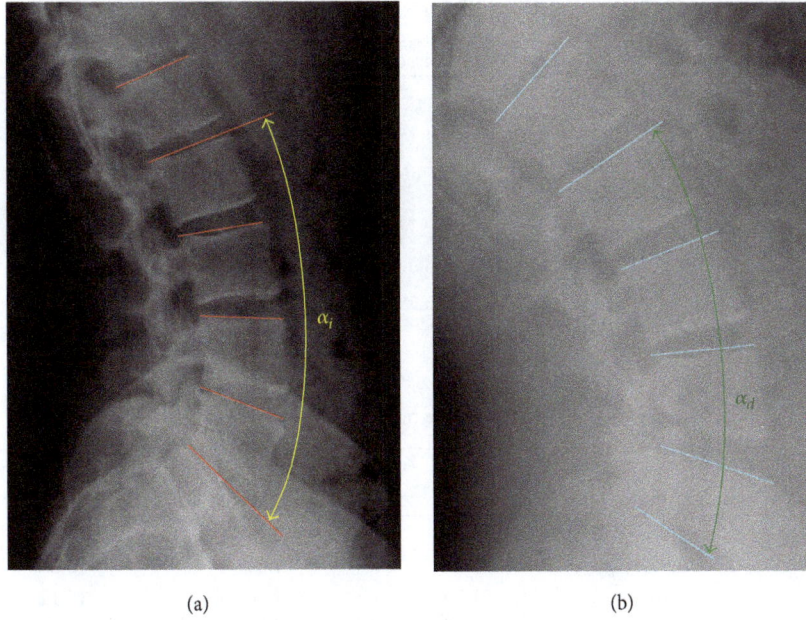

FIGURE 8: Measurements on radiological images (standing and extension).

(6) *sagittal plane:*

$$l_s = \frac{X_{s2} - X_{s1}}{L_s}, \qquad m_s = \frac{Y_{s2} - Y_{s1}}{L_s}, \qquad n_s = \frac{Z_{s2} - Z_{s1}}{L_s}. \tag{11}$$

The same applies for the sacrum. Then the relative angle with respect to the sacrum can be obtained by means of the scalar product, applying the above equations to every particular case as the following:

(1) flexo-extension (sagittal plane, $YZ$):

$$\cos \alpha_{\text{FE}} (Li - S) = m_s^{Li} m_s^S + n_s^{Li} n_s^S, \tag{12}$$

(2) lateral bending (frontal plane, $XZ$):

$$\cos \alpha_{\text{LB}} (Li - S) = l_f^{Li} l_f^S + n_f^{Li} n_f^S, \tag{13}$$

(3) axial rotation (horizontal plane, $XY$):

$$\cos \alpha_{\text{AR}} (Li - S) = l_f^{Li} l_f^S + m_f^{Li} m_f^S. \tag{14}$$

Computing the above values for every vertebra and sacrum, it is possible to determine their relative positions in space, both in the undeformed and deformed configurations, corresponding to the different analyzed movements.

In order to collect the radiological measurements which make it possible to contrast and validate the developed model, a group of 25 healthy volunteers, male individuals with an average age of 27.4 years, ranging from 23 to 33, and an average weight of 78.6 Kg, ranging from 72.1 to 81.7, with the corresponding informed consent were taken.

Two radiographic techniques have been used: (a) standing, starting from a neutral position and performing movements of flexion, extension, and lateral bending; (b) the radiographs of flexion and extension were repeated placing the individuals in sitting position with hips bent at 90° above the torso and knees also bent at 90°, with a dense, rubber, and foam device at the level of the abdomen. No significant differences between the values of flexion and extension were found with respect to those obtained in the standing position.

For the measurements on the radiological images, we proceed at a graphic level with the same methodology of comparing the relative positions of common points. Lines are depicted at the top of every vertebra, and then the final position after movement is compared with the equivalent line in the standing position for the different movements (Figure 8). The radiological monitoring of the torsion has not been performed due to the fact that reliable measurements cannot be obtained from frontal, dorsal, and/or lateral images as those used in the rest of movements. The in vivo study was used to verify the accuracy obtained for the movement of individual vertebrae. In fact, there is a lot of sets of values for muscle force that produce the global movement, but only one of them is coherent with all the individual movements.

In the case of disc degeneration, MRI can detect disc space narrowing, osteophyte formation, vacuum phenomena, and water content. The incompressibility is reduced due to nucleus dehydration, and the disc deformation implies some compressibility. From a mechanical point of view, two effects have to be taken into account: a loss of disc rise and a loss of tension in the ligaments, basically in the anterior and in the posterior ones. The degenerative process induces a certain degree of instability in the affected unit depending on the degree of degeneration. From the healthy model,

TABLE 2: Mechanical properties of degenerated disc.

| Material | Young modulus (MPa) | Poisson coefficient | Element type | Number of elements |
|---|---|---|---|---|
| Annulus fibrosus | 6.0 | 0.35 | Hexahedron | 8288 |
| Nucleus pulposus* | 1.3 | 0.4 | Tetrahedron | 14410 |
| Annulus fiber layers 1 | 36.0 | 0.3 | Truss** | 592 |
| Annulus fiber layers 2 | 40.8 | 0.3 | Truss** | 592 |
| Annulus fiber layers 3 | 45.5 | 0.3 | Truss** | 592 |
| Annulus fiber layers 4 | 50.3 | 0.3 | Truss** | 592 |
| Annulus fiber layers 5 | 55.0 | 0.3 | Truss** | 296 |

* Elastic material (compressible).
** Only tension.

TABLE 3: Results from the radiological measurements.

| Vertebra | Flexion (°) | | Extension (°) | | Lateral bending (°) | |
|---|---|---|---|---|---|---|
| | Mean | Standard deviation | Mean | Standard deviation | Mean | Standard deviation |
| L1 | 33.94 | 4.91 | 38.73 | 4.29 | 23.40 | 2.39 |
| L2 | 30.25 | 3.93 | 34.17 | 4.29 | 20.08 | 2.55 |
| L3 | 24.78 | 6.20 | 31.70 | 4.28 | 16.12 | 1.38 |
| L4 | 18.09 | 6.83 | 24.25 | 5.24 | 9.45 | 1.33 |
| L5 | 9.69 | 4.50 | 12.66 | 4.06 | 4.21 | 0.63 |

pathological conditions were simulated in the L5-S1 disc diminishing the nucleus compressibility and modifying the stiffness in the different elements according Table 2. In this case, a normalized moment of 15 m·N has been used for every movement except for axial rotation (6 m·N) acting on L1, according to the range used in the specialized literature. The objective is not to realize a sophisticated model for disc behaviour, as is done in specialized studies involving just a functional unit [35, 46, 47], but to analyze the influence of disc degeneration in the global mobility of the lumbar spine.

## 3. Results and Discussion

The results concerning radiological measurements are included in Table 3. The results obtained from the simulation model and from radiological measurements are depicted in Figure 9 for the four movements analysed. It can be seen that in every case a progressive movement of the vertebrae is produced as the distance to the sacrum increases, so that the global movements are increasing in the order L5 → L4 → L3 → L2 → L1. Concerning the radiological movements, the range of values obtained coincides with the results of the simulation by means of FE, as well as with the physiological values [50, 52].

Revising every movement, the evolution of the values obtained for flexion can be seen in Figure 9(a), compared to the radiological measurements and physiological values of reference. A maximum angle of 34.40° is achieved (L1), and an accurate correspondence can be observed with the radiological measurements (33.94 ± 4.91°) as well as a good approximation to the physiological values [50, 52].

The evolution of the values obtained for extension can be seen in Figure 9(b), comparing them again with the radiological measurements and physiological values of reference.

A maximum angle of 35.58° is achieved (L1), and a very good agreement with the radiological measurements (38.73±4.29°) as well as a good approximation to the physiological values can be observed. The results for the complete movement of flexion-extension are shown in Figure 9(c), with a whole flexion-extension angle of 69.98° (mean value of 72.67° in the radiological measurements). Logically a good degree of approximation is maintained with both the results of the radiological measurement and physiological values [50, 52], both in the global movement and in the ones corresponding to every vertebra.

In Figure 9(d) the values obtained for lateral bending and its evolution are shown. Once more the values are compared with the radiological measurements and with the physiological values of [50, 52] and show a very high degree of approximation again. A maximum angle of 19.33° was reached (23.40 ± 2.39° in the radiological measurements).

Finally, the values obtained for the movement of torsion are shown in Figure 9(e). In this case the values obtained by means of simulation are compared with the physiological values of [50, 52] and once again show a good concurrence. A maximum angle of 9.96° was reached in this movement.

Finally, concerning recent references, in Mosnier [57] are collected a lot of results, corresponding to different in vivo tests. In Figure 10 can be seen a comparison between those results and the values obtained in the present work. A good agreement is obtained for the different movements.

As for the tensional state, due to the anatomical accuracy of the model, precise stress distributions can be obtained for either part. So, as an example, Figure 11 shows the distribution of von Mises stress (MPa) in vertebrae and sacrum in the movement of torsion, as it relates to a movement which has been studied less than the rest of movements in the specialized literature. A progressive increase in

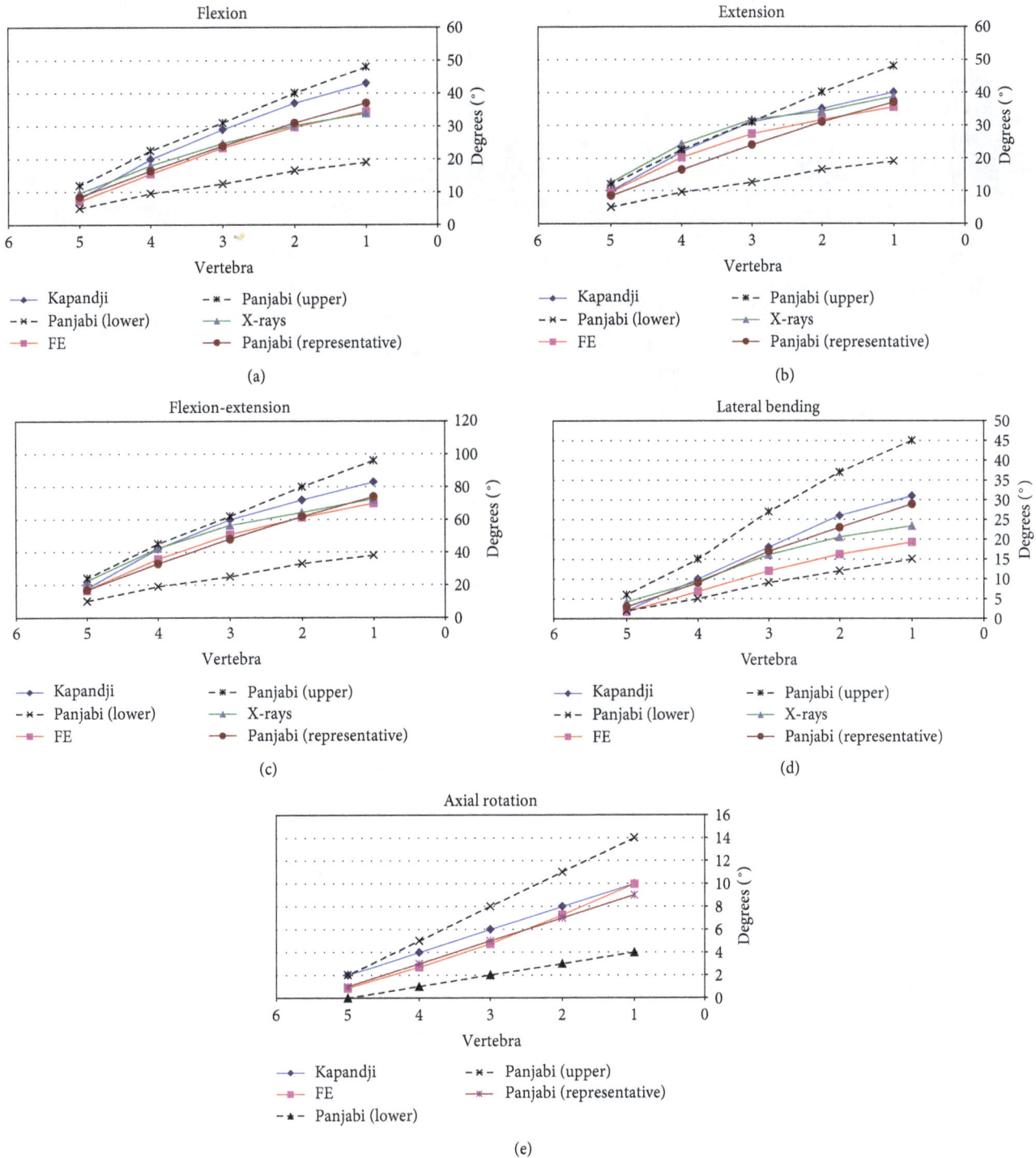

FIGURE 9: Comparison of angles in flexion, extension, flexion-extension, lateral bending, and rotation.

the tensional level in the order L5 → L4 → L3 → L2 → L1 is observed. The distribution of maximum shear tension (MPa) in disc L5-S1 in extension where the effect of shear is more marked is also shown. Some localized zones of maximum shear in the posterior zone of the annulus fibrosus are detected, with a noticeable tensional discontinuity between the annulus and the nucleus, as corresponds to materials with

very different rigidities. All the obtained values are according to the previously published ranges [18, 54, 58].

For the discs, in Figure 12 the stresses (tension) on the different fibers of the annulus fibrosus are shown. The blue-coloured fibers are situated in the zones of compression, hence they are not working. In the movement of flexion, the maximum tensions in the posterior fibers are reached,

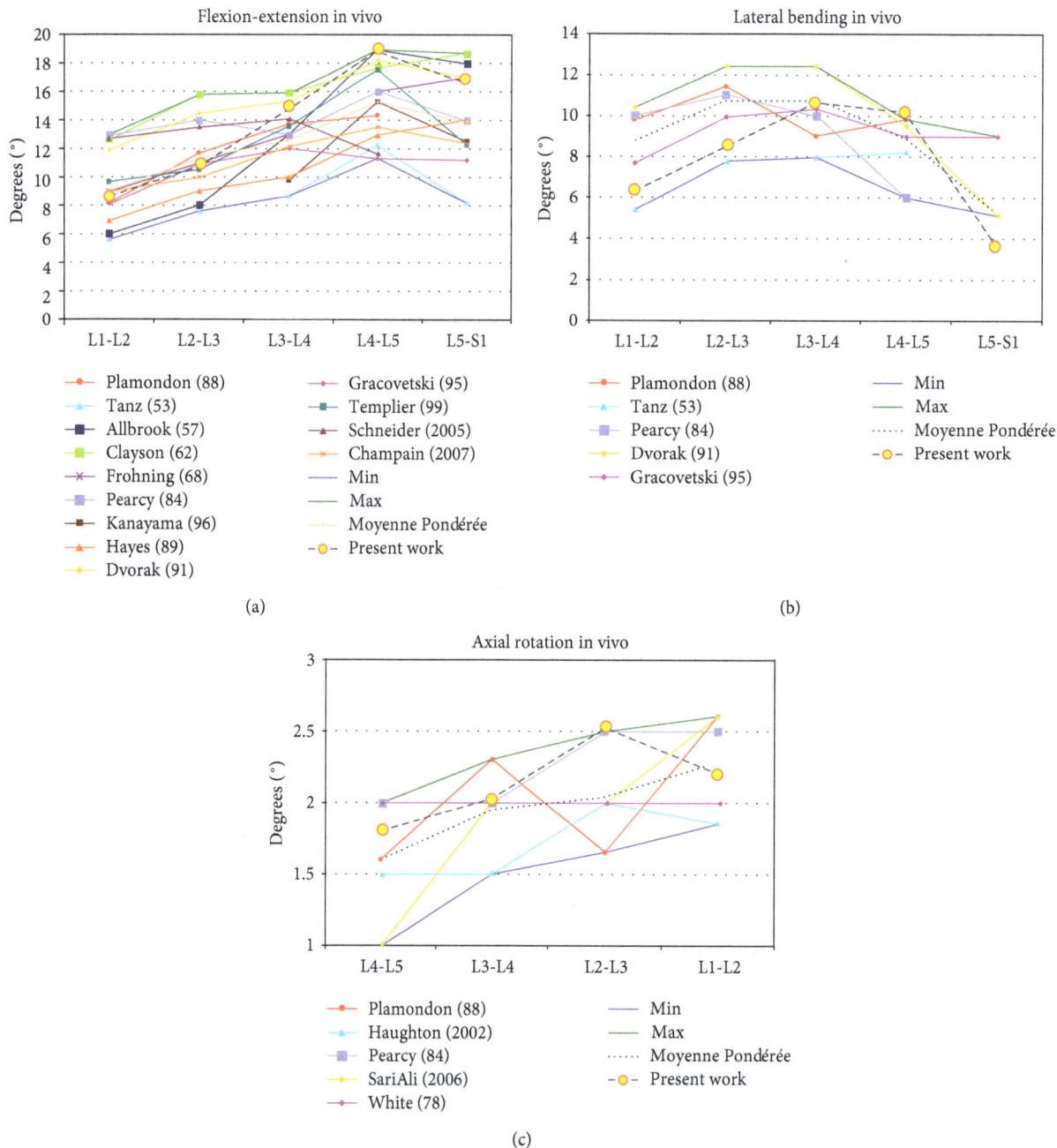

FIGURE 10: Comparison of angles in flexion-extension, lateral bending, and rotation with data from [57].

whilst in extension the maximum tensions correspond to the anterior fibers. In lateral bending the fibers on the opposite side to the inclination of the torso are loaded. Finally, in the movement of torsion, the fibers tensioned along the five discs form a helix, which corresponds to the optimum mechanism of resistance to torsion of any mechanical element. In the same way, the precise stress distribution for every component in the model (vertebrae, nucleus, annulus, and ligaments) can be obtained for every analyzed movement or even different combinations of the basic movements.

In the simulated pathologic conditions, a higher mobility is detected at every vertebral level when comparing with healthy conditions, according to [59]. So in Figure 13 a comparison of the deformed configurations for the different movements is presented, and in Figure 14 the numerical values for the rotation at different vertebral levels are included. Finally, in Figure 15 a comparative diagram shows the mobility differences between healthy and pathological conditions, with values of 19.4% for flexion, 23.3% for extension, 29.1% for lateral bending, and 10.3% for axial rotation.

Despite one can find in the literature previous validated models of the lumbar spine with a good agreement with experimental tests [54], the developed model incorporates

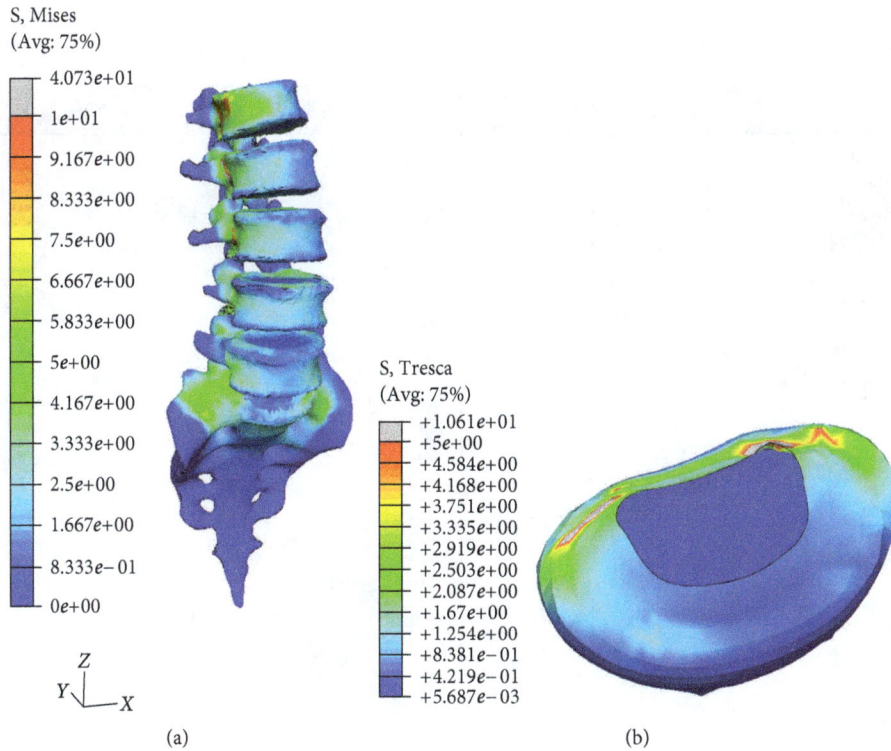

FIGURE 11: Stress distribution: (a) von Mises stress in vertebrae and sacrum (axial rotation) (MPa); (b) maximum shear stress in L5-S1disc (extension) (MPa).

improvements in some aspects. So a precise geometrical characterization (without simplifications) of all of the constituent elements (vertebrae, discs, ligaments, and cartilages) according to anatomical features is done. This allows a better characterization of the ligaments-apophyses interaction, avoiding the local effects produced by one-dimensional elements in the 3D models.

The model also provides a suitable definition concerning conditions of interaction between elements (vertebrae-discs interaction, vertebrae/discs-ligaments interaction, and contact between articular apophyses). This gives rise to a nonlinear behaviour of the whole model, with results that reliably reproduce those obtained in other studies. There are models in the literature much better in the properties characterization (porous materials, hypoelasticity, incompressible fluid, etc.), but such models concern to only one element (vertebra, disc) or a functional unit (two vertebrae and intervertebral disc) at most [30, 38–40].

The role of the fibres on the annulus fibrosus is essential in the biomechanical behaviour of the lumbar spine [60], its adequate modelling being fundamental. In the developed model, fibres have been added in great detail, respecting the distribution in layers, as well as the variable inclination from the interior to the exterior of the annulus (Figure 2), making it possible to obtain precise stresses distributions (Figures 11 and 12). This is very significant with regard to previous models [61–65], which excessively simplify the behaviour of the fibres.

Another important topic in the model is the anatomical accuracy of ligaments when comparing with previous works which simplify them to linear or nonlinear one-dimensional springs or truss type elements [30–34, 36] and then cannot obtain precise stress distributions or detect transverse displacements which can produce local instability.

Moreover, the model is capable of providing reliable results of stresses values in any of the elements which form the model. This differs to the existing models which are in general limited in this aspect, when the behaviour of some elements is simplified [30–34, 36]. This is essential at the time of analysing different pathologies and when making it possible to simulate the biomechanical repercussion of the fixations, since the clinical studies [66–68] suggest that the stress concentrations in the adjacent spaces can give rise to, in the medium and long term, new pathologies in these levels. Finally, the developed model makes it possible to obtain results in a wide range of each movement, reaching the usual anatomical maximum values.

The mobility of the lumbar spine has been studied, both in vivo and in vitro using different methods [8–25]. In the radiological measurements, it is difficult to get fixed references, due to the different degrees of rotation in each X-ray, focal distance, and position of the hip and pelvis. The same is applicable to the studies with video fluoroscope or computer monitor. In addition, the mobility measurements on the same individual can vary throughout the day [24]. The studies with CT are carried out in decubitus position and in

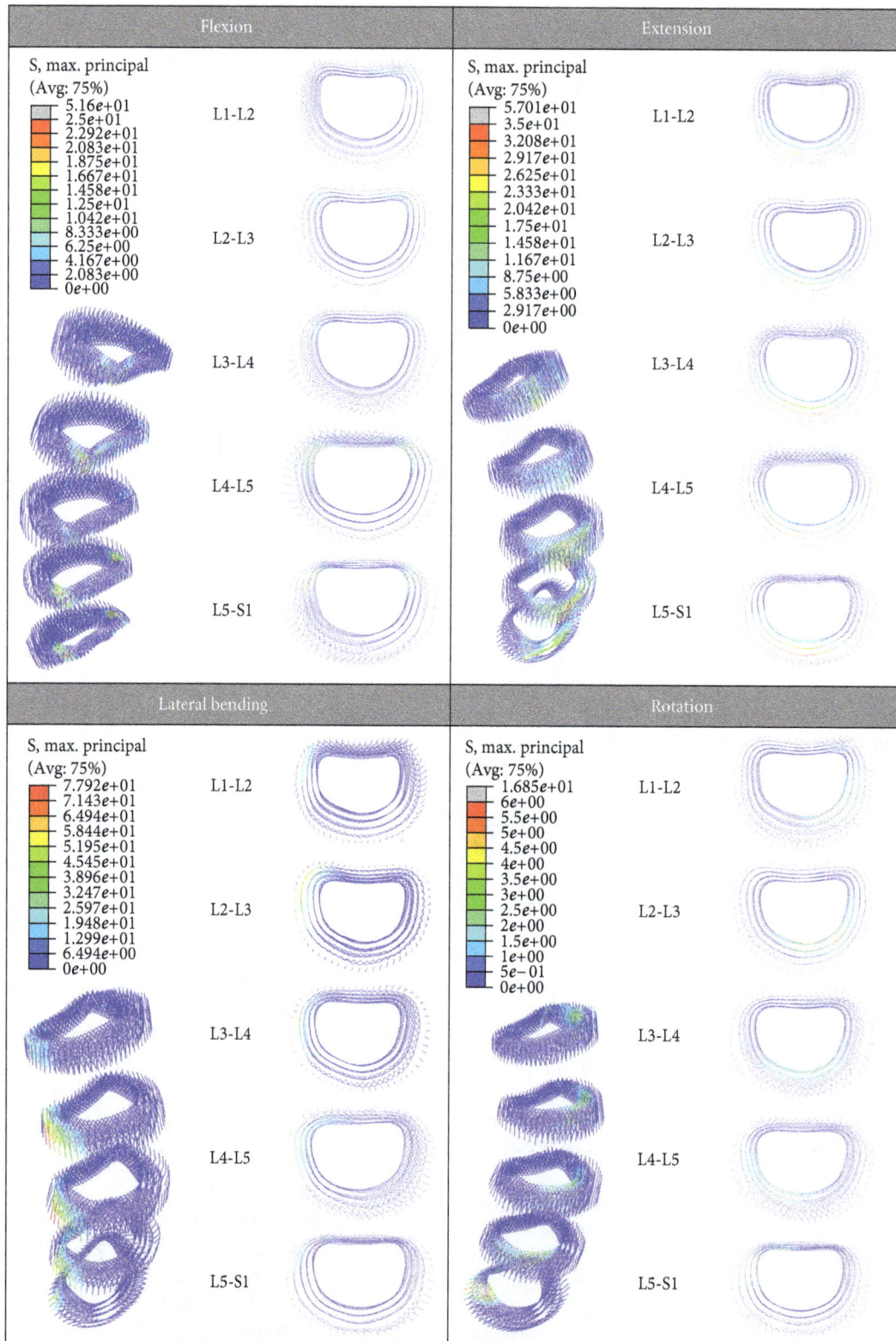

Figure 12: Stress distribution on the fibers of the annulus fibrosus (MPa).

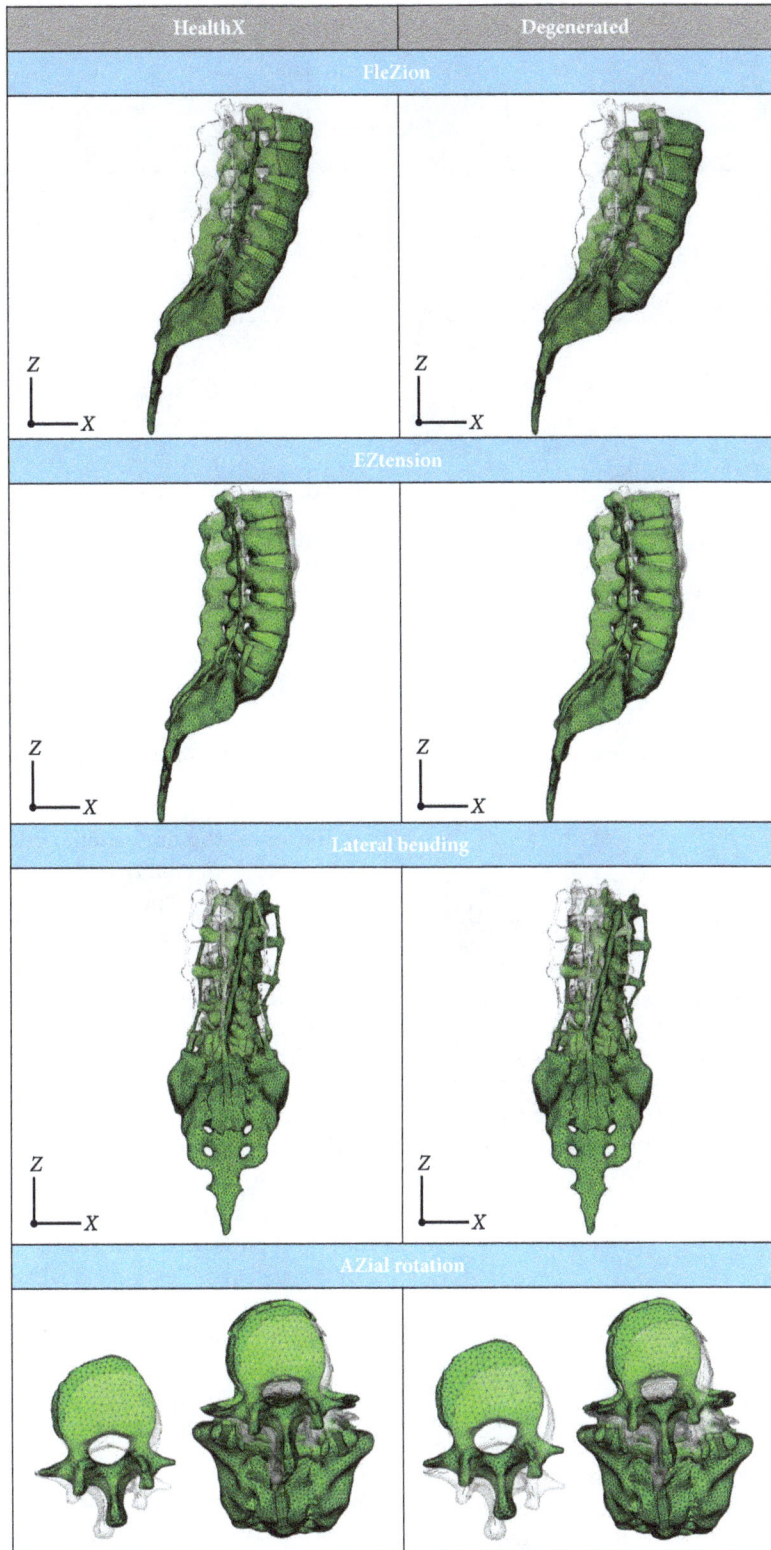

FIGURE 13: Deformed configuration: healthy model versus model with disc degeneration at L5-S1 level.

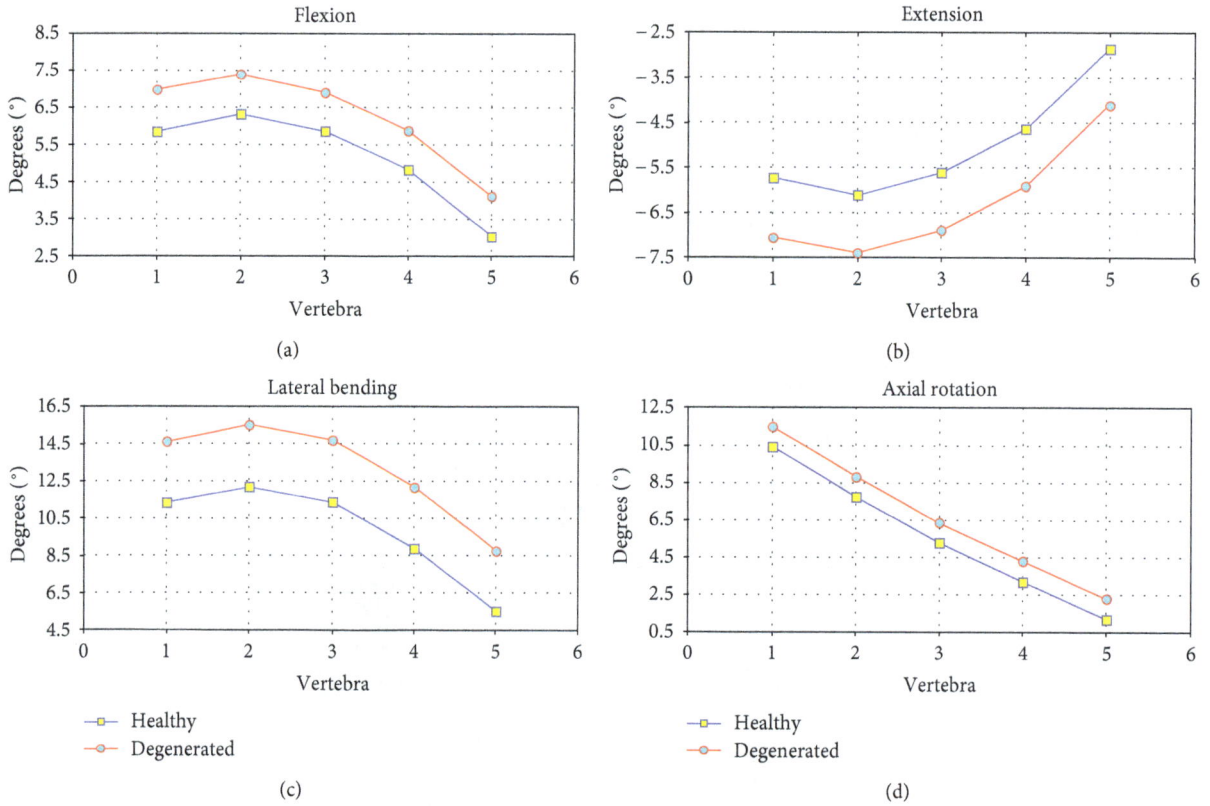

FIGURE 14: Mobility results. Comparison between healthy model and model with disc degeneration at L5-S1 level.

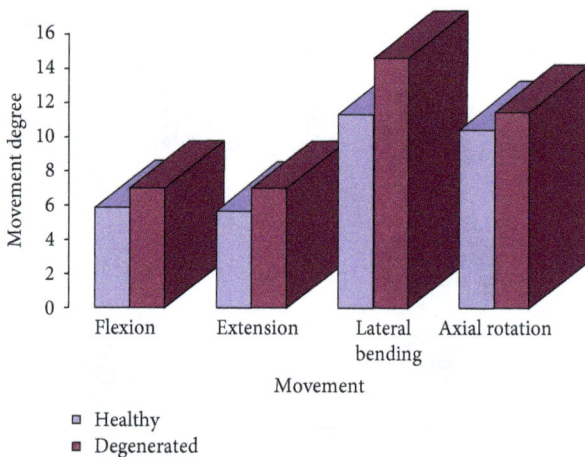

FIGURE 15: Global mobility. Difference of mobility between healthy model and model with disc degeneration at L5-S1 level.

wide range of movements, above all in rotation [17]; in the studies with MRI, there are limitations in the number of cases and in the range of mobility that this technique currently allows (a maximum flexion angle of 45°) [21–23]. The studies with corpse spines [8–11] are of little value, due to the loss of flexibility and range of movements. All of this leads to a great variability in the ranges of mobility in the different published works; in addition to the fact that mobility and biomechanics vary with age [18, 25, 69] and with the underlying pathology [21, 41].

The aim of the simulation models is to fulfill the requirements of reproducibility and versatility, with the advantage of being able to repeat the study as it is a noninvasive investigation, and the initial conditions are changeable. Some authors model one or two functional units [30, 34–39, 61–63], while others model the complete lumbar spine [31–33, 46, 54, 65]. An important geometrical simplification is present in most of the models, concerning mainly to ligaments (uniaxial models with spring or truss type elements) and annulus fibrosus (number and disposition of fiber layers) [30–34, 36]. Models in which the behaviour of the intervertebral disc is simulated in a more complex way [34, 35] only consider one or two functional units instead of the complete lumbar zone. This provides results at a local level, but they cannot be extrapolated to the level of global behaviour. Some works that are dedicated to specific problems exist [39, 40] but have not managed to integrate a complete model of the lumbar spine with nonlinear behaviour.

In the developed model, in flexion and extension a progressive movement of the vertebrae is produced as the distance to the sacrum increases, and the accumulated movements are increasing in the order L5 → L4 → L3 → L2 → L1. However, the segmental movement between two vertebrae is less in the segments nearest the sacrum (lower levels) (Figure 9(c)) according to other authors [32, 65]. The ranges obtained correspond with the average values of the radiological measurements carried out and with those of classic works [50, 52], in addition to up-to-date references

[19, 70]. However, opposite results are also referenced, both in classic studies [11, 13] and in more recent ones with electrogoniometer, spectrometry, and MRI [23]. This difference could be due to the fact that in these studies flexion is limited to 45°, because of the limitations of MRI devices. Besides, the sample is not numerous enough to establish general patterns.

In lateral bending, in relative terms greater mobility is observed in the intermediate levels and in the range of maximum values accepted in the more classical references [50, 52]. Mobility is less in the upper segments coinciding with dual fluoroscope studies and MRI [23]. A greater degree of accumulated movement is observed the higher the vertebrae are (Figure 9(d)), in concordance with [23].

In rotation, the comparison of twisting is more complex, as the study of this movement "in vivo" is much more difficult due to the difficulty of finding reference points. In the developed model the movements in torsion can be studied in a similar way to the rest of movements (Figure 9(e)). There is not a noticeable difference in the degrees of rotation of the different vertebrae in agreement with [23], and once again the maximum ranges are accepted in the classic references [50, 52]. Torsion has been studied in different situations and with different techniques: MRI [21–23], X-rays [12–15], three-dimensional television system [16], and CT [19]. In the majority of the studies the upper levels have a greater degree of mobility when assessing the rotation in supine position. According to [23], the different results in the torsion could be due to the different load conditions and to the position of the lumbar spine. This makes quantitative comparison difficult. Due to that dispersion of results in the different measurements, a former comparative analysis was carried out with classical references [50, 52] recognized by the majority of authors. Notwithstanding the aforementioned, in [57] a complete review of different in vivo works along more than fifty years is done, and it can be confirmed that the present results are close to the average values included in the review (Figure 10). In pathologic conditions, the obtained results agree with [56], showing an increase in the mobility at every vertebral level (Figure 15).

## 4. Conclusions

A complete three-dimensional FE model of the lumbar spine has been developed and verified, in to which all of the structures of the spine have been incorporated. This can be modified to reproduce the biomechanics in physiological and pathological conditions. Therefore making it possible to simulate the pathological conditions of hypermobility and lumbar segmental instability produced by disc degeneration, which is associated with pain of discogenic origin [14, 71, 72].

The developed model provides a valid tool for predicting the biomechanical behaviour of the lumbar spine in different conditions and is capable of reproducing a wide range of physiological movements. The model represents the first step for the analysis of the behavioural changes induced by different pathologies, allowing parametric studies for different ranges of disc degeneration.

## Abbreviations

FE:   Finite elements
3D:   Three-dimensional
TV:   Television
CT:   Computed tomography
NMR:  Nuclear magnetic resonance
L1:   First lumbar vertebrae
L2:   Second lumbar vertebrae
L3:   Third lumbar vertebrae
L4:   Fourth lumbar vertebrae
L5:   Fifth lumbar vertebrae
S1:   First sacral vertebrae
FEM:  Finite-element method
MRI:  Magnetic resonance imaging.

## Authors' Contributions

A. Herrera and J. R. Vela conceived the design of the radiological study and performed it, including the statistical treatment of data. E. Ibarz, Y. Más, J. Cegoñino, S. Puértolas and L. Gracia conceived and developed the finite element model and carried out all the simulations A. Herrera and L. Gracia coordinated the work between surgeons and engineers. All authors participated in the drawing up of the paper, and read and approved the final paper.

## Acknowledgments

This work has been partially financed through the project (a) Research Project PM100/2006, Aragón Regional Government, (b) Research Project 2006-0296, Foundation SECOT, and (c) Research Project ICS08/0333, Foundation Mutua Madrileña.

## References

[1] L. Manchikanti, V. Singh, S. Datta, S. P. Cohen, and J. A. Hirsch, "Comprehensive review of epidemiology, scope, and impact of spinal pain," *Pain Physician*, vol. 12, no. 4, pp. E35–E70, 2009.

[2] L. E. Griffith, R. P. Wells, H. S. Shannon, S. D. Walter, D. C. Cole, and S. Hogg-Johnson, "Developing common metrics of mechanical exposures across aetiological studies of low back pain in working populations for use in meta-analysis," *Occupational and Environmental Medicine*, vol. 65, no. 7, pp. 467–481, 2008.

[3] J. C. Iatridis, J. J. MacLean, P. J. Roughley, and M. Alini, "Effects of mechanical loading on intervertebral disc metabolism in vivo," *Journal of Bone and Joint Surgery*, vol. 88, no. 2, pp. 41–46, 2006.

[4] K. C. Wong, R. Y. Lee, and S. S. Yeung, "The association between back pain and trunk posture of workers in a special school for the severe handicaps," *BMC Musculoskeletal Disorders*, vol. 10, article no. 43, 2009.

[5] M. C. Battié, T. Videman, and E. Parent, "Lumbar disc degeneration: epidemiology and genetic influences," *Spine*, vol. 29, no. 23, pp. 2679–2690, 2004.

[6] M. C. Battié and T. Videman, "Lumbar disc degeneration: epidemiology and genetics," *Journal of Bone and Joint Surgery A*, vol. 88, no. 2, pp. 3–9, 2006.

[7] M. J. Yaszemski, A. A. White III, and M. M. Panjabi, "Biome-chanics of spine," in *Orthopaedic Knowledge Update: Spine 2*, D. F. Fardon and S. R. Garfin, Eds., pp. 17–26, AAOS, Rosemont, Ill, USA, 2002.

[8] M. M. Panjabi, K. Takata, and V. K. Goel, "Kinematics of lumbar intervertebral foramen," *Spine*, vol. 8, no. 4, pp. 348–357, 1983.

[9] I. Yamamoto, M. M. Panjabi, T. Crisco, and T. Oxland, "Three-dimensional movements of the whole lumbar spine and lumbosacral joint," *Spine*, vol. 14, no. 11, pp. 1256–1260, 1989.

[10] M. M. Panjabi, T. R. Oxland, I. Yamamoto, and J. J. Crisco, "Mechanical behavior of the human lumbar and lumbosacral spine as shown by three-dimensional load-displacement curves," *Journal of Bone and Joint Surgery A*, vol. 76, no. 3, pp. 413–424, 1994.

[11] R. C. Hilton, J. Ball, and R. T. Benn, "In-vitro mobility of the lumbar spine," *Annals of the Rheumatic Diseases*, vol. 38, no. 4, pp. 378–383, 1979.

[12] M. J. Pearcy and S. B. Tibrewal, "Axial rotation and lateral bending in the normal lumbar spine measured by three-dimensional radiography," *Spine*, vol. 9, no. 6, pp. 582–587, 1984.

[13] M. A. Hayes, T. C. Howard, C. R. Gruel, and J. A. Kopta, "Roentgenographic evaluation of lumbar spine flexion-extension in asymptomatic individuals," *Spine*, vol. 14, no. 3, pp. 327–331, 1989.

[14] J. M. Fritz, S. R. Piva, and J. D. Childs, "Accuracy of the clinical examination to predict radiographic instability of the lumbar spine," *European Spine Journal*, vol. 14, no. 8, pp. 743–750, 2005.

[15] T. Steffen, R. K. Rubin, H. G. Baramki, J. Antoniou, D. Marchesi, and M. Aebi, "A new technique for measuring lumbar segmental motion in vivo: method, accuracy, and preliminary results," *Spine*, vol. 22, no. 2, pp. 156–166, 1997.

[16] M. J. Pearcy, J. M. Gill, M. W. Whittle, and G. R. Johnson, "Dynamic back movement measured using a three-dimensional television system," *Journal of Biomechanics*, vol. 20, no. 10, pp. 943–949, 1987.

[17] R. S. Ochia, N. Inoue, S. M. Renner et al., "Three-dimensional in vivo measurement of lumbar spine segmental motion," *Spine*, vol. 31, no. 18, pp. 2073–2078, 2006.

[18] K. W. N. Wong, J. C. Y. Leong, M. K. Chan, K. D. K. Luk, and W. W. Lu, "The flexion-extension profile of lumbar spine in 100 healthy volunteers," *Spine*, vol. 29, no. 15, pp. 1636–1641, 2004.

[19] K. W. N. Wong, K. D. K. Luk, J. C. Y. Leong, S. F. Wong, and K. K. Y. Wong, "Continuous dynamic spinal motion analysis," *Spine*, vol. 31, no. 4, pp. 414–419, 2006.

[20] S. W. Lee, K. W. Wong, M. K. Chan, H. M. Yeung, J. L. Chiu, and J. C. Leong, "Development and validation of a new technique for assessing lumbar spine motion," *Spine*, vol. 27, no. 8, pp. E215–E220, 2002.

[21] V. M. Haughton, B. Rogers, M. E. Meyerand, and D. K. Resnick, "Measuring the axial rotation of lumbar vertebrae in vivo with MR imaging," *American Journal of Neuroradiology*, vol. 23, no. 7, pp. 1110–1116, 2002.

[22] K. Kulig, C. M. Powers, R. F. Landel et al., "Segmental lumbar mobility in individuals with low back pain: In vivo assessment during manual and self-imposed motion using dynamic MRI," *BMC Musculoskeletal Disorders*, vol. 8, article no. 8, 2007.

[23] G. Li, S. Wang, P. Passias, Q. Xia, G. Li, and K. Wood, "Segmental in vivo vertebral motion during functional human lumbar spine activities," *European Spine Journal*, vol. 18, no. 7, pp. 1013–1021, 2009.

[24] F. B. M. Ensink, P. M. M. Saur, K. Frese, D. Seeger, and J. Hildebrandt, "Lumbar range of motion: Influence of time of day and individual factors on measurements," *Spine*, vol. 21, no. 11, pp. 1339–1343, 1996.

[25] G. K. Fitzgerald, K. J. Wynveen, W. Rheault, and B. Rothschild, "Objective assessment with establishment of normal values for lumbar spinal range of motion," *Physical Therapy*, vol. 63, no. 11, pp. 1776–1781, 1983.

[26] L. H. Riley, J. C. Eck, H. Yoshida, Y. D. Koh, J. W. You, and T. H. Lim, "A biomechanical comparison of calf versus cadaver lumbar spine models," *Spine*, vol. 29, no. 11, pp. E217–E220, 2004.

[27] A. Kettler, L. Liakos, B. Haegele, and H. J. Wilke, "Are the spines of calf, pig and sheep suitable models for pre-clinical implant tests?" *European Spine Journal*, vol. 16, no. 12, pp. 2186–2192, 2007.

[28] M. J. Turner, R. W. Clough, H. C. Martin, and L. J. Topp, "Stiffness and deflection analysis of complex structures," *Journal of the Aeronautical Sciences*, vol. 23, no. 9, pp. 805–823, 1956.

[29] O. C. Zienkiewicz and Y. K. Cheung, *Finite Element Method in Structural & Continuum Mechanics*, McGraw-Hill, London, UK, 1967.

[30] G. Denozière and D. N. Ku, "Biomechanical comparison between fusion of two vertebrae and implantation of an artificial intervertebral disc," *Journal of Biomechanics*, vol. 39, no. 4, pp. 766–775, 2006.

[31] A. Rohlmann, T. Zander, and G. Bergmann, "Comparison of the biomechanical effects of posterior and anterior spine-stabilizing implants," *European Spine Journal*, vol. 14, no. 5, pp. 445–453, 2005.

[32] A. Rohlmann, N. K. Burra, T. Zander, and G. Bergmann, "Comparison of the effects of bilateral posterior dynamic and rigid fixation devices on the loads in the lumbar spine: a finite element analysis," *European Spine Journal*, vol. 16, no. 8, pp. 1223–1231, 2007.

[33] T. Zander, A. Rohlmann, N. K. Burra, and G. Bergmann, "Effect of a posterior dynamic implant adjacent to a rigid spinal fixator," *Clinical Biomechanics*, vol. 21, no. 8, pp. 767–774, 2006.

[34] A. Fantigrossi, F. Galbusera, M. T. Raimondi, M. Sassi, and M. Fornari, "Biomechanical analysis of cages for posterior lumbar interbody fusion," *Medical Engineering and Physics*, vol. 29, no. 1, pp. 101–109, 2007.

[35] A. Rohlmann, T. Zander, H. Schmidt, H. J. Wilke, and G. Bergmann, "Analysis of the influence of disc degeneration on the mechanical behaviour of a lumbar motion segment using the finite element method," *Journal of Biomechanics*, vol. 39, no. 13, pp. 2484–2490, 2006.

[36] A. Glema, T. Lodygowski, W. Kakol, M. Ogurkowska, and M. Wierszycki, "Modeling of intervertebral discs in the numerical analysis of spinal segment," in *Proceedings of the European Congress on Computational Methods in Applied Sciences and Engineering (ECCOMAS '04)*, July 2004.

[37] J. Noailly, D. Lacroix, and J. A. Planell, "The mechanical significance of the lumbar spine components—a finite element stress analysis," in *Proceedings of the Summer Bioengineering Conference*, June 2003.

[38] T. Pitzen, F. Geisler, D. Matthis et al., "A finite element model for predicting the biomechanical behaviour of the human lumbar spine," *Control Engineering Practice*, vol. 10, no. 1, pp. 83–90, 2002.

[39] A. Boccaccio, P. Vena, D. Gastaldi, G. Franzoso, R. Pietrabissa, and C. Pappalettere, "Finite element analysis of cancellous bone failure in the vertebral body of healthy and osteoporotic subjects," *Proceedings of the Institution of Mechanical Engineers H*, vol. 222, no. 7, pp. 1023–1036, 2008.

[40] A. Polikeit, L. P. Nolte, and S. J. Ferguson, "Simulated influence of osteoporosis and disc degeneration on the load transfer in a lumbar functional spinal unit," *Journal of Biomechanics*, vol. 37, no. 7, pp. 1061–1069, 2004.

[41] A. Fujiwara, T. H. Lim, H. S. An et al., "The effect of disc degeneration and facet joint osteoarthritis on the segmental flexibility of the lumbar spine," *Spine*, vol. 25, no. 23, pp. 3036–3044, 2000.

[42] R. E. Thompson, M. J. Pearcy, and T. M. Barker, "The mechanical effects of intervertebral disc lesions," *Clinical Biomechanics*, vol. 19, no. 5, pp. 448–455, 2004.

[43] J. P. G. Urban and S. Roberts, "Degeneration of the intervertebral disc," *Arthritis Research and Therapy*, vol. 5, no. 3, pp. 120–130, 2003.

[44] H. S. An and K. Masuda, "Relevance of in vitro and in vivo models for intervertebral disc degeneration," *Journal of Bone and Joint Surgery A*, vol. 88, no. 2, pp. 88–94, 2006.

[45] R. N. Natarajan, J. R. Williams, and G. B. J. Andersson, "Recent advances in analytical modeling of lumbar disc degeneration," *Spine*, vol. 29, no. 23, pp. 2733–2741, 2004.

[46] H. Schmidt, A. Kettler, A. Rohlmann, L. Claes, and H. J. Wilke, "The risk of disc prolapses with complex loading in different degrees of disc degeneration—a finite element analysis," *Clinical Biomechanics*, vol. 22, no. 9, pp. 988–998, 2007.

[47] J. P. Little, C. J. Adam, J. H. Evans, G. J. Pettet, and M. J. Pearcy, "Nonlinear finite element analysis of anular lesions in the L4/5 intervertebral disc," *Journal of Biomechanics*, vol. 40, no. 12, pp. 2744–2751, 2007.

[48] L. M. Ruberté, R. N. Natarajan, and G. B. Andersson, "Influence of single-level lumbar degenerative disc disease on the behavior of the adjacent segments—a finite element model study," *Journal of Biomechanics*, vol. 42, no. 3, pp. 341–348, 2009.

[49] R. N. Natarajan, J. R. Williams, and G. B. J. Andersson, "Modeling changes in intervertebral disc mechanics with degeneration," *Journal of Bone and Joint Surgery A*, vol. 88, no. 2, pp. 36–40, 2006.

[50] A. A. White and M. M. Panjabi, *Clinical Biomechanics of the Spine*, Lippincott, Philadelphia, Pa, USA, 1990.

[51] SIEMENS, http://www.plm.automation.siemens.com/.

[52] I. A. Kapandji, *The Physiology of the Joints*, Churchill Livingstone, New York, NY, USA, 2008.

[53] Dassault Systèmes, http://www.3ds.com/.

[54] A. Rohlmann, L. Bauer, T. Zander, G. Bergmann, and H. J. Wilke, "Determination of trunk muscle forces for flexion and extension by using a validated finite element model of the lumbar spine and measured in vivo data," *Journal of Biomechanics*, vol. 39, no. 6, pp. 981–989, 2006.

[55] N. Arjmand, D. Gagnon, A. Plamondon, A. Shirazi-Adl, and C. Larivière, "Comparison of trunk muscle forces and spinal loads estimated by two biomechanical models," *Clinical Biomechanics*, vol. 24, no. 7, pp. 533–541, 2009.

[56] E. Ibarz, *Finite element simulation of the biomechanical behaviour of the lumbar spine. Application to the study of degenerative pathologies and the evaluation of fixation systems [Ph.D. thesis]*, University of Zaragoza, Department of Mechanical Engineering, 2010.

[57] T. Mosnier, *Contribution à l'analyse biomécanique et à l'évaluation des implants rachidiens [Ph.D. thesis]*, ENSAM, Paris, France, 2008.

[58] N. A. Langrana, S. P. Kale, W. T. Edwards, C. K. Lee, and K. J. Kopacz, "Measurement and analyses of the effects of adjacent end plate curvatures on vertebral stresses," *Spine Journal*, vol. 6, no. 3, pp. 267–278, 2006.

[59] K. Kulig, C. M. Powers, R. F. Landel et al., "Segmental lumbar mobility in individuals with low back pain: in vivo assessment during manual and self-imposed motion using dynamic MRI," *BMC Musculoskeletal Disorders*, vol. 8, article no. 8, 2007.

[60] L. J. Smith and N. L. Fazzalari, "The elastic fibre network of the human lumbar anulus fibrosus: architecture, mechanical function and potential role in the progression of intervertebral disc degeneration," *European Spine Journal*, vol. 18, no. 4, pp. 439–448, 2009.

[61] P. Vena, G. Franzoso, D. Gastaldi, R. Contro, and V. Dallolio, "A finite element model of the L4-L5 spinal motion segment: biomechanical compatibility of an interspinous device," *Computer Methods in Biomechanics and Biomedical Engineering*, vol. 8, no. 1, pp. 7–16, 2005.

[62] T. H. Lim, J. G. Kim, A. Fujiwara et al., "Biomechanical evaluation of diagonal fixation in pedicle screw instrumentation," *Spine*, vol. 26, no. 22, pp. 2498–2503, 2001.

[63] A. P. Dooris, V. K. Goel, N. M. Grosland, L. G. Gilbertson, and D. G. Wilder, "Load-sharing between anterior and posterior elements in a lumbar motion segment implanted with an artificial disc," *Spine*, vol. 26, no. 6, pp. E122–E129, 2001.

[64] J. L. Wang, M. Parnianpour, A. Shirazi-Adl, and A. E. Engin, "Viscoelastic finite-element analysis of a lumbar motion segment in combined compression and sagittal flexion: effect of loading rate," *Spine*, vol. 25, no. 3, pp. 310–318, 2000.

[65] F. Ezquerro, A. Simón, M. Prado, and A. Pérez, "Combination of finite element modeling and optimization for the study of lumbar spine biomechanics considering the 3D thorax-pelvis orientation," *Medical Engineering and Physics*, vol. 26, no. 1, pp. 11–22, 2004.

[66] N. Miyakoshi, E. Abe, Y. Shimada, K. Okuyama, T. Suzuki, and K. Sato, "Outcome of one-level posterior lumbar interbody Fusion for spondylolisthesis and postoperative intervertebral disc degeneration adjacent to the fusion," *Spine*, vol. 25, no. 14, pp. 1837–1842, 2000.

[67] G. Ghiselli, J. C. Wang, N. N. Bhatia, W. K. Hsu, and E. G. Dawson, "Adjacent segment degeneration in the lumbar spine," *Journal of Bone and Joint Surgery A*, vol. 86, no. 7, pp. 1497–1503, 2004.

[68] T. L. Schulte, F. Leistra, V. Bullmann et al., "Disc height reduction in adjacent segments and clinical outcome 10 years after lumbar 360° fusion," *European Spine Journal*, vol. 16, no. 12, pp. 2152–2158, 2007.

[69] M. S. Sullivan, C. E. Dickinson, and J. D. G. Troup, "The influence of age and gender on lumbar spine sagittal plane range of motion: a study of 1126 healthy subjects," *Spine*, vol. 19, no. 6, pp. 682–686, 1994.

[70] E. Ulucam and B. S. Cigali, "Measurement of normal lumbar spine range of motion in the college-aged Turkish population using a 3D ultrasound-based motion analysis system," *Trakya Universitesi Tip Fakultesi Dergisi*, vol. 26, no. 1, pp. 29–35, 2009.

[71] J. H. Abbott, B. McCane, P. Herbison, G. Moginie, C. Chapple, and T. Hogarty, "Lumbar segmental instability: a criterion-related validity study of manual therapy assessment," *BMC Musculoskeletal Disorders*, vol. 6, article no. 56, 2005.

[72] J. H. Abbott, J. M. Fritz, B. McCane et al., "Lumbar segmental mobility disorders: comparison of two methods of defining abnormal displacement kinematics in a cohort of patients with non-specific mechanical low back pain," *BMC Musculoskeletal Disorders*, vol. 7, article no. 45, 2006.

# Assessment of Genotoxic and Cytotoxic Hazards in Brain and Bone Marrow Cells of Newborn Rats Exposed to Extremely Low-Frequency Magnetic Field

**Monira M. Rageh, Reem H. EL-Gebaly, and Nihal S. El-Bialy**

*Department of Biophysics, Faculty of Science, Cairo University, Giza 12013, Egypt*

Correspondence should be addressed to Monira M. Rageh, monirarageh@yahoo.com

Academic Editor: Brynn Levy

The present study aimed to evaluate the association between whole body exposure to extremely low frequency magnetic field (ELF-MF) and genotoxic , cytotoxic hazards in brain and bone marrow cells of newborn rats. Newborn rats (10 days after delivery) were exposed continuously to 50 Hz, 0.5 mT for 30 days. The control group was treated as the exposed one with the sole difference that the rats were not exposed to magnetic field. Comet assay was used to quantify the level of DNA damage in isolated brain cells. Also bone marrow cells were flushed out to assess micronucleus induction and mitotic index. Spectrophotometric methods were used to measure the level of malondialdehyde (MDA) and the activity of glutathione (GSH) and superoxide dismutase (SOD). The results showed a significant increase in the mean tail moment indicating DNA damage in exposed group ($P < 0.01, 0.001, 0.0001$). Moreover ELF-MF exposure induced a significant ($P < 0.01, 0.001$) four folds increase in the induction of micronucleus and about three folds increase in mitotic index ($P < 0.0001$). Additionally newborn rats exposed to ELF-MF showed significant higher levels of MDA and SOD ($P < 0.05$). Meanwhile ELF-MF failed to alter the activity of GSH. In conclusion, the present study suggests an association between DNA damage and ELF-MF exposure in newborn rats.

## 1. Introduction

Considerable attention has been focused on the effects of the electromagnetic field (EMF) due to its wide-ranging use in everyday life. Wertheimer and Leeper [1] showed a possible relationship between electrical power lines and childhood cancer. Other studies have been carried out to investigate the biological effects of extremely low-frequency electromagnetic fields (ELF-EMFs). However, the results obtained are contradictory, and a comparison between them is difficult, because of the many differences in exposure parameters (periodicity of the exposure, flux intensity, and endpoint are investigated) [2].

Nevertheless, there is an increase in the number of studies detecting the genotoxic effects upon exposure to ELF-MF. Lai and Singh [3, 4] and Svedenstål and coworkers [5, 6] have reported that whole body exposure to power frequency of 10, 100, and 500 $\mu$T MF can result in DNA single- and double-strand breaks (DSBs) in the brains of rodents. Such

DNA double-strand breaks are considered as the most potent form of DNA damage which can induce genomic instability. Unrepaired DSBs cause cell cycle arrest and apoptosis. Moreover, insufficient or incorrect repair of DSBs can lead to carcinogenesis due to translocations, inversions, or deletions. Other studies conducted on mouse m5S cells and human cells detected a significant dose-dependent increase of chromatid-type chromosomal aberrations at 5, 50, and 400 mT [7]. Also Pasquini et al. [8] observed an increased frequency of micronuclei in Jurkat cells exposed for 24 h to 50 Hz and 5 mT magnetic field.

Ager and Radul [9] studied the effects of ELF-MF on mitotic recombination in *Saccharomyce scerevisiae* and found no effect. Antonopoulos et al. [10] found that incubation of peripheral blood culture upon exposure to an electromagnetic field (EMF, 50 Hz, and 5 mT) leads to stimulation of cell proliferation in lymphocytes but has no influence on the frequencies sister chromatid exchange.

Exposure to magneticfield (MF) induced changes in the activity of some enzymes involved in the antioxidant system, and the formed damages are due to an imbalance between the activities of an oxidant agent and the antioxidant system within the cell. Although oxygen is required for many important aerobic cellular reactions, it may undergo electron transfer reactions, which generate highly reactive membrane-toxic intermediates such as superoxide, hydrogen peroxide, or hydroxyl radical. The observed cytotoxicity is related to reactive oxygen species (ROS), hydrogen peroxide ($H_2O_2$), and hydroxyl radical which are produced by various factors. These reactive species cause oxidative damage to cell membrane, increase in oxygen radical's production, and permit leakage of enzymes leading to organ damage [11, 12].

Living cells have evolved numerous defense mechanisms to neutralize the harmful effects of free radicals [13, 14]. The antioxidant defense system includes enzymes such as superoxide dismutase (SOD) and other scavengers such as glutathione (GSH). Di Loreto et al. [15] investigated the effects of exposures to two different 50 Hz sinusoidal ELF-MFs intensities (0.1 and 1 mT) in maturing rat cortical neurons' major antioxidative enzymatic and nonenzymatic cellular protection systems, membrane peroxidative damage, as well as growth factor, and cytokine expression pattern. The results showed that ELF-MFs affected positively the cell viability and concomitantly reduced the levels of apoptotic death in rat neuronal primary cultures, with no significant effects on the main antioxidative defenses.

Therefore, the present study was designed to evaluate the effect of continuous exposure to 50 Hz and 0.5 mT MF on the brain and bone marrow of newborn rats by measuring some parameters indicative of oxidative stress (lipid peroxidation MDA, superoxide dismutase SOD, and glutathione GSH) and on oxidative DNA damage that subsequently cause injure in the brain and bone marrow cells.

## 2. Materials and Methods

*2.1. Animals.* Animals were housed and maintained according to guidelines for the Care and Use of Laboratory Animals 1996 [16] and approved by the Animal Ethics Committee at Cairo University. Briefly, timed pregnant Wistar rats were purchased from the National Center of Researches in Egypt and gave birth at the animal house of Cairo University. They were maintained for one week in the laboratory for adaptation. Adult female rats with their newborn were housed in plastic cages with free access to water and standard chow diet. They were also maintained in a controlled environment (humidity 50% ± 5 and temperature 25°C) with 12 h light cycle. All animal groups were housed in clean first hand cages under standard condition in a separate laboratory which belongs to animal care unit.

*2.2. Magnetic Field Exposure.* The exposure was performed using a magnet with a fixed magnetic field value of 0.5 mT ± 0.025. The magnetic field was generated by a solenoid carrying current of 18 ampere at 50 Hz from the main supply (220–230 Volt) via a variac (made in Yugoslavia). The

FIGURE 1: Experimental setup of the exposure system.

magnet consisted of a coil with 320 turns made of electrically insulated 0.8 mm copper wire. The coil was wounded around a copper cylinder of 2 mm thickness, 40 cm diameter, and 40 cm length. The cylinder wall was earthed to eliminate the electric field (Figure 1). The magnetic field was measured at different locations to find out the most homogenous zone inside the solenoid core. This was done using Gauss/Tesla meter with probe T-4048 manufactured by Bell Technologies Inc. (Orlando, FL, USA).

Plastic cages ($40 \times 25 \times 20$ cm), containing 7 adult female rats and their newborns ($n = 60$), were placed in the middle of the exposure chamber prior to MF exposure. When the newborns reached 10 days of age, they were randomly divided into two groups of 30 each: one control (sham) and one exposed. The latter group has been exposed to 50 Hz, 0.5 mT ± 0.025 MF for 30 days 24 h per day. The control group was treated like the exposed group with the sole difference that it was not exposed to magnetic field. The two groups were treated equally considering light and food. The temperature and humidity were monitored continuously throughout the experimental period. This ensures that the control and the exposed animals were maintained in the same condition. During the experimental period, cleaning and changing water and food were done to all animals two times daily. The field was switched off during cleaning the cages.

After 30 days of exposure, the two groups of newborn rats were sacrificed by decapitation. Ten newborn rats from each group were used for micronucleus test, comet assay, and histopathological examination. Another 10 newborn rats were used for mitotic index determination and lastly 10 newborn rats were maintained for oxidative stress evaluation.

*2.3. Biochemical Estimation.* Ten newborn rats from both control and exposed groups were sacrificed by decapitation, and the brains were dissected and prepared for biochemical analyses. Superoxide dismutase (SOD) assay relies on the ability of the enzyme to inhibit the phenazine

TABLE 1: The levels of MDA ($\mu$mol/L), GSH (mg/dl) and SOD (U/g tissue) in the brain of control newborn rats and those exposed to 50 Hz 0.5 mT MF for 30 days 24 h/day. Each value was expressed as mean of 10 rats $\pm$ standard error.

| Parameters | Control group | Exposed group |
|---|---|---|
| MDA ($\mu$mol/L) | 0.113 $\pm$ 0.017 | 0.302 $\pm$ 0.021* |
| GSH (mg/dl) | 0.033 $\pm$ 0.003 | 0.030 $\pm$ 0.000 |
| SOD (U/g tissue) | 182.33 $\pm$ 8.192 | 263.33 $\pm$ 12.019* |

*Significant difference in comparison with the corresponding control at $\alpha = 0.05$ ($P < 0.05$).

methosulphate-mediated reduction of nitroblue tetrazolium dye and measured at 560 nm [17]. Glutathione (GSH) assay is based on the reduction of 2-nitrobenzoic acid with glutathione to produce a yellow compound measured at 405 nm [18]. Lipid peroxidation was measured by estimating the amount of malondialdehyde (MDA) according to Uchiyama and Mihara [19], and the assay depends on the colorimetric reaction of MDA with thiobarbituric acid giving a pink complex that can be measured at 532 nm.

### 2.4. Histopathological Examination.

The brains of newborn rats from control and exposed groups were dissected, removed, and fixed in 10% neutral formalin. Then, they were embedded in paraffin blocks, sectioned and stained with hematoxylin and eosin (H&E).

All brain tissue sections were examined using light microscope (CX31 Olympus microscope, Tokyo, Japan) connected with digital camera (Canon).

### 2.5. Comet Assay (Single Cell Gel Electrophoresis).

Comet assay (single cell gel electrophoresis) is considered as a rapid, simple, visual, and sensitive technique to assess DNA fragmentation typical for toxic DNA damage and early stage of apoptosis [20, 21]. The comet assay was performed under alkaline conditions (PH > 13) according to the method developed by Singh et al. and Tice et al. [22, 23]. Briefly, a small piece of brain tissues ($n = 10$) from each group were placed in 1 mL cold HBSS containing 20 mM EDTA (ethylenediaminetetraacetic acid)/10% DMSO (dimethylsulfoxide, Qualigens, CPW59). The tissues were minced into fine pieces and let settled. 5 $\mu$L of aliquot was mixed with 70 $\mu$L of 0.7% low melting point (LMP) agarose (Sigma, A9414). This agarose was prepared in $Ca^{2+}$, $Mg^{2+}$ free PBS (phosphate buffered saline, HiMedia, TS1006) at 37°C and placed on a microscope slide, which was already covered with a thin layer of 0.5% normal melting point (NMP) agarose (HiMedia.RM273). After cooling at 4°C for 5 min, slides were covered with a third layer of LMP agarose. After solidification at 4°C for 5 min, slides were immersed in freshly prepared cold lysis solution (2.5 M NaCl, 1 mM $Na_2$EDTA, 10 mM tris base, pH 10, with 1% Triton X-100 and 10% DMSO added just before use) at 4°C for at least 1 h. Following lyses, slides were placed in a horizontal gel electrophoresis unit and incubated in fresh alkaline electrophoresis buffer (1 mM $Na_2$EDTA, 300 mM NaOH, pH 13). Electrophoresis was conducted for 30 min at 24 V (~0.74 V/cm) and 300 mA

at 4°C. Then, the slides were immersed in neutralized buffer (0.4 MTris-HCl, pH 7.5) and gently washed three times for 5 min at 4°C. All the above procedures were performed under dimmed light to prevent the occurrence of additional DNA damage. Comets were visualized by 80 $\mu$L, 1X ethidium bromide staining (Sigma E-8751) and examined at 400x magnification using a fluorescent microscope.

Comet 5 image analysis software developed by Kinetic Imaging, Ltd. (Liverpool, UK) linked to a CCD camera was used to assess the quantitative and qualitative extent of DNA damage in the cells by measuring the length of DNA migration and the percentage of migrated DNA. Finally, the program calculates tail moment. In all the samples, 100 cells were analyzed and classified into 5 types (0–4) depending on their tail moment. Type 0 represents the cells without visible damage, while cells of type 4 have total degradation of DNA (long, broad tail, and poorly visible head of the comet). Types 1, 2, and 3 represent the symptoms of increasing DNA damage. To calculate the extent of DNA damage, three types of the comet: numbers 2, 3, and 4 were selected.

### 2.6. Micronucleus Test.

Bone marrow slides for micronucleus assay from 10 newborn rats of each group were prepared and stained according to the method described by Schmidt [24] using the modifications of Agarwal and Chauhan [25]. Bone marrow was flushed out from tibias using 1 mL fetal calf serum and centrifuged at 2000 ×g for 10 min. The supernatant was discarded. Evenly spread bone marrow smears were stained using the May-Grunwald and Giemsa protocol. Slides were scored at a magnification of 1000x using light microscope (CX31 Olympus microscope, Tokyo, Japan). 2000 polychromatic erythrocytes per animal were scored, and the number of micronucleated polychromatic erythrocytes (MNPCE) was determined. In addition, the number of polychromatic erythrocytes (PCEs) was counted in fields that contained 100 cells (mature and immature) to determine the score of PCE and normochromatic erythrocytes (NCEs).

### 2.7. Mitotic Index Determination.

Slides were prepared according to the method described by Adler [26] with some modification. Briefly, 10 newborn rats (from control and exposed groups) were injected intraperitoneally (i.p.) with colchicine (2 mg/Kg) 2 hours prior to tissue sampling. Bone marrow cells were collected from the tibia by flushing in KCl (0.075 M, at 37°C) and incubated at 37°C for 25 min. Material was centrifuged at 2000 ×g for 10 min, and fixed in acetomethanol (acetic-acid : ethanol. 1 : 3, v/v). Centrifugation and fixation (in the cold) were repeated five times at an interval of 20 min. The material was resuspended in a small volume of the fixative, dropped onto chilled slides, lame dried, and stained the following day in 5% buffered Giemsa (pH 6.8). Slides were scored at a magnification of 1000x using light microscope (CX31 Olympus microscope, Tokyo, Japan). At least 1000 cells were examined for each rat and the number of dividing cells including late prophase and metaphase was determined. The mitotic activity is expressed

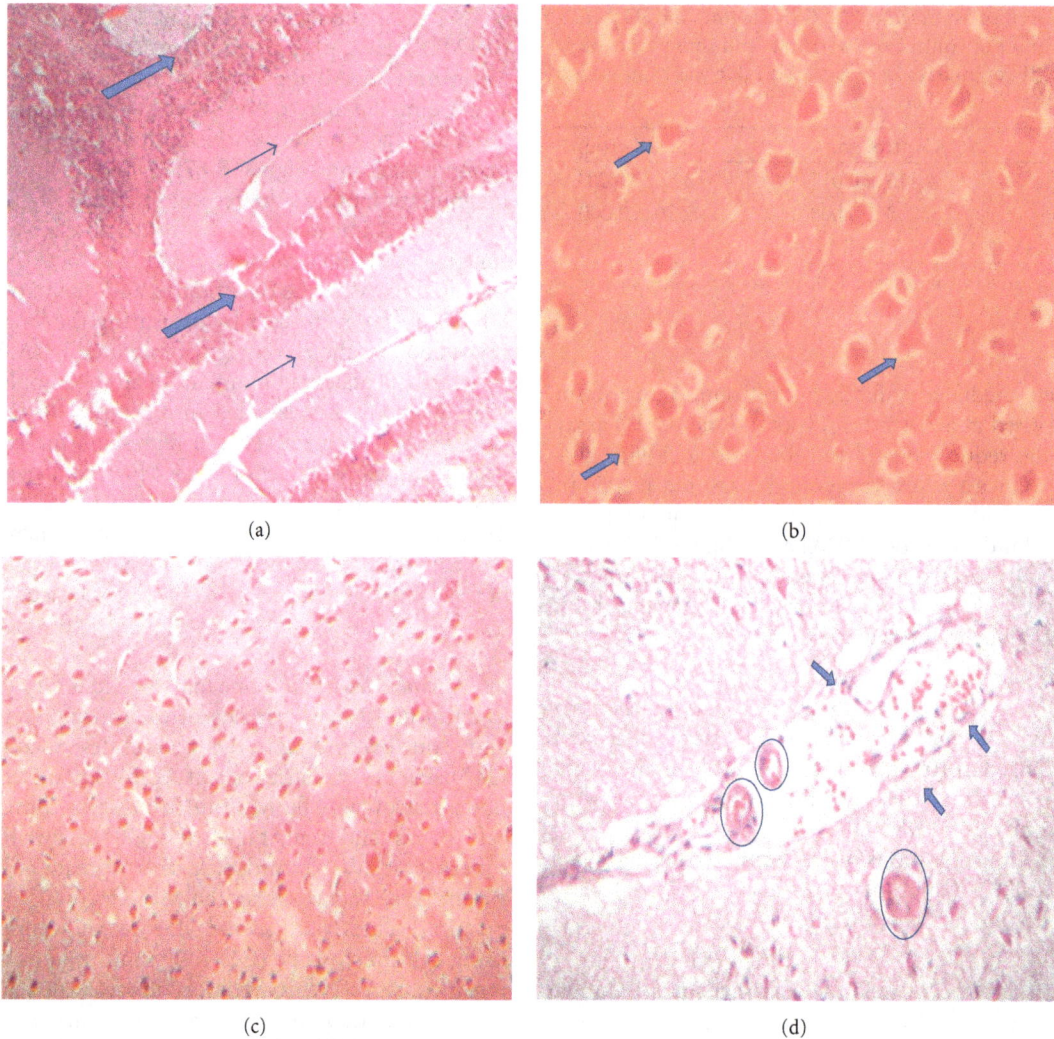

FIGURE 2: Brain of control group (a and b). (a) The neuroepithelium (bold arrow) is of normal cellularity and architecture, with the supportive astroglial tissue (thin arrow) at ×100 (H and E). (b) The astro-glial tissue is of normal cellularity, showing pyramidal astro-glial cells, with indistinct cell boundaries, having fibrillary cytoplasmic processes (arrows) at ×400 (H and E). Brain of exposed group (c and d). (c) The field shows marked astro-glial hypercellularity (i.e., reactive gliosis) at ×200 (H and E). (d) The astro-glial tissue shows marked hypercellularity (i.e., secondary reactive gliosis), with a diffused vascular congestion (arrows) accompanied by extravasation of RBCs, as well as partial obliteration of lumen with thickened hypertrophied fibrous vessel wall (encircled) at ×300 (H and E).

TABLE 2: DNA damage in the brain of newborn rats after whole body exposure to 50 Hz and 0.5 mT MF for 30 days 24 h/day, assessed by comet assay. Each value was expressed as mean of 10 rats ± standard error.

| Groups | Grade of damage | | | | | |
|---|---|---|---|---|---|---|
| | Type (0) | Type (1) | Type (2) | Type (3) | Type (4) | Types (2 + 3 + 4) |
| Control | $49.3 \pm 0.67$ | $28.7 \pm 2.196$ | $15.3 \pm 0.33$ | $6.7 \pm 0.33$ | 0 | $22 \pm 0.57$ |
| Exposed | $17.3 \pm 0.67^{**}$ | $43.3 \pm 2.31^{*}$ | $19.7 \pm 0.33^{**}$ | $13.7 \pm 0.88^{**}$ | $6 \pm 0.58^{***}$ | $39.3 \pm 1.76^{**}$ |

*,**,***Significant difference in comparison with the corresponding control at $\alpha = 0.01$, 0.001, and 0.0001, respectively.

by the mitotic index (MI), which is the number of dividing cells in 1000 cells per rat.

*2.8. Statistical Analysis.* All statistical analysis were executed by the aid of Statistical Package for the Social Science (SPSS) version 20. In order to compare between various variables (control and exposed), student *t*-test was used at $\alpha = 0.05$, 0.01, 0.001, and 0.0001.

## 3. Results

The measured malondialdehyde (MDA), glutathione (GSH), and super oxidedismutase (SOD) levels in brain cells of

Assessment of Genotoxic and Cytotoxic Hazards in Brain and Bone Marrow Cells of Newborn Rats Exposed to Extremely Low-Frequency Magnetic Field

111

(a)

(b)

(c)

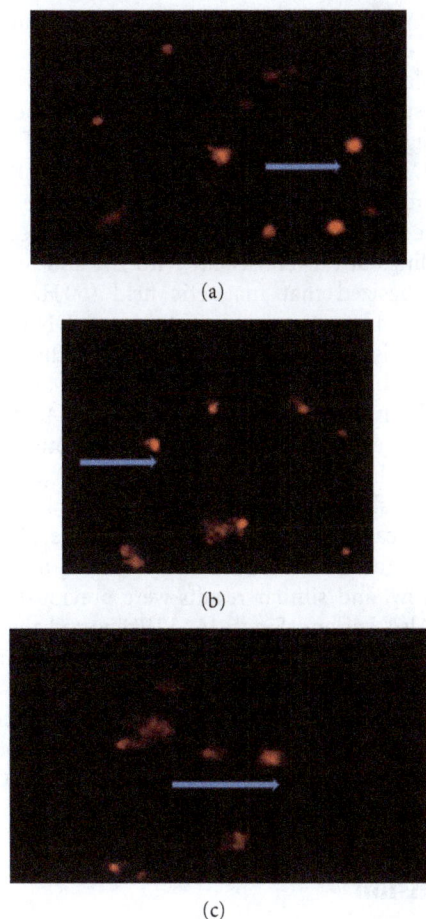

FIGURE 3: Evaluation of DNA damage induced by 50 Hz and 0.5 mT MF for 30 days 24 h/day in the brain cells of newborn rats processed by comet assay. (a) represents cells without visible DNA damage type (0), (b) represents cells with symptom of increasing DNA damage types (1, 2, and 3), and (c) represents cells with a major DNA damage (type 4).

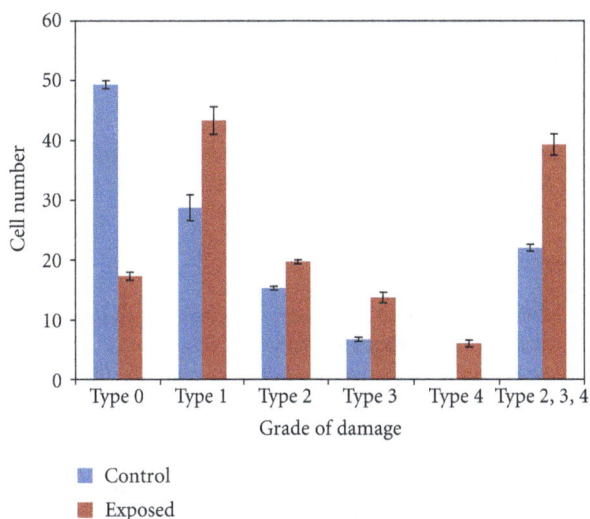

FIGURE 4: The level of DNA damage in brain cells of control newborn rats and those exposed to 50 Hz and 0.5 mT MF for 30 days 24 h/day. Assessed by comet assay. Mean ± S.E. ($P < 0.05$).

TABLE 3: Micronucleus induction (MNPCE), PCE, and NCE in bone marrow cells of control newborn rats and those exposed to 50 Hz and 0.5 mT MF for 30 days 24 h/day. Each value was expressed as mean of 10 rats ± standard error.

| Groups | NCE/100 | PCE/100 | MNPCE/2000 |
|--------|---------|---------|------------|
| Control | 60.67 ± 1.53 | 39.33 ± 1.53 | 17.6 ± 0.58 |
| Exposed | 56.33 ± 0.58* | 43.67 ± 0.58* | 72 ± 3.61** |

*,** Significant difference in comparison with the corresponding control at $\alpha = 0.01$ and 0.001, respectively.

TABLE 4: Mitotic index in bone marrow cells of control newborn rats and those exposed to 50 Hz and 0.5 mT MF for 30 days 24 h/day. Each value was expressed as mean of 10 rats ± standard error.

| Group | MI/1000 cells |
|-------|---------------|
| Control | 26 ± 3 |
| Exposed | 65 ± 2.65* |

*Significant difference in comparison with the corresponding control at $\alpha = 0.0001$.

newborn rats for the two groups were listed in Table 1. For exposed group when compared with control one, a significant increase in the MDA and SOD levels was observed ($P < 0.05$). In contrast ELF-MF failed to alter GSH level.

Brain section of control group showed fibrillar matrix and normal vasculature. No necrodegenerative or atrophic changes appeared. Also, there are no vascular or inflammatory lesions with the absence of gliosis (i.e., no astroglial proliferation) Figures 2(a) and 2(b). Meanwhile Figures 2(c) and 2(d) showed the brain section of exposed group revealing mild atrophy and degeneration within the glia. Moreover, reactive gliosis appeared accompanied by mild vascular congestion.

The levels of DNA damage in brain cells of exposed newborn rats showed a significant increase ($P < 0.01, 0.001, 0.0001$) compared to control one (Table 2 and Figures 3 and 4). For type (0) the data revealed that about 50% of brain cells did not exhibit any DNA damage in control group compared to 17% for exposed one. Meanwhile in type (4) about 6% of brain cells showed complete DNA damage in exposed group relative to 0% for control one. The total percent of DNA damage in brain cells represented by types (2, 3 and 4) showed two-fold increase for exposed group with respect to control one.

Table 3 and Figure 5 showed the frequencies of MNPCE, PCE, and NCE in bone marrow cells of newborn rats for both control and exposed groups. The results showed a significant increase in the formation of PCE and MNPCE, respectively, ($P < 0.01$ and $P < 0.001$) for exposed group compared to control one.

The results of the mitotic index (used to evaluate cell cycle kinetics) are summarized in Table 4 and Figure 6. The results showed that MI of bone marrow cells exposed to ELF-MF was increased significantly compared to the control one ($P < 0.0001$).

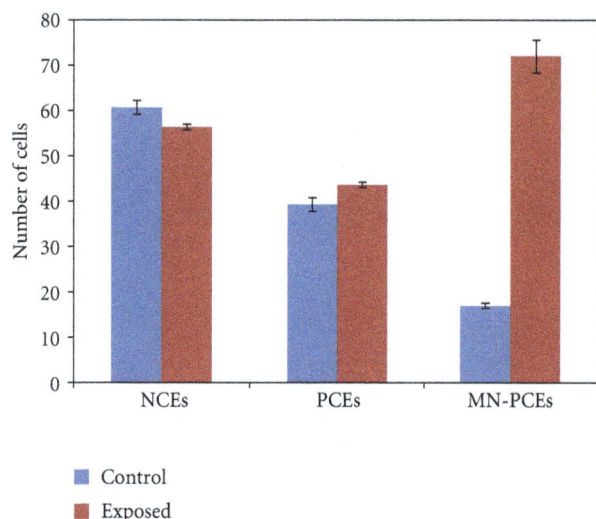

FIGURE 5: Micronucleus induction, NCE, PCE, and MN-PCE, in bone marrow cells of control newborn rats and those exposed to 50 Hz and 0.5 mT MF for 30 days 24 h/day. Mean ± S.E. ($P < 0.01$, 0.001) for PCE and MNPCE, respectively.

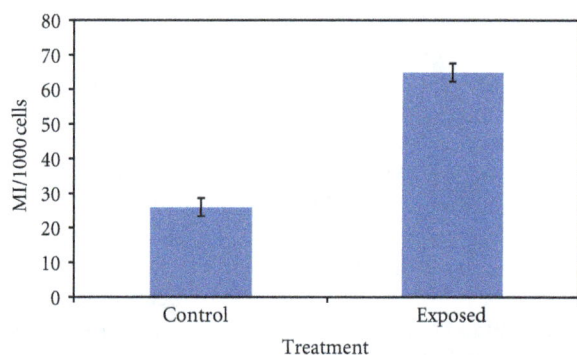

FIGURE 6: Mitotic index in bone marrow cells of control newborn rats and those exposed to 50 Hz and 0.5 mT MF for 30 days 24 h/day. Mean ± S.E. ($P < 0.001$).

## 4. Discussion

Health concerns about exposure to ELF-MF are particularly significant in children due to its potential harmful outcomes. It is generally accepted that ELF-MF is unable to transfer energy to cells in sufficient amounts that may cause direct cytotoxic and genotoxic effects. However, it is possible that certain cellular processes such as free radicals production and/or activity are altered by exposure to ELF-MF [27]. Thus, antioxidant activity, DNA damage, micronucleus induction, and cellular damage are probably induced by indirect mechanisms. In agreement with previous literature [28, 29], our results showed a significant increase in the main antioxidant enzymatic activities in the brain cells of exposed newborn rats which was considered as an evidence for occurrence of a physiological disturbance. Such increase of MDA and SOD levels means the presence of lipid peroxidation and oxidative stress. Accordingly, ELF-MF may induce oxygen free radicals which can evoke an

interaction with membrane lipids, proteins, and nucleic acids leading to extensive tissue damage that was confirmed by histopathological examination (Table 1 and Figure 2).

The results emphasized that exposing newborn rats to ELF-MF significantly increases the number of cells with damaged DNA types (2, 3, and 4) compared to the control one (Figures 3, 4 and Table 2). This damage might be due to the excess of oxygen derived from free radicals. These findings are in consistent with Lai and Singh [3, 4] who hypothesized that magnetic field (60 Hz, 0.1, 0.25, and 0.5 mT, 24 hr exposure) could damage DNA by radical pairs mechanism that cause an increase in the number of single strand breaks in brain cells of the rat. Supportive to this hypothesis is their observation that DNA breaks were blocked by free radicals scavengers (melatonin and N-t-butyl-$\alpha$ phenylnitrone).

Free radical activity and DNA breaks may be considered as the main causes of the significant increase (Table 3 and Figure 5) in the micronucleus induction of newborn rats bone marrow, and similar results were previously reported [30, 31]. Also Lai and Singh [4, 32] reported that reactive oxygen species (ROS) interact with the spindle apparatus and cause micronucleus induction.

In the current study, the significant increase in mitotic index (Table 4 and Figure 6) measurements may be attributed to the enhancement effect of ELF-MF on Ca$^{++}$ ions flux throughout the cell membrane [33].

## 5. Conclusion

The current study showed that continuous exposure of newborn rats to 50 Hz and 0.5 mT MF for 30 days 24 h/day led to cytogenetic and genotoxic hazards. Our results strongly demonstrate that chronic exposure of newborn rats can induce serious DNA damage that may lead to genomic instability, leading to carcinogenesis. Future studies about the health risks of ELF-MFs should be extended to evaluate the other physiological disturbances that might occur due to daily environmental exposure.

## Conflict of Interests

The authors report no conflict of interests. The authors alone are responsible for the content and writing of the paper.

## References

[1] N. Wertheimer and E. Leeper, "Original contributions. Electrical wiring configurations and childhood cancer," *American Journal of Epidemiology*, vol. 109, no. 3, pp. 273–284, 1979.

[2] H. Berg, "Problems of weak electromagnetic field effects in cell biology," *Bioelectrochemistry and Bioenergetics*, vol. 48, no. 2, pp. 355–360, 1999.

[3] H. Lai and N. P. Singh, "Acute exposure to a 60 Hz magnetic field increases DNA strand breaks in rat brain cells," *Bioelectromagnetics*, vol. 18, no. 2, pp. 156–165, 1997.

[4] H. Lai and N. P. Singh, "Melatonin and N-tert-butyl-$\alpha$-phenylnitrone block 60 Hz magnetic field-induced DNA single

Assessment of Genotoxic and Cytotoxic Hazards in Brain and Bone Marrow Cells of Newborn Rats Exposed to
Extremely Low-Frequency Magnetic Field

113

and double strand breaks in rat brain cells," *Journal of Pineal Research*, vol. 22, no. 3, pp. 152–162, 1997.

[5] B. M. Svedenstål, K. J. Johanson, and K. Hansson Mild, "DNA damage induced in brain cells of CBA mice exposed to magnetic fields," *In Vivo*, vol. 13, no. 6, pp. 551–552, 1999.

[6] B. M. Svedenstål, K. J. Johanson, M. O. Mattsson, and L. E. Paulsson, "DNA damage, cell kinetics and ODC activities studied in CBA mice exposed to electromagnetic fields generated by transmission lines," *In Vivo*, vol. 13, no. 6, pp. 507–513, 1999.

[7] H. Yaguchi, M. Yoshida, G. R. Ding, K. Shingu, and J. Miyakoshi, "Increased chromatid-type chromosomal aberrations in mouse m5S cells exposed to power-line frequency magnetic fields," *International Journal of Radiation Biology*, vol. 76, no. 12, pp. 1677–1684, 2000.

[8] R. Pasquini, M. Villarini, G. Scassellati Sforzolini, C. Fatigoni, and M. Moretti, "Micronucleus induction in cells co-exposed in vitro to 50 Hz magnetic field and benzene, 1,4-benzenediol (hydroquinone) or 1,2,4-benzenetriol," *Toxicology in Vitro*, vol. 17, no. 5-6, pp. 581–586, 2003.

[9] D. D. Ager and J. A. Radul, "Effect of 60 Hz magnetic fields on ultraviolet light-induced mutation and mitotic recombination in Saccharomyces cerevisiae," *Mutation Research*, vol. 283, no. 4, pp. 279–286, 1992.

[10] A. Antonopoulos, B. Yang, A. Stamm, W. D. Heller, and G. Obe, "Cytological effects of 50 Hz electromagnetic fields on human lymphocytes in vitro," *Mutation Research*, vol. 346, no. 3, pp. 151–157, 1995.

[11] S. Amara, H. Abdelmelek, C. Garrel et al., "Influence of static magnetic field on cadmium toxicity: Study of oxidative stress and DNA damage in rat tissues," *Journal of Trace Elements in Medicine and Biology*, vol. 20, no. 4, pp. 263–269, 2006.

[12] J. Everse and N. Hsia, "The toxicities of native and modified hemoglobins," *Free Radical Biology and Medicine*, vol. 22, no. 6, pp. 1075–1099, 1997.

[13] B. A. Freeman and J. D. Crapo, "Biology of disease. Free radicals and tissue injury," *Laboratory Investigation*, vol. 47, no. 5, pp. 412–426, 1982.

[14] N. Hogg and B. Kalyanaraman, "Nitric oxide and lipid peroxidation," *Biochimica et Biophysica Acta*, vol. 1411, no. 2-3, pp. 378–384, 1999.

[15] S. Di Loreto, S. Falone, V. Caracciolo et al., "Fifty hertz extremely low-frequency magnetic field exposure elicits redox and trophic response in rat-cortical neurons," *Journal of Cellular Physiology*, vol. 219, no. 2, pp. 334–343, 2009.

[16] National Research Council, *Guide for the Care and Use of Laboratory Animals*, National Academy Press, Washington, DC, USA, 1996.

[17] M. Nishikimi, N. Appaji Rao, and K. Yagi, "The occurrence of superoxide anion in the reaction of reduced phenazine methosulfate and molecular oxygen," *Biochemical and Biophysical Research Communications*, vol. 46, no. 2, pp. 849–854, 1972.

[18] E. Beautler, O. Duron, and B. M. Kelly, "Improved method for the determination of blood glutathione.," *The Journal of Laboratory and Clinical Medicine*, vol. 61, pp. 882–888, 1963.

[19] M. Uchiyama and M. Mihara, "Determination of malonaldehyde precursor in tissues by thiobarbituric acid test," *Analytical Biochemistry*, vol. 86, no. 1, pp. 271–278, 1978.

[20] P. Moller, L. E. Knudsen, S. Loft, and H. Wallin, "The comet assay as a rapid test in biomonitoring occupational exposure to DNA-damaging agents and effect of confounding factors," *Cancer Epidemiology Biomarkers and Prevention*, vol. 9, no. 10, pp. 1005–1015, 2000.

[21] W. M. Awara, S. H. El-Nabi, and M. El-Gohary, "Assessment of vinyl chloride-induced DNA damage in lymphocytes of plastic industry workers using a single-cell gel electrophoresis technique," *Toxicology*, vol. 128, no. 1, pp. 9–16, 1998.

[22] N. P. Singh, M. T. McCoy, R. R. Tice, and E. L. Schneider, "A simple technique for quantitation of low levels of DNA damage in individual cells," *Experimental Cell Research*, vol. 175, no. 1, pp. 184–191, 1988.

[23] R. Tice, E. Agurell, D. Anderson et al., "Single cell gel/comet assay: guidelines for in vitro and in vivo genetic toxicology testing," *Environmental and Molecular Mutagenesis*, vol. 35, no. 3, pp. 206–221, 2000.

[24] W. Schmidt, "The micronucleus test for cytogenetic analysis," in *Chemical Mutagens, Principles and Methods for Their Detection*, A. Hollaender, Ed., pp. 31–53, Plenum Press, New York, NY, USA, 1976.

[25] D. K. Agarwal and L. K. S. Chauhan, "An improved chemical substitute for fetal calf serum for the micronucleus test," *Biotechnic and Histochemistry*, vol. 68, no. 4, pp. 187–188, 1993.

[26] I. Adler, "Cytogenetic tests in mammals," in *Mutagenicity Testing, A Practical Approach*, S. Venitt and J. M. Parry, Eds., pp. 275–306, IRL Press, Oxford, UK, 1984.

[27] B. Brocklehurst and K. A. McLauchlan, "Free radical mechanism for the effects of environmental electromagnetic fields on biological systems," *International Journal of Radiation Biology*, vol. 69, no. 1, pp. 3–24, 1996.

[28] B. Zheng, G. Yao, L. Xie, Y. Lin, D. Q. H. Lu, and Chiang, "Effects of 50 Hz magnetic fields on lipid peroxydation and antioxidase activities in brain tissue of mice," in *Proceedings of the 2nd World Congress for Electricity and Magnetism in Biology and Medicine*, 1997.

[29] T. B. Kryston, A. B. Georgiev, P. Pissis, and A. G. Georgakilas, "Role of oxidative stress and DNA damage in human carcinogenesis," *Mutation Research*, vol. 711, no. 1-2, pp. 193–201, 2011.

[30] I. Udroiu, M. Cristaldi, L. A. Ieradi, A. Bedini, L. Giuliani, and C. Tanzarella, "Clastogenicity and aneuploidy in newborn and adult mice exposed to 50 Hz magnetic fields," *International Journal of Radiation Biology*, vol. 82, no. 8, pp. 561–567, 2006.

[31] L. Dominici, M. Villarini, C. Fatigoni, S. Monarca, and M. Moretti, "Genotoxic hazard evaluation in welders occupationally exposed to extremely low-frequency magnetic fields (ELF-MF)," *International Journal of Hygiene and Environmental Health*, vol. 215, no. 1, pp. 68–75, 2011.

[32] H. Lai and N. P. Singh, "Magnetic field-induced DNA strand breaks in brain cells of the rat," *Environmental Health Perspectives*, vol. 112, no. 6, pp. 687–694, 2004.

[33] A. Lacy-Hulbert, J. C. Metcalfe, and R. Hesketh, "Biological responses to electromagnetic fields," *FASEB Journal*, vol. 12, no. 6, pp. 395–420, 1998.

# Relationship between Anthropometric Factors, Gender, and Balance under Unstable Conditions in Young Adults

Júlia Maria D'Andréa Greve,[1] Mutlu Cuğ,[2] Deniz Dülgeroğlu,[3]
Guilherme Carlos Brech,[1] and Angelica Castilho Alonso[1]

[1] *Laboratory of Movement Studies (LEM), Institute of Orthopedics and Traumatology (IOT), Hospital das Clínicas (HC),*
 *School of Medicine, University of São Paulo, 05403-010 São Paulo, SP, Brazil*
[2] *Physical Education and Sports Department, Cumhuriyet University, 58140 Sivas, Turkey*
[3] *Diskapi Yildirim Beyazit Education and Research Hospital, Physical Medicine and Rehabilitation Clinic, Ankara, Turkey*

Correspondence should be addressed to Angelica Castilho Alonso; angelicacastilho@msn.com

Academic Editor: Giuseppe Spinella

The objective of this study was to evaluate the relationship between the anthropometric factors of height, body mass, body mass index and postural balance and to compare the balance indices between genders in the upright standing position, in healthy adult subjects under conditions of instability. Forty individuals were subjected to functional tests of body stability using the Biodex Balance System, and the resulting indices were correlated with body mass, height, and body mass index, and also compared between genders. Body mass was the main anthropometric factor that influenced variations in postural balance, with a high correlation between groups and with all variables. A linear regression analysis showed that body mass associated with BMI explained 66% of the overall stability, and body mass explained 59% of the anteroposterior stability index and 65% of the mediolateral stability index. In the female group, body mass explained 72% of the overall balance, 66% of the anteroposterior, and 76% of the medio-lateral stability index. Increased body mass requires greater movements to maintain postural balance. Height and BMI presented moderate correlations with balance. Women showed less movement than men on the Biodex Balance System.

## 1. Introduction

Balance is defined as the ability to maintain the body's center of mass within the base of support. The balance is maintained through the movement of body weight in different directions with safety, speed (response time), and coordination. Balance is dynamic and requires constant adjustments to adapt to external perturbations, through the use of vision, muscle activity, articular positioning and proprioception, and the vestibular system, all acting in concert [1, 2].

Balance evaluation tests that simulate functional activities are the most appropriate type of test to determine the contributions of the musculoskeletal, vestibular, and visual systems. Systems of maintaining postural balance can be affected by lesions, musculoskeletal, or neurological limitations, anthropometric factors, aging, use of medications, physical conditioning, and specific training (e.g., high impact sports), as well as extrinsic factors such as the type of shoes and the type of ground [3].

Many balance assessment methods exist, ranging from simple observation, clinical tests, scales, and posturographic measurements, to integrated assessment systems of greater complexity. All of them have advantages and limitations and can demonstrate different results with multiple interpretations, and this is exacerbated by the lack of consensus regarding which individual characteristics (especially anthropometric factors) should to be controlled for, so that quantitative evaluations can be considered reliable. This lack of consensus impedes the use of such tests in clinical practice as a safe tool for assessing the risk of falls and the results from therapeutic interventions [3–6].

Cote et al. [7] reported that balance measurements should be controlled in order to avoid errors in analyzing the results and highlighted anthropometric factors as important in this type of evaluation. Studies seeking normative data for controlling the variables that might influence balance assessments are needed in the scientific literature, since there is still no consensus regarding the influence of anthropometric measurements on postural balance.

The objective of this study was to evaluate the relationship between the anthropometric factors of height, body mass, body mass index (BMI), and postural balance and to compare the balance indices between genders in the upright standing position, in healthy adult subjects under conditions of instability.

## 2. Method

This study was conducted in Diskapi Yildirim Beyazit Education and Research Hospital, Physical Medicine and Rehabilitation Clinic, Ankara, Turkey in partnership with the Laboratory of Movement Studies, University of São Paulo Medical School, Brazil. The study was approved by the METU School of Natural and Applied Science Ethics Committee (no. 00.00/126/78-1597).

This was a descriptive observational, cross-sectional study, without intervention. Forty volunteers (twenty-five males and fifteen females) were evaluated. The mean age was $21.7 \pm 1.4$ years (20–27), with a mean body mass of $65.7 \pm 12.4$ kg (43–94), mean height of 170 cm (157–185), and mean BMI of $22.5 \pm 3.5$ kg/m$^2$ (16.1–31.4). For the males, the mean age was $21.5 \pm 1.4$ years (20–27), with a mean body mass of $71.2 \pm 9.1$ kg (57–90), mean height of $173.8 \pm 0.1$ cm (161–185) and mean BMI of $23.5 \pm 2.8$ kg/m$^2$ (18.8–29.3). For the females, the mean age was $21.8 \pm 1.1$ years (20–24), with a mean body mass of $56.5 \pm 12.1$ kg (43–94), mean height of $164 \pm 0.5$ cm (157–173), and mean BMI of $20.8 \pm 3.9$ kg/m$^2$ (16.1–31.4).

The inclusion criteria were as follows: (a) signing of a free and informed consent statement; (b) age between 20 and 30 years; (c) no physical activity for a minimum of six months (according to the physical activity readiness questionnaire); (d) absence of neurological, cardiovascular, metabolic, rheumatic or vestibular diseases; (e) no injuries or previous surgery on the legs; and (f) absence of knee or ankle clinical instability. Individuals who required more than three attempts to accomplish the test were excluded from the study, but none of the individuals met this exclusion criterion.

All the volunteers answered a questionnaire requesting personal information, in order to identify the inclusion criteria. Thirty-seven volunteers (92.5%) had a dominant right leg (kicking side), and three (7.5%) had a dominant left leg.

The same evaluator performed all the anthropometric measurements (body mass, height, and BMI) and the balance test. The balance test was performed using the Biodex Balance System (BBS) (Neurocom International, Inc. Clackamas, OR, USA) with the level 2 stability protocol, which allows an inclination of up to 20° in the horizontal plane in all directions. Stability varies according to the resistance level equipment (from level 8—more stable, to level 1—less stable) [1, 7]. Three stability indices were calculated as follows: antero-posterior stability index—represented the variance of foot platform displacement in degrees, from level, for motion in the sagittal plane. Mediolateral stability index- represented the variance of foot platform displacement in degrees, from level, for motion in the frontal plane and overall stability index (sum of the first two)—represented the variance of foot platform displacement in degrees in all motions during a test. It was the angular excursion of the patient's center of gravity. A high number was indicative of a lot of movement during a test with static measures; it was the angular excursion of the patient's center of gravity. The arithmetic means of the results were calculated from the three tests. The subjects were tested with their eyes open at all times.

*2.1. Positioning.* The subjects were asked to step onto the BSS platform and take a comfortable bipedal standing position, while maintaining slight flexion of the knees (15°), looking straight ahead, and folding the arms across the chest. The subjects were tested barefoot at all times.

The platform was released, and the patients were instructed to balance themselves, keeping the indicator at the center of the target on the screen. When the patient was capable of keeping the indicator in the center of the target on the screen (balanced position) without hand support, the foot position was recorded using the platform rail.

*2.2. Test.* Once the subjects had been positioned, they were instructed not to move their feet until the end of each measurement. Changes were recorded in relation to the center of the platform. Two 20-second measurements were made separated by one-minute intervals. The result was the arithmetic mean of the two measurements, which was supplied automatically by the equipment.

*2.3. Statistical Analysis.* The Kolmogorov-Smirnov test was used to analyze the gender distribution and the continuous variable in the parametric and nonparametric tests. As the data presented normal distribution, the paired Student's *t*-test was used. The SPSS 20.0 software for Windows was used, adopting a level of significance of $P \leq 0.05$.

Pearson's coefficient was used to evaluate the correlation between anteroposterior, medio-lateral and general stability (sum of the first two) and the anthropometric measurements (body mass, height, and BMI) of each subject. The correlation was high when $R \geq 0.7$, moderate when $0.5 \leq R < 0.7$, low when $0.1 \leq R \leq 0.5$, and null when $R = 0$.

To analyze the linear regression model, all the variables that presented $P \leq 0.20$ in the correlation coefficient analysis were selected. These were then placed in order from lowest to highest $P$ value. The multiple modeling processes followed the stepwise forward selection method, in which the variables were added one by one, according to their ranking. The variables that resulted in $P \leq 0.05$ remained in the model.

## 3. Results

The females presented better stability indices (general, antero-posterior and medio-lateral) than the males (Table 1).

The correlation analysis between the anthropometric variables and postural balance in the whole group and divided according to gender is described in Table 2.

Regression analysis on the anthropometric variables in relation to postural balance in the whole group and divided according to gender is described in Table 3.

## 4. Discussion

Identifying factors that influence balance can help improve the accuracy of diagnosis and quality of treatment and rehabilitation (indication of specific exercises) and is fundamental for preventing falls and incapacities [5, 8]. The BBS is a reliable and reproducible method for evaluating antero-posterior and medio-lateral movements from the body's center of mass that can be made while maintaining postural stability [9].

Anthropometric variables influence the stability limits of the organism and can affect the motor strategies relating to balance control [5]. Some anthropometric variables, such as body mass, are directly related to postural balance; however, the majority of works evaluate specific groups, such as obese adolescents [4], the elderly [1, 10, 11], and athletes [12]. Yet few studies have reported on individuals with normal body mass or overweight who were subjected to conditions of instability.

The male group demonstrated stronger correlations for overall, antero-posterior, and medio-lateral stability index with BMI. Women had moderate correlations for all variables with BMI. In the whole group evaluation, the correlations between BMI and the stability indexes were also moderate in regression analysis associated with body mass, which explained 66% of overall balance. Greve et al. [6] showed that in young adult males, the higher the BMI, worst postural balance, needing more postural adjustments to maintain balance in single leg stance. Thus, their data were similar to the current study. The women of this study had lower BMI and body mass than the men, but had similar height, which may have contributed to reducing the strength of correlation and regression with BMI. Other studies conducted on individuals with normal or slightly higher than normal BMI have shown low correlations between body mass and balance [13, 14]. The differences between these studies are likely related to the evaluation techniques. It seems that in situations of instability, body mass presented greater impairment of balance.

The male group had stronger correlations for overall, antero-posterior and medio-lateral stability with body mass. Women had moderate correlations for all variables with body mass. In the whole group evaluation, the correlations between body mass and the stability indexes were also stronger. The difference between genders can slightly alter the behavior of the inverted pendulum model. The inverted pendulum model should be carefully applied to study postural control. In order to avoid any misunderstanding during the analysis

TABLE 1: Comparison between general, anteroposterior, and medi-olateral stability indexes (mean and SD) distributed according gender.

| Stability index | Male (N = 25) Mean (SD) | Female (N = 15) Mean (SD) | P |
|---|---|---|---|
| Overall balance | 6.6 (±2.8) | 3.7 (±2.7) | 0.003* |
| Anteroposterior | 4.9 (±2.0) | 2.9 (±2.0) | 0.004* |
| Mediolateral | 4.5 (±2.0) | 2.5 (±1.9) | 0.004* |

*$P \leq 0.05$—Student's $t$-test.

TABLE 2: Correlation ($r$ value) between the general, anteroposterior, and mediolateral stability indexes and height (cm), body mass (kg), and BMI (kg/m$^2$).

| Stability index | Height R ($P$ value) | Body mass R ($P$ value) | BMI R ($P$ value) |
|---|---|---|---|
| General (N = 40) | | | |
| Overall balance | 0.624 (0.000)* | 0.808 (0.000)* | 0.647 (0.000)* |
| Anteroposterior | 0.598 (0.000)* | 0.779 (0.000)* | 0.627 (0.000)* |
| Mediolateral | 0.631 (0.000)* | 0.813 (0.000)* | 0.650 (0.000)* |
| Male (N = 25) | | | |
| Overall balance | 0.423 (0.117) | 0.864 (0.000)* | 0.804 (0.000)* |
| Anteroposterior | 0.408 (0.132) | 0.829 (0.000)* | 0.767 (0.000)* |
| Mediolateral | 0.449 (0.094) | 0.885 (0.000)* | 0.822 (0.000)* |
| Female (N = 15) | | | |
| Overall balance | 0.530 (0.000)* | 0.680 (0.000)* | 0.411 (0.041)* |
| Anteroposterior | 0.488 (0.013)* | 0.636 (0.001)* | 0.391 (0.053)* |
| Mediolateral | 0.534 (0.006)* | 0.688 (0.000)* | 0.415 (0.039)* |

*$P \leq 0.05$—Pearson's coefficient.
General group: all the anthropometric variables presented moderate to strong positive correlations with the postural balance variables.
Male group: the variables of body mass and BMI presented strong positive correlations with the postural balance variables.
Female group: all the anthropometric variables presented weak to moderate positive correlations with the postural balance variables.

of the inverted pendulum model applied in the quiet standing posture, it is necessary to consider the mass distribution in the calculus for the position of the center of mass [5, 13, 15].

The majority of studies indicate that there was a direct relationship between obesity and increased postural instability, as evaluated by means of various tools and methods [4, 8, 14–17]. In the present study, body mass presented a high correlation with the stability indices; that is, there was a need for greater movements to maintain postural balance. This finding was similar to that of Ledin and Odkvist [8] who demonstrated that a 20% increase in body mass reduced the ability to make adjustments in response to external perturbations in the orthostatic position, with a consequent increase in postural instability. Hue et al. [18] found that body mass was responsible for more than 50% of balance at speed and Chiari et al. [13] demonstrated a strong correlation between body mass, antero-posterior movements, and the area of detachment. In both of these studies, a force platform was used in the evaluations. Other authors have reported that greater postural adjustments are necessary to maintain an

TABLE 3: Linear regression analysis between postural balance and anthropometric variables.

| Stability index | Height $\beta$ (P value) | Body mass $\beta$ (P value) | BMI $\beta$ (P value) | Adjusted $r^2$ |
|---|---|---|---|---|
| General (N = 40) | | | | |
| Overall balance | | 0.292 (<0.001) | −0.365 (0.058) | 0.668 |
| Anteroposterior | — | 0.140 (<0.001) | — | 0.596 |
| Mediolateral | — | 0.143 (<0.001) | — | 0.652 |
| Male (N = 25) | | | | |
| Overall balance | 15.003 (0.046) | 0.175 (0.001) | — | 0.512 |
| Anteroposterior | — | 0.143 (0.001) | — | 0.378 |
| Mediolateral | 10.900 (0.042) | 0.128 (0.001) | — | 0.526 |
| Female (N = 15) | | | | |
| Overall balance | — | 0.192 (<0.001) | — | 0.727 |
| Anteroposterior | — | 0.136 (<0.001) | — | 0.663 |
| Mediolateral | — | 0.136 (<0.001) | — | 0.767 |

$r^2$ = linear regression coefficient.
General group: body mass associated with BMI explained 66% of the stability index for overall balance, and body mass explained 59% of the anteroposterior stability index and 65% of the mediolateral stability index.
Male group: body mass associated with height explained 51% of the overall balance, body mass explained 37% of the anteroposterior stability index, and body mass associated with height explained 52% of the mediolateral stability index.
Female group: body mass explained 72% of the overall balance, 66% of the anteroposterior stability index, and 76% of the mediolateral stability index.

erect posture when there is a buildup of adipose tissue, thus causing a reduction in balance and an increase in injuries and falls [4, 8, 17].

Due to the high degree of correlation between balance and body mass, we can safely infer that the mechanical factor of body mass inertia requires greater musculoskeletal force to balance it against the force of gravity, and therefore, to maintain balance. Obese individuals require greater movement from the center of gravity to remain in the orthostatic position. This study showed that only a high body mass can contribute towards decreasing the balance and occurrences of falls in situations of instability.

The male and female groups, and the group overall, had moderate correlations for all variables with height. In the regression analysis for the male group, height explained over half of the overall and medio-lateral stability index, associated with body mass. Ku et al. [19] found that BMI has an impact on postural control in both the bipedal and unipodal stance. These findings corroborate the data in the literature. There is a consensus that the greater the height is, the worse the balance. Berger et al. [20] and Alonso et al. [21] stated that ankle displacements and the response of the gastrocnemius increased with increasing height. Allard et al. [22] and Lee and Lin [23] reported that ectomorph (lanky) individuals present greater postural sway than do endomorph or mesomorph individuals, and they attributed this to the higher position of the center of mass. Other studies have found that body stability is inversely related to the height of the center of gravity and that, for this reason, posturography measurements are affected by individuals' anthropometric characteristics [5, 18].

Chiari et al. [13] and Kejonen et al. [5] suggested that height and body mass could be affected by the total load of movements that occur at the top of the "inverted pendulum," in which when standing upright on two feet, the pivot is the ankle joint and the support base is the interface between the body and the ground (geometry). These authors recommended that for this reason, these variables should always be evaluated as a set.

Since the correlations found were similar, it was not possible to determine which movements were the greatest instability factors: medio-lateral (movements of the pelvis and lower limbs) or antero-posterior (movements of the trunk).

Comparing the genders, we saw that women showed less movement on the BBS than men did, and these findings were similar to those of Rozzi et al. [24] who evaluated basketball and American football players using the same equipment. In another study involving children, it was observed that girls presented better postural balance than boys [23].

This could be due to anthropometric factors (greater in men), but other factors such as neuromuscular (flexibility) and neurophysiologic (processing of inferences), as well as the habit of using higher heels, may also account for the differences.

The BBS is a simple system that enables rapid evaluation and, when used with certain criteria, it may be of assistance for quantifying alterations in balance. It can also be used for dysfunction training.

These results safely suggest that incorporating the evaluation of body composition in patients with equal BMI can help understand and use this correlation in order to prevent falls and other incapacities of obese patients.

## 5. Conclusion

Increased body mass required greater movement to maintain postural balance. Height and BMI presented moderate correlations with balance. Women showed less movement than men on the BBS.

## Acknowledgments

We wish to confirm that there are no known conflicts of interest associated with this publication and there has been no significant financial support for this work that could have influenced its outcome. The study was conducted in Diskapi Yildirim Beyazit Education and Research Hospital, Physical Medicine and Rehabilitation Clinic, Ankara, Turkey in partnership with the Laboratory of Movement Studies, University of São Paulo Medical School, Brazil. The study was approved by the METU School of Natural and Applied Science Ethics Committee (no. 00.00/126/78-1597).

## References

[1] J. M. Prado, T. A. Stoffregen, and M. Duarte, "Postural sway during dual tasks in young and elderly adults," *Gerontology*, vol. 53, no. 5, pp. 274–281, 2007.

[2] A. C. Alonso, J. M. D. A. Greve, and G. L. Camanho, "Evaluating the center of gravity of dislocations in soccer players with and without reconstruction of the anterior cruciate ligament using a balance platform," *Clinics*, vol. 64, no. 3, pp. 163–170, 2009.

[3] A. C. Alonso, G. C. Brech, A. M. Bourquin, and J. M. D. A. Greve, "The influence of lower-limb dominance on postural balance," *Sao Paulo Medical Journal*, vol. 129, no. 6, pp. 410–413, 2011.

[4] B. McGraw, B. A. McClenaghan, H. G. Williams, J. Dickerson, and D. S. Ward, "Gait and postural stability in obese and nonobese prepubertal boys," *Archives of Physical Medicine and Rehabilitation*, vol. 81, no. 4, pp. 484–489, 2000.

[5] P. kejonen, K. Kauranen, and H. Vanharanta, "The relationship between anthropometric factors and body-balancing movements in postural balance," *Archives of Physical Medicine and Rehabilitation*, vol. 84, no. 1, pp. 17–22, 2003.

[6] J. M. D. A. Greve, A. C. Alonso, A. C. P. G. Bordini, and G. L. Camanho, "Correlation between body mass index and postural balance," *Clinics*, vol. 62, no. 6, pp. 717–720, 2007.

[7] K. P. Cote, M. E. Brunet, B. M. Gansneder, and S. J. Shultz, "Effects of pronated and supinated foot postures on static and dynamic postural stability," *Journal of Athletic Training*, vol. 40, no. 1, pp. 41–46, 2005.

[8] T. Ledin and L. M. Odkvist, "Effects of increased inertial load in dynamic and randomized perturbed posturography," *Acta Oto-Laryngologica*, vol. 113, no. 3, pp. 249–252, 1993.

[9] W. J. C. Cachupe, B. Shifflett, L. Kahanov, and E. H. Wughalter, "Reliability of biodex balance system measures," *Measurement in Physical Education and Exercise Science*, vol. 5, no. 2, pp. 97–108, 2001.

[10] M. R. M. Mainenti, E. C. Rodrigues, J. F. Oliveira, A. S. Ferreira, C. M. Dias, and A. L. S. Silva, "Adiposity and postural balance control: correlations between bioelectrical impedance and stabilometric signals in elderly Brazilian women," *Clinics*, vol. 66, no. 9, pp. 1513–1518, 2011.

[11] J. Swanenburg, E. D. de Bruin, K. Favero, D. Uebelhart, and T. Mulder, "The reliability of postural balance measures in single and dual tasking in elderly fallers and non-fallers," *BMC Musculoskeletal Disorders*, vol. 9, article 162, 2008.

[12] A. C. Alonso, E. Bronzatto-Filho, G. C. Brech, and F. V. Moscoli, "Estudo comparativo do equilíbrio postural entre atletas de judô e indivíduos sedentários," *Revista Brasileira de Biomecânica*, vol. 9, no. 17, pp. 130–137, 2008.

[13] L. Chiari, L. Rocchi, and A. Cappello, "Stabilometric parameters are affected by anthropometry and foot placement," *Clinical Biomechanics*, vol. 17, no. 9-10, pp. 666–677, 2002.

[14] R. Molikova, M. Bezdickova, K. Langova et al., "The relationship between morphological indicators of human body and posture," *Biomedical Papers of the Medical Faculty of the University Palacký, Olomouc, Czechoslovakia*, vol. 150, no. 2, pp. 261–265, 2006.

[15] I. D. Loram, H. Gollee, M. Lakie, and P. J. Gawthrop, "Human control of an inverted pendulum: is continuous control necessary? Is intermittent control effective? Is intermittent control physiological?" *Journal of Physiology*, vol. 589, no. 2, pp. 307–324, 2011.

[16] A. Goulding, I. E. Jones, R. W. Taylor, J. M. Piggot, and D. Taylor, "Dynamic and static tests of balance and postural sway in boys: effects of previous wrist bone fractures and high adiposity," *Gait and Posture*, vol. 17, no. 2, pp. 136–141, 2003.

[17] F. Berrigan, M. Simoneau, A. Tremblay, O. Hue, and N. Teasdale, "Influence of obesity on accurate and rapid arm movement performed from a standing posture," *International Journal of Obesity*, vol. 30, no. 12, pp. 1750–1757, 2006.

[18] O. Hue, M. Simoneau, J. Marcotte et al., "Body weight is a strong predictor of postural stability," *Gait and Posture*, vol. 26, no. 1, pp. 32–38, 2007.

[19] P. X. Ku, N. A. Abu Osman, A. Yusof, and W. A. W. Abas, "Biomechanical evaluation of the relationship between postural control and body mass index," *Journal of Biomechanics*, vol. 45, no. 9, pp. 1638–1642, 2012.

[20] W. Berger, M. Trippel, M. Discher, and V. Dietz, "Influence of subjects' height on the stabilization of posture," *Acta Oto-Laryngologica*, vol. 112, no. 1, pp. 22–30, 1992.

[21] A. C. Alonso, N. M. S. Luna, L. Mochizuki, F. Barbieri, S. Santos, and J. M. D. A. Greve, "The influence of anthropometric factors on postural balance: the relationship between body composition and posturographic measurements in young adults," *Clinics*, vol. 67, no. 12, pp. 1433–1441, 2012.

[22] P. Allard, M. L. Nault, S. Hinse, R. LeBlanc, and H. Labelle, "Relationship between morphologic somatotypes and standing posture equilibrium," *Annals of Human Biology*, vol. 28, no. 6, pp. 624–633, 2001.

[23] A. J. Y. Lee and W. H. Lin, "The influence of gender and somatotype on single-leg upright standing postural stability in children," *Journal of Applied Biomechanics*, vol. 23, no. 3, pp. 173–179, 2007.

[24] S. L. Rozzi, S. M. Lephart, W. S. Gear, and F. H. Fu, "Knee joint laxity and neuromuscular characteristics of male and female soccer and basketball players," *American Journal of Sports Medicine*, vol. 27, no. 3, pp. 312–319, 1999.

# Interplay of Biomechanical, Energetic, Coordinative, and Muscular Factors in a 200 m Front Crawl Swim

**Pedro Figueiredo,**[1,2] **David R. Pendergast,**[3]
**João Paulo Vilas-Boas,**[1,4] **and Ricardo J. Fernandes**[1,4]

[1] *Centre of Research, Education, Innovation and Intervention in Sport, Faculty of Sport, University of Porto, Rua Dr. Plácido Costa 91, 4200-450 Porto, Portugal*
[2] *Higher Education Institute of Maia (ISMAI), Avenida Carlos Oliveira Campos, 4475-690 Maia, Portugal*
[3] *Center for Research and Education in Special Environments, Department of Physiology and Biophysics, University at Buffalo, 3435 Main Street, Buffalo, NY 14214, USA*
[4] *Porto Biomechanics Laboratory, University of Porto, Rua Dr. Plácido Costa 91, 4200-450 Porto, Portugal*

Correspondence should be addressed to Pedro Figueiredo; spafg@vodafone.pt

Academic Editor: Francisco Miró

This study aimed to determine the relative contribution of selected biomechanical, energetic, coordinative, and muscular factors for the 200 m front crawl and each of its four laps. Ten swimmers performed a 200 m front crawl swim, as well as 50, 100, and 150 m at the 200 m pace. Biomechanical, energetic, coordinative, and muscular factors were assessed during the 200 m swim. Multiple linear regression analysis was used to identify the weight of the factors to the performance. For each lap, the contributions to the 200 m performance were 17.6, 21.1, 18.4, and 7.6% for stroke length, 16.1, 18.7, 32.1, and 3.2% for stroke rate, 11.2, 13.2, 6.8, and 5.7% for intracycle velocity variation in $x$, 9.7, 7.5, 1.3, and 5.4% for intracycle velocity variation in $y$, 17.8, 10.5, 2.0, and 6.4% for propelling efficiency, 4.5, 5.8, 10.9, and 23.7% for total energy expenditure, 10.1, 5.1, 8.3, and 23.7% for interarm coordination, 9.0, 6.2, 8.5, and 5.5% for muscular activity amplitude, and 3.9, 11.9, 11.8, and 18.7% for muscular frequency). The relative contribution of the factors was closely related to the task constraints, especially fatigue, as the major changes occurred from the first to the last lap.

## 1. Introduction

The goal of competitive swimming is to perform the race distance as fast as possible, for that swimmers must achieve their highest average velocity for that distance. Swimming velocity ($\bar{v}$) is the product of the stroke rate (SR) and the distance moved through the water with each complete stroke cycle (SL) [1] and can be expressed as

$$\bar{v} = \text{SR} \times \text{SL}. \tag{1}$$

For the same $\bar{v}$, several combinations of SR and SL are possible and are a result of modifications of the time spent in different phases of the stroke cycle (interarm coordination), which can be measure in front crawl with the index of coordination (IdC; [2–4]). However, swimmers do not move at a constant velocity within each stroke cycle, and variations in the action of the arms, legs, and trunk result in intermittent application of force and lead to variations in the swimming velocity around the mean velocity within each stroke cycle. These intermittent movements and resultant variations in velocity increase the work done by the swimmer [5], compared to swimming at a constant velocity. The average velocity attained by the swimmer results from the average of the instantaneous velocity, resulting from intracycle velocity variation (IVV):

$$\bar{v} = v_{\text{constant}} + \Delta v(t). \tag{2}$$

In addition to these factors, maximal swimming velocity ($\bar{v}_{\max}$) depends on the maximal metabolic power of the swimmers ($\dot{E}_{\text{tot-max}}$) and on their energy cost of locomotion ($C$):

$$\overline{v}_{\max} = \frac{\dot{E}_{\text{tot-max}}}{C}, \quad (3)$$

where $\dot{E}_{\text{tot-max}}$ can be computed based on measures/estimates of the aerobic, anaerobic lactic, and anaerobic alactic energy contributions and $C$ (i.e., the amount of metabolic energy spent to cover one unit of distance, KJ·m$^{-1}$). The $C$ depends on biomechanical factors such as the mechanical efficiency ($\eta_m$), the propelling efficiency ($\eta_p$), and the mechanical work to overcome hydrodynamic resistance ($W_d$):

$$C = \frac{W_d}{\left(\eta_p \times \eta_m\right)}. \quad (4)$$

To assess $W_d$ several methods have been proposed; however there is no agreement on the most valid method [6–8], and thus it remains difficult to determine active drag during a competitive event while preserving the ecology of the movement. On the other hand, propelling efficiency includes work done against drag and is defined as the ratio of useful mechanical work ($W_d$) to total mechanical work ($W_{\text{tot}}$):

$$\eta_p = \frac{W_d}{W_{\text{tot}}}, \quad (5)$$

where in aquatic environments $W_d$ is lower than $W_{\text{tot}}$, since a fraction of the work produced by the contracting muscles is used to accelerate a variable amount of water backwards (wasted work) [9] and for the internal work [10]. The $\eta_p$ includes $W_{\text{tot}}$ and is dependent on the swimmers' technique and is velocity-dependent and affected by fatigue. In addition, mechanical efficiency is related to how muscles produce the mechanical work needed to sustain a given speed [10, 11]. Muscle efficiency arises from the range of either their force/length and/or force/speed relationships. Relations between force and iEMG have been used to estimate different efficiencies. Also, it has been suggested that the reduction in electrical efficiency with fatigue indicated that more motor units were recruited to generate the same amount of force compared with the nonfatigued muscle [12, 13]. However, the diagnostic value of the time domain analysis (iEMG) in muscle fatigue evaluation is considered to be more limited than that of the frequency domain analysis (Freq; [14]). So, to minimize the metabolic cost of high performance activities, the limbs must generate large power outputs while the muscles perform work at high efficiencies.

As described above, theoretical models have been developed that attempt to explain the influence of various factors on performance. In spite of the fact that velocity is common to the theoretical approaches, they cannot be combined due to incompatibility of terms and units. This has led to attempts at practical approaches, relating swimming performance to different anthropometrical, physiological, and biomechanical parameters [15–18]. This kind of research can be developed by comparing different competitive level swimmers, employing the neural network, computing cluster analysis, or developing

statistical models from the swimmer's profile [19]. However, these studies have not theorized/assessed swimming performance completely using a biophysical approach, particularly at high swimming speeds [19–21]. The 200 m swim and freestyle swimming are the dominant competitive events and thus of great interest. Therefore, the purpose of this study was to determine the relative contribution of selected biomechanical (SL, SR, horizontal IVV, vertical IVV, $\eta_p$), energetic ($\dot{E}_{\text{tot}}$), coordinative (IdC), and muscular factors (iEMG and Freq) for the 200 m front crawl performance and each of its four laps. The approach used, in the absent of an appropriate theoretical approach, was a multivariate analysis of the important factors among those listed above that would account for the average swimming velocity in a 200 m front crawl swim and its component lengths, in well-trained swimmers. It was hypothesized that the biomechanical and energetic factors would be most important, with the coordinative and muscular factors also playing an important, but lesser, role.

## 2. Methods

*2.1. Subjects.* Ten well-trained swimmers ($21.6 \pm 2.4$ yr) who were specialists in the 200 m front crawl event participated in this study. Height, arm span, body mass, and percentage of adipose tissue were $185.2 \pm 6.8$ cm, $188.7 \pm 8.4$ cm, $76.4 \pm 6.1$ kg, and $10.1 \pm 1.8\%$, respectively. The subjects had an average of $11.9 \pm 3.5$ yrs of competitive experience. Their performances in the 200 m front crawl were $109.3 \pm 2.1$ s, which correspond to a mean velocity that represents $91.6 \pm 2.1\%$ of the mean velocity of the short course pool world record for men. The protocol was approved by the local ethics committee and followed the rules of the Declaration of Helsinki (2000). Swimmers were informed of the procedure, the potential risks involved, and the benefits of the study and then gave a written consent to participate. During the testing period, subjects were asked to adapt the intensity and the total volume of training to avoid stressful training programs. Swimmers' practiced with and were accustomed to all procedures, particularly swimming with the snorkel used for measurement of $\dot{V}O_2$.

*2.2. Experimental Procedures.* All tests were conducted in a 25 m indoor pool and each subject swam alone in the middle lane, avoiding pacing or drafting effects. Following a warm-up that consisted of a self-selected swim of about 1000 m, including some swimming with the snorkel, swimmers performed a 200 m maximum effort front crawl swim after a push start and using open turns without a glide. They were instructed to replicate their pacing and strategy used in competition. After 90 min of active rest, swimmers performed a 50 m front crawl test and twenty-four hours later a 150 m and a 100 m tests, with 90 min active rest interval between them. Together 50, 100, and 150 m tests were at the same swimming speed as in the previous 200 m paced by a visual light pacing system placed in the bottom of the pool. The pacing lights led the swimmers as the lights progressed down the pool with a flash every 5 m (TAR 1.1, GBK-Electronics, Aveiro, Portugal).

### 2.3. Data Collection and Analysis

*2.3.1. Biomechanical Factors.* Each swimmer's performance was recorded with a total of six stationary and synchronized video cameras (Sony, DCR-HC42E, Tokyo, Japan), four below and two above the water. The calibration set-up, accuracy, and reliability procedures have been previously described in detail [22]. The twenty-one landmarks videoed (Zatsiorsky's model adapted by [23]) that define the three-dimensional position and orientation of the head, torso, upper arms, forearms, hands, thighs, shanks, and feet were manually digitized at 50 Hz using a commercial software package (Ariel Performance Analysis System, Ariel Dynamics, Inc., USA). The Direct Linear Transformation Algorithm [24] was used for three-dimensional reconstruction and a digital low-pass filter at 6 Hz was used to smooth the data.

*2.3.2. Stroking Parameters.* One complete stroke cycle (defined as the period between the instant of entry of one hand to the next instant of entry of the same hand) for each of the 50 m laps of the 200 m front crawl was analyzed. From these data, the center of mass position as a function of time was computed. The mean velocity ($\bar{v}$) was calculated by dividing the horizontal displacement of the center of mass in one stroke cycle over its total duration. Additionally, the horizontal distance travelled by the center of mass during the stroke cycle was used to determine the stroke length (SL). The stroke rate (SR) was determined as the inverse of the time (seconds) to complete one stroke cycle, which was then multiplied by 60 to yield units of strokes per minute.

*2.3.3. Intracycle Velocity Variation.* To determine and analyze the whole body centre of mass' IVV in the $x$, $y$, and $z$ axes of motion, the coefficient of variation (CV = SD/mean) was computed as previously suggested [19, 22, 25].

*2.3.4. Propelling Efficiency.* Propelling efficiency ($\eta_p$) was calculated from the computed 3D hand velocity as the sum of the instantaneous 3D velocity of the right and left hand combined during the underwater phase of the stroke (3Du). The $\eta_p$ was calculated from the ratio of the speed of the center of mass to the 3D mean hand velocity ($\eta_p = \bar{v}/3Du$), since this ratio represents the theoretical efficiency in all fluid machines and has been used in swimming [18, 26].

### 2.4. Energetic Factors

*2.4.1. Total Energy Expenditure and Energy Cost of Swimming (C).* Oxygen uptake ($\dot{V}O_2$) was recorded by means of the K4b$^2$ telemetric gas exchange system (COSMED, Rome, Italy), during the 200 m front crawl test. This equipment was connected to the swimmer by a low hydrodynamic resistance respiratory snorkel and valve system. This system was previously validated and widely used [15, 25, 26]. Expired gas concentrations were measured breath-by-breath

and averaged every 5 s, to get the $\dot{V}O_2$ used in subsequent calculations. Net $\dot{V}O_2$ was calculated by subtracting the resting $\dot{V}O_2$ from the steady state $\dot{V}O_2$ measured during swimming. Before, and after, the 50, 100, 150, and 200 m tests, capillary blood samples (5 $\mu$L) were collected from the ear lobe to assess rest and postexercise blood lactate (La$_b$) using a portable lactate analyzer (Lactate Pro, ARKRAY, Inc.). Lactate was measured at 1, 3, 5, and 7 min after test, and the peak value was used for further analysis.

Since the 200 m front crawl energy contribution is supplied from the three energy sources [26–28], $\dot{E}_{tot}$ was calculated for each 50 m lap (for review see [28]):

$$\dot{E}_{tot} = \dot{V}O_2 + \beta\dot{L}a_b + PCr\left(1 - e^{-t/\tau}\right), \tag{6}$$

where $\dot{E}_{tot}$ is the total energy expenditure, $\dot{V}O_2$ is the aerobic contribution (calculated from the time integral of the net $\dot{V}O_2$ versus time), $\beta\dot{L}a_b$ is the net accumulation of lactate after exercise, $\beta$ is the energy equivalent for lactate accumulation in blood (2.7 mL O$_2\cdot$mM$^{-1}\cdot$kg$^{-1}$), PCr is the alactic contribution, $t$ is the time duration, and $\tau$ is the time constant of PCr splitting at work onset (23.4 s). The contribution of each energy pathway was calculated for each lap, and on the basis of these data, $\dot{E}_{tot}$ was computed and $C$ was calculated as the ratio between $\dot{E}$ and $\bar{v}$.

### 2.5. Coordinative Factors

*2.5.1. Index of Coordination.* The calculation of the index of coordination (IdC) requires the identification of key points in the stroke cycle [2, 4], specifically, (A) entry and catch of the hand in the water, (B) pull in the water, (C) push in the water, and (D) recovery out of the water. Each phase, within the stroke cycle, was determined from the swimmer's horizontal ($x$) and vertical ($y$) displacement of the hand noting the time corresponding to start and end of these phases for two arm stroke cycles previously digitized.

The IdC was calculated as the time gap between the propulsion (pull and push phases) of the two arms and expressed as a percentage of the duration of the complete arm-stroke cycle (sum of the propulsive and nonpropulsive phases (catch and exit phases)) [3, 29, 30]. IdC was the mean of IdC left and IdC right.

*2.6. Muscular Factors.* The EMG signals of eight muscles (flexor carpi radialis, biceps brachii, triceps brachii, pectoralis major, upper trapezius, rectus femoris, biceps femoris, and tibialis anterior), which have been shown to have high activity during front crawl swimming [31, 32], were recorded simultaneously from the right side of the body using bipolar (interelectrode distance of 2.0 cm) Ag–AgCl circular surface electrodes. The skin of the swimmer was shaved and cleaned with alcohol and the electrodes with preamplifiers placed in line with the muscle's fibre orientation on the surface in the midpoint of the contracted muscle belly according to international standards [33] and covered with an adhesive bandage (OPSITE FLEXIFIX) [34, 35]. A reference electrode was attached to the body's patella. All cables were fixed to the skin by adhesive tape to minimize artifacts during swimming.

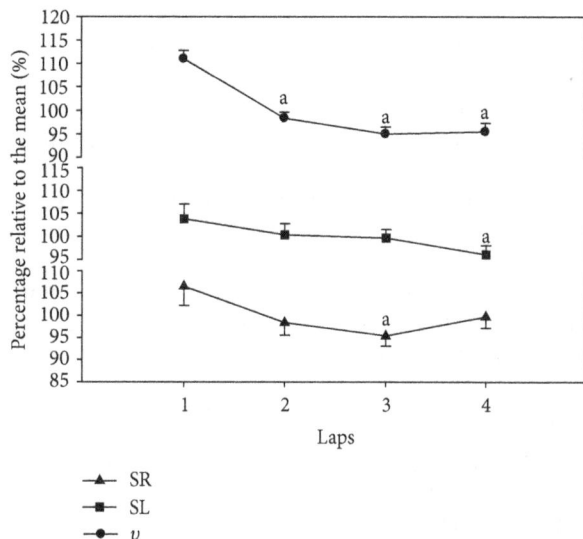

FIGURE 1: Mean (SE) values expressed as a percentage of the mean value for the 200 m front crawl for velocity ($v$), stroke length (SL), and stroke rate (SR) are plotted as a function of the 50 m laps. [a]Significantly different from the 1st lap.

Additionally, swimmers wore a total body coverage swimsuit (Fastskin, Speedo) to cover the electrodes and recording wires. The total gain of the amplifier was set at 1100 times with a common mode rejection ratio of 110. The data were sampled at 1000 Hz with a 16-bit analog to digital conversion and recording system (BIOPAC System, Inc.) and stored on a computer for later analysis. An electronic flashlight signal synchronized with an electronic trigger marked simultaneously the video and EMG recordings, respectively, to synchronize EMG and video recordings. The EMG data analysis was performed using the MATLAB 2008a software environment (MathWorks Inc., Natick, Massachusetts, USA).

### 2.6.1. iEMG. 
Raw EMG signals were band-passed (8–500 Hz), rectified to obtain the full wave signals, and smoothed with a 4th order Butterworth filter (10 Hz) for the linear envelope. The integration of the rectified EMG was calculated, per unit of time, to eliminate the stoke cycle duration effect (iEMG/T) and normalized to the maximum iEMG observed (signal was partitioned in 40 ms windows to identify the maximal iEMG) [36]. All iEMG values from the measured muscles taken in the mid-pool section for each 50 m were averaged. In addition, the average iEMG values of all 8 muscles were added together (iEMG) and used to represent the total electrical activity of swimming.

### 2.6.2. Frequency Analysis. 
For the frequency analysis (Freq), spectral indices were calculated [37] and averaged. Spectral indices were obtained for each stroke, defined by video analysis, in the mid-section of the pool for each 50 m lap and they were averaged for each muscle. The spectral indices for each muscle were then averaged to determine the Freq factor used to represent spectral muscle information. Spectral indices have been shown to most accurately detect changes

in muscle power during dynamic contractions [38], and their increases indicate fatigue [37, 38].

### 2.6.3. Statistical Analysis. 
Mean (SD) computations for descriptive analysis were obtained for all variables (normal Gaussian distribution of the data was verified by the Shapiro-Wilk's test). A one-way repeated measures ANOVA was used to compare each factor along the 200 m. When a significant $F$-value was achieved, Bonferroni post-hoc procedures were performed to locate the pairwise differences between the means. All the statistical analysis was performed using STATA 10.1 (StataCorp, USA), with the level of significance set at 0.05. The effect size ($f$) for each variable was calculated in accordance with Cohen [39] to measure the magnitude of difference.

### 2.7. Modeling of Performance. 
As described in the Introduction, absence of a theoretical model to combine the factors that contribute to swimming performance, a multiple linear regression was used to identify the relative contributions of factors that are associated with swimming performance. These, among the previous defined, factors are biomechanical (SL, SR, IVV$x$, IVV$y$, $\eta_p$), energetic ($\dot{E}_{tot}$), coordinative (IdC), and muscular (iEMG and Freq). This analysis was carried out for the 200 m front crawl velocity and then repeated for the velocities of each of the component 50 m laps to examine and compare the relative contribution of the factors in each segment of the swim. A common general multiple linear regression analysis was used to identify the weight of the factors identified as contributing to 200 m swim velocity and attaining 100% of the variance of the performance. The equation used for all the models tested was

$$\bar{v} = \text{constant} + k\text{SL} + k\text{SR} + k\text{IVV}x + k\text{IVV}y$$
$$+ k\eta_p + k\dot{E}_{tot} + k\text{IdC} + k\text{iEMG} + k\text{Freq}, \quad (7)$$

where $\bar{v}$ is the mean swimming velocity for the 200 m or the mean velocity of each 50 m lap that equals the sum of the model' constant with the factors, stroke length, stroke rate, intracycle velocity variation ($x$ and $y$), propulsive efficiency, total energy expenditure, index of coordination, muscular activation, and spectral indices weighted by their specific beta coefficients ($k$). Both $C$ and IVV$z$ were not used in the model to limit the number of factors and they were reflected in $\dot{E}_{tot}$ and $\eta_p$ or IVV$x$, IVV$y$, respectively. To better express the relative importance of the factor, the weights of the regression were converted to standardized regression coefficients (beta weights).

## 3. Results

Mean velocity for the total 200 m front crawl was 1.41 ($\pm$0.04) m$\cdot$s$^{-1}$. Figure 1 shows the data for the average velocity of each 50 m lap, along with the observed stroke rate and stroke length, expressed as a percentage of their mean for the 200 m swim. The velocity in the first lap was faster than the average velocity but decreased below the average in the second lap, after which it remained constant ($F_{3,27} =$

TABLE 1: The beta coefficients ($k$) determined to identify the importance of the factors included in the multiple linear regression models computed for the mean for the overall 200 m performance, as well as for each individual performance lap.

|  |  | SL | SR | IVV$x$ | IVV$y$ | $\eta_p$ | $\dot{E}_{\text{tot}}$ | IdC | iEMG | Freq | Constant |
|---|---|---|---|---|---|---|---|---|---|---|---|
| 200 m performance | Lap 1 | −1.10 | −0.04 | 4.04 | −1.40 | 11.55 | 0.01 | 0.05 | −0.32 | 0.01 | 0.89 |
|  | Lap 2 | 5.90 | 0.26 | −14.15 | −3.52 | −28.08 | 0.04 | −0.08 | −0.64 | −0.04 | −2.12 |
|  | Lap 3 | 0.20 | 0.02 | 0.25 | −0.02 | −0.16 | 0.002 | −0.006 | 0.03 | 0.001 | 0.05 |
|  | Lap 4 | −0.13 | −0.002 | 0.21 | 0.12 | −0.97 | 0.005 | −0.02 | −0.04 | 0.002 | 1.04 |
| Each 50 m performance | Lap 1 | −6.52 | −0.25 | 19.87 | −8.02 | 62.06 | 0.03 | 0.26 | −1.44 | 0.02 | 4.29 |
|  | Lap 2 | 1.32 | 0.07 | −1.65 | −0.52 | −2.72 | 0.004 | −0.01 | −0.08 | −0.01 | −1.75 |
|  | Lap 3 | 0.63 | 0.05 | 0.35 | −0.18 | −0.49 | −0.002 | −0.01 | 0.05 | −0.001 | −1.59 |
|  | Lap 4 | 0.50 | 0.03 | −0.15 | −0.03 | −0.12 | 0.001 | −0.004 | −0.01 | −0.0001 | −0.92 |

24.72, $P < 0.001$, $f = 1.27$). Swimming velocity is the product of SR and SL, and they both decreased concomitantly with velocity (Figure 1). SR had a mean value for the 200 m of 38.41 (±3.05) cycles·min$^{-1}$ but decreased across the swim, reaching a statistical difference after the third lap ($F_{3,27} = 5.08$, $P = 0.006$, $f = 0.38$). SL decreased below the mean for the 200 m of 2.20 (±0.14) m in lap 3 but reached significance only in the last lap ($F_{3,27} = 4.55$, $P = 0.01$, $f = 0.33$).

Figure 2 shows the four groups of factors identified as contributing to the 200 m front crawl swim (i.e., biomechanical, energetic, coordinative, and muscular). Biomechanical factor IVV ($x$, $y$, and $z$) (Figure 2(a)) mean values for the 200 m were 0.22 (0.03), 0.76 (0.08), and 0.83 (0.03), respectively. A stable pattern over the 50 m laps was observed (IVV$x$: $F_{3,27} = 1.60$, $P = 0.21$, $f = 0.18$; IVV$y$: $F_{3,27} = 0.82$, $P = 0.49$, $f = 0.00$; IVV$z$: $F_{3,27} = 2.18$, $P = 0.13$, $f = 0.24$). Another biomechanical factor, $\eta_p$, presented a mean value over the four laps of 0.42 (0.02) (Figure 2(a)), however, showed a significant reduction in the 4th 50 m lap ($F_{3,27} = 6.64$, $P = 0.002$, $f = 0.41$). Energetic factors, $\dot{E}_{\text{tot}}$ ($F_{3,27} = 19.58$, $P < 0.001$, $f = 0.63$) and C ($F_{3,27} = 19.77$, $P < 0.001$, $f = 0.63$) (Figure 2(b)), showed significant changes for the 50 m laps, with a mean of 80.11 (7.97) mml O$_2$·kg$^{-1}$·min$^{-1}$ and 1.60 (0.16) KJ·m$^{-1}$, respectively. The coordinative factor, IdC, presented a mean value of −14.94 (2.15)% (Figure 2(c)) and showed a significant increase in the 4th 50 m lap ($F_{3,27} = 4.09$, $P = 0.02$, $f = 0.34$). The two muscular factors, Freq ($F_{3,27} = 30.40$, $P < 0.001$, $f = 0.89$) and iEMG (Figure 2(d)), showed a significant increase ($F_{3,27} = 4.22$, $P = 0.01$, $f = 0.22$), in the last 50 m lap and the mean values were 1.97e$^{-14}$ (0.22e$^{-14}$) and 1.76 (±0.37), respectively.

The beta coefficients for all factors are presented in Table 1, for their contribution in the four laps to the 200 m velocity (upper half) and to the average velocity in each 50 m lap (lower half). Standardized coefficients from the multiple linear regression model showed that the contributions of the first and last 50 m laps velocity to the mean 200 m velocity were higher (26.1 and 30.8%, resp.) than the contributions of the second and third laps (21.7 and 21.4%, resp.) of the 200 m front crawl. The model had an $F_{4,5} = 339.159$, $P < 0.001$, $R^2 = 0.996$, and adjusted $R^2 = 0.993$ for these factors. These data are consistent with the changes in velocity shown in Figure 1.

The biomechanical factors showed a great importance, manly the SL and SR (Figure 3) to the overall performance of the 200 m front crawl (16.2% and 17.5%, resp.). However, their contribution decreased in the final lap (from 17.6% and 16.1% to 7.6% and 3.2%, resp.). The SR had a very high contribution in the third 50 m lap (32.1%); concomitant with this, there was a great decrease in the contribution of the other biomechanical factors (6.7% for IVV$x$, 1.3% for IVV$y$, and 2.0% for $\eta_p$), with the IVV$y$ and $R^2$ factors increasing afterwards (5.4% and 6.4%, resp.). The $\dot{E}_{\text{tot}}$ contribution increases continually during the four laps (4.5%, 5.8%, 10.9% and 23.7%), while the IdC factor shows a "U" pattern with a large contribution at the beginning (10.1%), a decrease in the middle (5.1%), and then increase at the end of the swim (23.7% for the fourth lap). Relative to the muscular parameters (iEMG and Freq), iEMG appears to be quite stable (ranging from 5.5 to 9.0%), with only small oscillations, while the contribution of Freq increased over the length of the swim (from 3.9% in the first lap to 18.7% in the last lap).

In Figure 4 the contributions of the relative importance of the factors used in the analysis for the average velocity in each lap individually are showed. The biomechanical factors (SL, SR, IVV$x$, IVV$y$, and $\eta_p$) had a higher contribution (81.1%) than the energetic ($\dot{E}_{\text{tot}}$, 3.9%), coordinative (IdC, 5.5%), and muscular (iEMG and Freq, 9.5%) factors. Within all the analyzed factors SL and SR showed the highest contribution (26.4% and 34.6%, respectively) the remaining ones (IVV$x$, IVV$y$, $\eta_p$, $\dot{E}_{\text{tot}}$, IdC, iEMG and Freq) had a similar contributions (ranging from 3.8 to 6.9%). It should be noted that SL and SR are related mathematically with the $v$. However, the contribution of each of these two factors for each 50 m lap performance showed that SL contribution decreased in the third lap (from 27.9% to 24.8%), in spite of its increase tendency over the four laps (from 20.0% in the first lap to 33.1% in the last lap), while the SR increased throughout the entire 200 m swim (from 17.6% to 49.4%). All the other factors used in the model showed a tendency to decrease their contribution from the beginning until the end of the swim, as the contributions of SL and SR increase.

## 4. Discussion

Although previous studies have evaluated the role of biomechanical [1, 40, 41], energetic [26, 27, 42], muscular

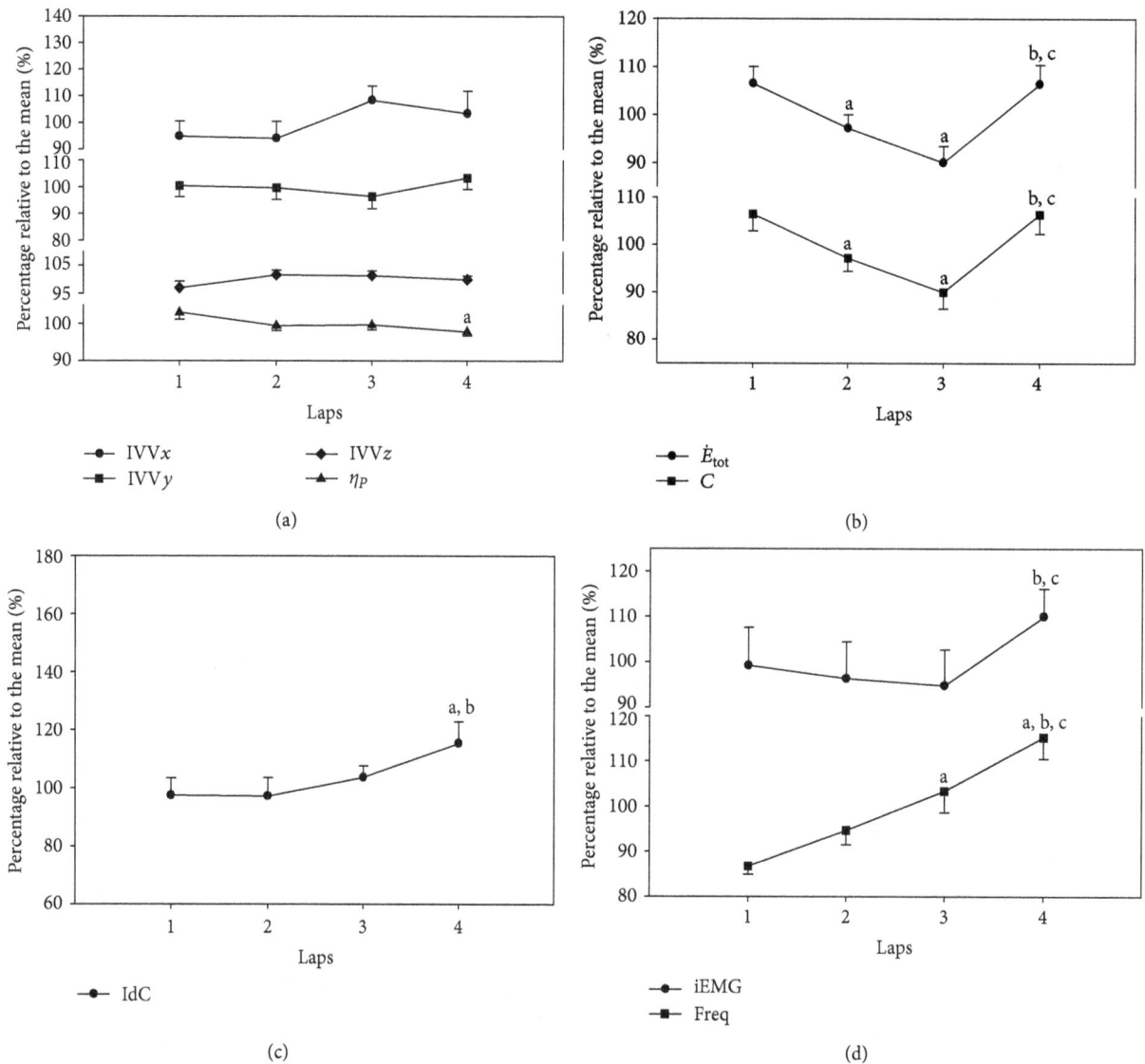

FIGURE 2: Mean (±SE) values for the percentage of the 200 m front crawl mean value for the (i) biomechanical factors: IVV for $x$, $y$, and $z$ axes and $\eta_p$ (a); (ii) energetic factors $\dot{E}_{tot}$ and $C$ (b); (ii) coordinative factor: IdC (c) and, (d) muscular factors: iEMG and Freq (d) for the 200 m front crawl event. [a,b,c] Significantly different from the 1st, 2nd, and 3rd laps, respectively.

[32, 35, 43], or coordinative [2, 4, 29] factors on the performance and others developed models to predict performance combining several factors [44], we are unaware of a study that examined their combined interactive effects as was performed in this study. The regression analysis performed was not intended to predict performance but to determine the contribution of the important factors to it. For the mean velocity of 1.41 m·s$^{-1}$, the biomechanical, energetic, coordinative, and muscular factors were 58.1%, 11.2%, 18.9%, and 11.8%, respectively, with SL and SR factors explaining 33.7% of the 200 m mean velocity. A decrease in velocity during the second 50 m lap was observed, and then velocity was constant. Although the patterns were different, SL and SR decreased from the first 50 m and together accounted for the decrease in velocity. These changes in SL and SR are in agreement with previous studies [1, 29], showing the increase

on the last lap of the SR to compensate for the SL decrease, in an attempt to maintain the velocity as high as possible. Also, the velocities that account for the major contribution to the overall performance of the 200 m front crawl were the first and last lap velocities, suggesting two important stages during this particular event. On the first lap, the highest velocities are achieved and on the last lap the consequences of fatigue were felt, and although velocity was constant, the contribution of the factors determining it changed. Among the 50 m laps, the contribution of biomechanical, energetic, coordinative, and muscular factors was on average 81.1%, 3.9%, 9.5%, and 5.5%, respectively, and 61% of the biomechanical contribution was attributed to the SL and SR.

4.1. Biomechanical Factors. Stability in the IVV ($x$, $y$, and $z$) was observed over the four laps, as previously reported by

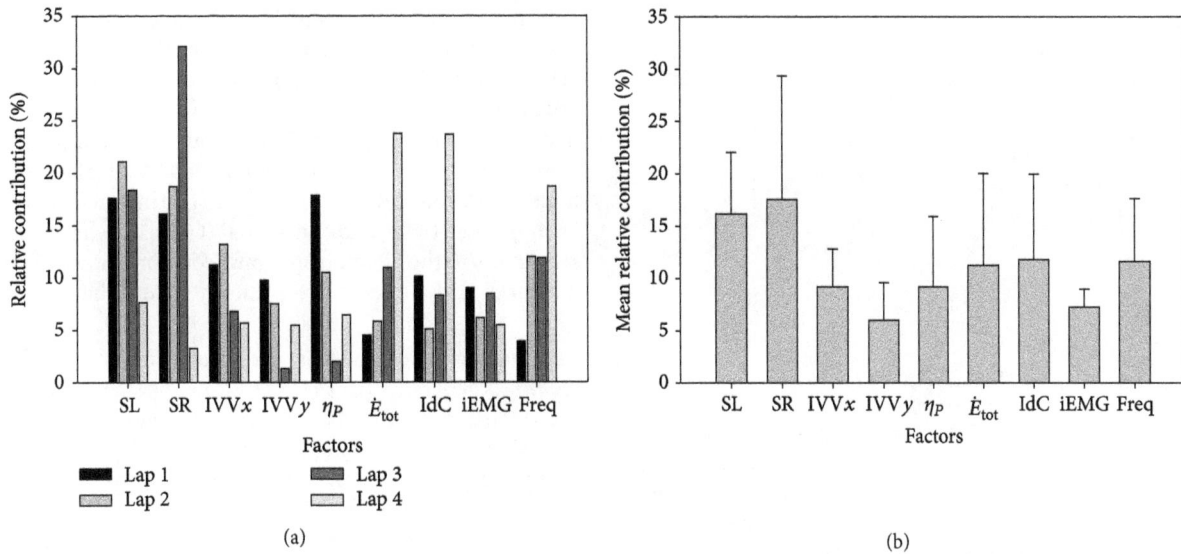

FIGURE 3: The percentage of the contributions of each factor in each lap for the 200 m swim performance (a) and mean percentages for all laps (b).

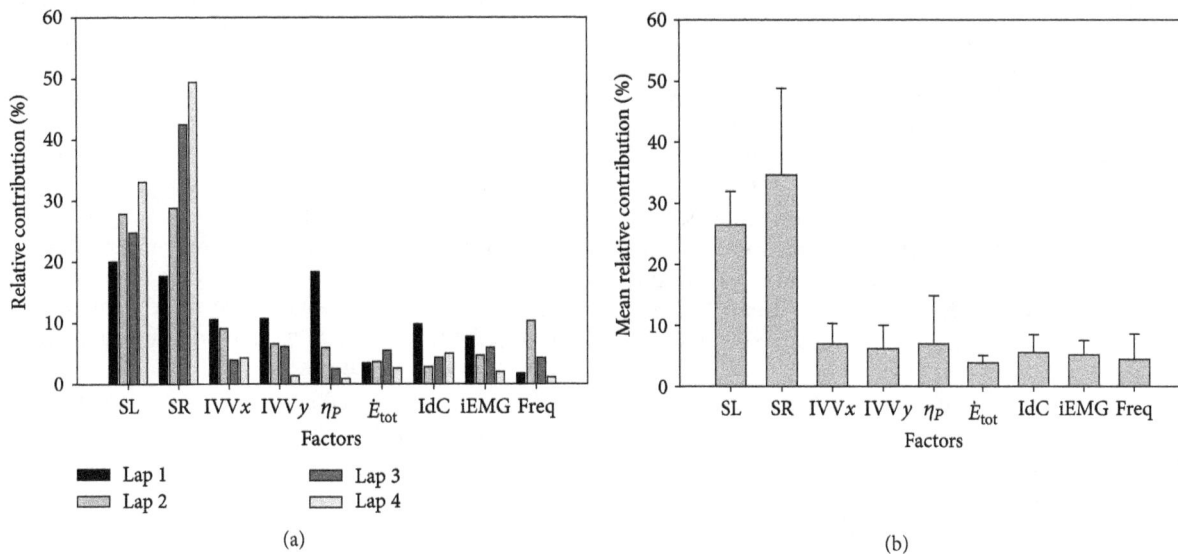

FIGURE 4: The relative contributions of the factors for the 50 m laps performances (a) and mean percentages for all laps (b) of the 200 m front crawl.

Psycharakis et al. [40] and Alberty et al. [29]. IVV stability seems to be related with a coordinative adaptation of the upper limbs, as IdC changes along the effort, as well as the SR, mainly in the last 50 m lap [45, 46]. IVV $(x, y)$ accounted for 15.2% of the variability of the 200 m swim and 13.2% for the 50 m laps. In spite of the stability of IVVx, the $\eta_p$ decreased in the last lap likely due to fatigue, as fatigue has been shown to evolve during the 200 m front crawl [26, 29]. Also, it indicates a reduction in stroke technique quality; at the end of effort [47], higher lactate accumulation occurs [48], as well as neuromuscular fatigue [49]. As a result, $\eta_p$ accounted for 9.2% of the variability of the 200 m swim and on average 6.9% for the 50 m laps individual performance.

4.2. Energetic Factors. $\dot{E}_{tot}$ and $C$ decreased over the second and third 50 m laps, concomitant with the velocity decrease. However, taking into account the determinants of $C$ (the hydrodynamic resistance and the propelling efficiency) and since $\eta_p$ decreased due to the development of fatigue, $C$, and thus $\dot{E}_{tot}$, increased in the last lap, which is in agreement with previous studies [26, 50]. $\dot{E}_{tot}$ accounted for 11.2% of the variability of the 200 m swim and on average 3.9% for the 50 m laps.

4.3. Muscular Factors. The assessed muscular factors revealed in spite of swimming at maximum effort that the

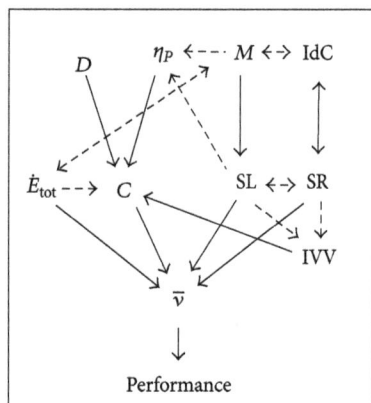

FIGURE 5: The relationship between biomechanical, energetic, coordinative, and muscular factors to performance in competitive swimming. $C$: energy cost, $\dot{E}_{tot}$: energy expenditure, IVV: intracycle velocity variation of the center of mass, IdC: index of coordination, SR: stroke rate, SL: stroke length, $\bar{v}$: mean swimming velocity, $D$: hydrodynamic drag, $\eta_p$: propelling efficiency, and $M$: muscular activation and frequency.

observed muscles were involved at a submaximum level, as amplitude increased and frequency decreased (i.e., increase in the spectral indices), as previously reported for amplitude [43] and frequency [51] for a $4 \times 50$ m test simulating the 200 m front crawl, and also shown for the 100 m swim [32]. Similar results were observed in other sports activities [52, 53]. Most of these studies interpreted the increase in the EMG activity amplitude as increased motor units recruitment and increased motor units synchronization, as well as the decrease in muscle fiber conduction velocity, due to an accumulation of metabolic products. The iEMG and Freq factors contributed by 7.3% and 11.6%, respectively, to the variance of the 200 m swim and on average by 5.1% and 4.4%, respectively, to the 50 m laps.

### 4.4. Coordinative Factors.
As velocity and the SL-SR ratio changed, interarm coordination adapted, with an increase in IdC in the final stages of the 200 m event. This observation is consistent with the development of fatigue as reported previously [29, 30]. Interlimb coordination is adapted, as an optimization mechanism to obtain as much speed as possible in face of constraints imposed [54], showing that an effective front crawl swimming technique must be sufficiently flexible and adaptable [55]. This factor (IdC) accounted for 18.9% of the variance of the 200 m swim performance and on average for 5.5% of the 50 m laps.

### 4.5. Interplay among Factors.
A theoretical framework for the interaction of the biomechanical, energetic, coordinative, and muscular factors is presented in Figure 5 and used in the subsequent discussion.

The biomechanical factors had the highest contribution to the 200 m front crawl and also to each 50 m lap mean velocities, where together they accounted for up to 33.7% and 61.0%, respectively. These contributions are understandable, as the product of two of these factors (SL and SR)

determines swimming velocity [1]. The contributions of SL and SR to the total performance are very important to achieve high velocities however their contributions decreased during the swim, which suggested that several other factors had increased importance in determining the last 50 m lap velocity (see Figure 5). The contribution of SL showed a higher contribution than SR in the last 50 m. This observation is supported by a study of Craig et al. [1], where the best swimmers in the 200 m front crawl could maintain higher SL distances at the end of the event, in spite of having similar SRs.

Changes in SL and SR are associated with IVV$x$ and IVV$y$ (see Figure 5); however, the latter showed a stable pattern over the 200 m. In spite of its stability, IVV$x$ showed a decreased contribution over the length of the swim. IVV$y$'s contribution decreased even more than IVV$x$ in the third lap, and then it increased in the fourth lap. Relative to the individual lap performances, $\eta_p$ had similar mean contributions to velocity as IVV$x$ and IVV$y$, and all of them decreased from the beginning to the end of the swim. IVV$x$'s and SL contributions to the 200 m performance showed a similar pattern, which could be explained by the increased time between propulsive phases as SL decreases and SR increases [2–4]. This change is also associated with a decrease in $\eta_p$ (see Figure 5) [9, 18, 26] and increase in the IVV$x$ and $\dot{E}_{tot}$ [56]. This can be explained as a smaller IVV will lead to a lower energy cost, for example, if two swimmers swim at equal mean velocity but the IVV = 0 in swimmer 1 and in swimmer 2 IVV > 0, then mean power of swimmer 1 will be $v^3$ but in swimmer 2 it will be $> v^3$, as $\dot{E}_{tot}$ has the same relation with $v$ [17, 27]. Supporting this concept, it was found that swimming with hand paddles, which increases $\eta_p$ and SL [57], results in decrease of IVV$x$ and increase of IdC [58]. On the other hand, IVV$y$ and $\eta_p$ showed a contribution pattern that was inverse of that of SR.

IVV$y$ can be linked to the medial-lateral hand movements that account for vertical displacement changes suggesting great importance of the sideways movements during the stroke's propulsive actions, which have been highlighted by previous studies (for review see [59]) and are decreased with higher SRs. In addition, as $\eta_p$ is SL-related [18, 26], its contribution to the variance in performance decreases when the SR contribution increases. The similar pattern observed for IVV$y$ and $\eta_p$ seems to confirm the possible link between IVV$y$ and the sideways hand pattern motion, which resulted in a high $\eta_p$. This may also account for the larger contribution of $\eta_p$ than IVV$y$. Propelling efficiency's decreasing contribution to the laps performance might be linked to reduced muscles force production during the stroke due to fatigue. It is likely that a reduced muscle force production occurs, as indicated by the changes in EMG factors, and the swimmers became unable to sustain the initial SL [1, 60], as observed in this study. The spectral indices (Freq) have been suggested as one of the first indicators of fatigue [37, 38]. The SL and $\eta_p$ decreases shown in this study are likely the result of fatigue developing toward the end of the 200 m swim.

As the biomechanical factors show a decreased contribution to the variance of the 200 m in each 50 m swim performance between the first and the last laps, other factor's contributions must increase (see Figure 5). This was the case for the energetic and coordinative factors. Over the 50 m laps, the contribution of $\dot{E}_{tot}$ to the overall performance increased; thus the swimmer's capacity to deliver higher energy expenditure became more important over the 200 m. Swimmers can have the same time splits for the 50, 100, and 150 m, but if $\dot{E}_{tot}$ cannot be increased to match the increase in C in the last 50 m, velocity cannot be sustained. The contribution of $\dot{E}_{tot}$ in the three final laps is similar to that of IdC, which could be explained by the swimmer naturally adopting a movement pattern to minimize his metabolic energy expenditure [61, 62].

The reduction in C, and thus $\dot{E}_{tot}$, in laps 2 and 3 may involve the process of self-optimization [62] which occurs to overcome the constraints imposed, in this case by the fatigue task constraint [3]. The IdC factor had the inverse pattern of contribution to performance in the first three laps compared to that of SL. As indicated by previous studies, based on the dynamical theories of motor organization, stroke rate is the first determinant of motor organization in swimming [3] and it has an inverse relation with SL [4, 60]. As SR and IdC are associated with each other (see Figure 5), IdC had a higher contribution to first 50 m lap, as was the case in overall performance analysis. This is likely due to the direct relationship between IdC and velocity that has been suggested [2, 4]. After the first 50 m, the contribution of IdC starts to decrease, as a result of the decrease in velocity and SL, until the development of fatigue, which resulted in an increase in the contribution of the SR to a greater extent than SL. When strokes are closer to one another, or overlap, this has the effect of increasing the average propulsive force while the mean force per stroke is maintained [30]. These changes in stroke patterns increase the contribution of IdC in the latter stages of the 200 m swim, which also has previously been shown [29].

The increased $\dot{E}_{tot}$ contribution to overall performance reported in this study is related to the changes in the balance of the three energy pathways (aerobic, anaerobic lactic, and anaerobic alactic) as a function of time as previously reported [26–28]. The increase in contribution of the anaerobic lactate contribution and resultant lactate accumulation by the end of the effort [26] contribute to the explanation of the decrease in SL and $\eta_p$. These changes are consistent with the deterioration of stroke mechanics observed by other authors [50, 60]. The reduced SL and increased Freq are associated with muscle fatigue most likely brought about by high lactate levels and reduced muscle glycogen [63]. This conclusion is supported by the suggestion that the increase in blood lactate concentration may change the stroking strategy significantly [60] and thus IVV $y$ and $\eta_p$. These deviations from the optimal combination of SL and SR result in a significant increase in energetic demand (see Figure 5), suggesting that minimizing energy cost may be an important factor contributing to cadence determination in cyclical forms of locomotion [62]. Supporting this, swimmers preferred to swim front crawl at the lowest SR (or the longest SL) that does not require an increase in oxygen uptake [64], as a significant decrease in the preferred SR, for example, determines the decline in time

limit exercise duration [65], which might be caused by an unusual muscular recruitment.

The increase in $\dot{E}_{tot}$, particularly of anaerobic lactic contribution, in the final lap due to muscle fatigue is generally (although not exclusively) attributed to the reduced muscular fibre conduction velocity, which is causally related to a decrease in the pH [66]. Although pH was not directly measured, high values of blood lactate concentration collected after the 200 m swim [26, 27, 67] implied a significant pH decrease during swimming. As muscles fatigue, power output is reduced during the swim [41], as is the case for SL. Since the SL is an index of propelling efficiency [18, 59], $\eta_p$ should decrease, as was observed in this study. The resultant deterioration of stroke mechanics in fatigued subjects is expected to lead to a progressive increase in the energy cost of swimming (see Figure 5), as was observed in this study. However, to maintain the total mechanical power output as Craig et al. [1] have shown for races of 200 m and longer, the distance per stroke tends to decrease as fatigue develops and SR has to increase to compensate to maintain the speed constant, or if SR and $\dot{E}_{tot}$ cannot be increased velocity decreases, which happens in this study in the second lap. In addition, increases in muscle activity can lead to decreases in efficiency (see Figure 5) with no increase in power output if the muscle coordination is inappropriate [11]. Muscle coordination changes due to fatigue in swimming have been shown [35].

In the first lap, the contribution of the SL is higher than the SR, but in the last lap SR is greater suggesting fatigue in the last lap, which is supported by the EMG data (see Figure 5). The muscular factor iEMG (amplitude analysis) has a tendency to decrease its contribution to the overall performance of the 200 m during subsequent 50 m laps. In the first 50 m lap, the highest contribution of the iEMG over the 200 m could be associated with the high velocity and also a higher contribution of the SL, which is linked to higher force production (see Figure 5) [68]. This would also be associated with a higher power output and velocity, as was the case for the first lap and concomitant with the high contribution observed. On the third lap, the iEMG contribution increases after its decrease in the second lap, which might explain the decrease in the absolute value of SR in this particular lap. This is supported by the higher SR in this lap that was associated with higher EMG activity [69] and its increase in contribution. Also, additional recruitment or increased synchronization of muscle fibers as a result of submaximal fatigue [70] most likely explains the reduced contribution in the last 50 m. If the velocity is an indicator of the power output and it was stable in the last three laps, mechanical efficiency and concomitant efficiency of the electrical activity was decreased, as the iEMG increased, if the electrical efficiency the ratio force to iEMG was considered [12, 13].

In spite of these associations described above, the relationship between iEMG and force is not linear and the diagnostic value of the time domain analysis (iEMG) in muscle fatigue evaluation is considered to be more limited than that of the frequency domain analysis (Freq) [14]. Freq showed a higher contribution to the 200 m swim than iEMG in the mean values and for the second, third, and fourth laps. These higher contributions might be explained by the $\dot{E}_{tot}$ absolute values and contributions, as $\dot{E}_{tot}$ absolute value is higher on the first lap, because of the higher velocity when swimmers are not fatigued. However, after the first lap velocity starts to decline, as did $\dot{E}_{tot}$, and a statistical stability during the second and third laps is being maintained. Freq's contribution increases during these two laps. As swimmers reach the fourth lap, Freq increases and SL decreases suggesting the presence of fatigue; $\dot{E}_{tot}$ increases in both absolute value and contribution to velocity in spite of the constant velocity. The contribution of the increased Freq over the swim distance attained a similar contribution to the overall performance as did the energetic and coordinative factors.

For the mean velocity in each lap, both iEMG and Freq present a similar mean contribution; however their pattern of change over the laps is different. The iEMG has its highest contribution on the first lap, whereas Freq has a small contribution. However, Freq is higher, and iEMG lower, in the last lap. This can suggest that at the beginning of the effort higher muscular activation is needed to recruit more fast-twitch muscle fibers and achieve the higher SR at this stage. In the second lapes, the contribution of Freq surpasses that of iEMG, and after this, it decreases constantly until the end of the 200 m effort. The decreased contribution of iEMG is contrary to the increase in absolute values relative to the mean value for 200 m. This pattern of changes is similar to the decrease in spectral parameters that indicate the evolvement of fatigue. As higher $\eta_p$ requires higher effective application of propulsive force, the decrease of the contribution of iEMG might be associated with a decrease in the contribution of $\eta_p$ and be associated with muscle fatigue.

Notwithstanding the results and discussion, as well as the combined interactive effects of performance influencing factors on several research fields in well-trained swimmers, the approach used has some limitations that have to be acknowledged. The regression analysis was not intended to predict performance, only to determine the contribution of the factors, and the variables used represent discrete and extremely important outcomes, each of them for the understanding of the swimming performance and aquatic human locomotion. The relation between the number of variables and subjects evaluated was poor, which may influence the results of the analysis performed, over- or underestimating the contribution of the factors.

## 5. Conclusion

The swimmers in this study had the highest velocity in the first lap of the 200 m swim. The factors contributing to this were a balance of SL, SR, $\eta_p$, IVV$x$, IdC, and iEMG, denoting particular importance for the biomechanical factors (SL, SR, and $\eta_p$), as this first lap is done comfortable enough, without fatigue constraints. From the second through the fourth lap, although the velocity was similar, dynamical changes occurred in the importance of the contributing factors, especially in the fourth lap. In this last, the contributions of Freq and IdC were high and suggest fatigue of the muscles used in swimming, resulting in a high contribution of $\dot{E}_{tot}$ and

lower contribution of $\eta_p$. These data may suggest swimming at a uniform velocity, to avoid the effects of fatigue, and/or training to increase $\dot{E}_{tot}$ and muscular endurance.

## Acknowledgments

This investigation was supported by Grants of the Portuguese Science and Technology Foundation (SFRH/BD/38462/2007) (PTDC/DES/101224/2008—FCOMP-01-0124-FEDER-009577).

## References

[1] A. B. Craig Jr., P. L. Skehan, J. A. Pawelczyk, and W. L. Boomer, "Velocity, stroke rate, and distance per stroke during elite swimming competition," *Medicine and Science in Sports and Exercise*, vol. 17, no. 6, pp. 625–634, 1985.

[2] D. Chollet, S. Chalies, and J. C. Chatard, "A new index of coordination for the crawl: description and usefulness," *International Journal of Sports Medicine*, vol. 21, no. 1, pp. 54–59, 2000.

[3] L. Seifert, D. Chollet, and A. Rouard, "Swimming constraints and arm coordination," *Human Movement Science*, vol. 26, no. 1, pp. 68–86, 2007.

[4] L. Seifert, H. M. Toussaint, M. Alberty, C. Schnitzler, and D. Chollet, "Arm coordination, power, and swim efficiency in national and regional front crawl swimmers," *Human Movement Science*, vol. 29, no. 3, pp. 426–439, 2010.

[5] L. D'Acquisto, D. Ed, and D. Costill, "Relationship between intracyclic linear body velocity fluctuations, power and sprint breaststroke performance," *Journal of Swimming Research*, vol. 13, pp. 8–14, 1998.

[6] H. M. Toussaint, P. E. Roos, and S. Kolmogorov, "The determination of drag in front crawl swimming," *Journal of Biomechanics*, vol. 37, no. 11, pp. 1655–1663, 2004.

[7] P. Zamparo, G. Gatta, D. Pendergast, and C. Capelli, "Active and passive drag: the role of trunk incline," *European Journal of Applied Physiology*, vol. 106, no. 2, pp. 195–205, 2009.

[8] S. V. Kolmogorov and O. A. Duplishcheva, "Active drag, useful mechanical power output and hydrodynamic force coefficient in different swimming strokes at maximal velocity," *Journal of Biomechanics*, vol. 25, no. 3, pp. 311–318, 1992.

[9] H. M. Toussaint, W. Knops, G. de Groot, and A. P. Hollander, "The mechanical efficiency of front crawl swimming," *Medicine and Science in Sports and Exercise*, vol. 22, no. 3, pp. 402–408, 1990.

[10] P. Zamparo, D. R. Pendergast, B. Termin, and A. E. Minetti, "How fins affect the economy and efficiency of human swimming," *Journal of Experimental Biology*, vol. 205, part 17, pp. 2665–2676, 2002.

[11] J. M. Wakeling, O. M. Blake, and H. K. Chan, "Muscle coordination is key to the power output and mechanical efficiency of limb movements," *Journal of Experimental Biology*, vol. 213, no. 3, pp. 487–492, 2010.

[12] T. I. Arabadzhiev, V. G. Dimitrov, N. A. Dimitrova, and G. V. Dimitrov, "Interpretation of EMG integral or RMS and estimates of "neuromuscular efficiency" can be misleading in fatiguing contraction," *Journal of Electromyography and Kinesiology*, vol. 20, no. 2, pp. 223–232, 2010.

[13] M. R. Deschenes, J. A. Giles, R. W. McCoy, J. S. Volek, A. L. Gomez, and W. J. Kraemer, "Neural factors account for strength

decrements observed after short-term muscle unloading," *The American Journal of Physiology*, vol. 282, no. 2, pp. R578–R583, 2002.

[14] R. Merletti, A. Rainoldi, and D. Farina, "Myoelectric manifestations of muscle fatigue," in *Electromyography*, R. Merletti and F. A. Parker, Eds., IEEE Press, Piscataway, NJ, USA, 2004.

[15] R. J. Fernandes, K. L. Keskinen, P. Colaço et al., "Time limit at V·O$_{2max}$ velocity in elite crawl swimmers," *International Journal of Sports Medicine*, vol. 29, no. 2, pp. 145–150, 2008.

[16] D. Pendergast, P. Zamparo, P. E. di Prampero et al., "Energy balance of human locomotion in water," *European Journal of Applied Physiology*, vol. 90, no. 3-4, pp. 377–386, 2003.

[17] H. M. Toussaint and A. P. Hollander, "Energetics of competitive swimming: implications for training programmes," *Sports Medicine*, vol. 18, no. 6, pp. 384–405, 1994.

[18] P. Zamparo, D. R. Pendergast, J. Mollendorf, A. Termin, and A. E. Minetti, "An energy balance of front crawl," *European Journal of Applied Physiology*, vol. 94, no. 1-2, pp. 134–144, 2005.

[19] T. M. Barbosa, J. A. Bragada, V. M. Reis, D. A. Marinho, C. Carvalho, and A. J. Silva, "Energetics and biomechanics as determining factors of swimming performance: Updating the state of the art," *Journal of Science and Medicine in Sport*, vol. 13, no. 2, pp. 262–269, 2010.

[20] D. R. Pendergast, C. Capelli, A. Craig et al., "Biophysics in swimming," in *Proceedings of the 10th International Symposium of Biomechanics and Medicine in Swimming*, pp. 185–189, Porto, Portugal, 2006.

[21] J. P. Vilas-Boas, "Biomechanics and medicine in swimming, past, present and future," in *Biomechanics and Medicine in Swimming XI*, K. L. Kjendlie, R. K. Stallman, and J. Cabri, Eds., pp. 11–19, Norwegian School of Sport Science, Oslo, Norway, 2010.

[22] P. Figueiredo, J. P. Vilas-Boas, J. Maia, P. Gonçalves, and R. J. Fernandes, "Does the hip reflect the centre of mass swimming kinematics?" *International Journal of Sports Medicine*, vol. 30, no. 11, pp. 779–781, 2009.

[23] P. de Leva, "Adjustments to zatsiorsky-seluyanov's segment inertia parameters," *Journal of Biomechanics*, vol. 29, no. 9, pp. 1223–1230, 1996.

[24] Y. Abdel-Aziz and H. Karara, "Direct linear transformation: from comparator coordinates into object coordinates in close range photogrammetry," in *Proceedings of the Symposium on Close-Range Photogrammetry*, pp. 1–18, American Society of Photogrammetry, Falls Church, Va, UA, 1971.

[25] T. M. Barbosa, K. L. Keskinen, R. Fernandes, P. Colaço, A. B. Lima, and J. P. Vilas-Boas, "Energy cost and intracyclic variation of the velocity of the centre of mass in butterfly stroke," *European Journal of Applied Physiology*, vol. 93, no. 5-6, pp. 519–523, 2005.

[26] P. Figueiredo, P. Zamparo, A. Sousa, J. P. Vilas-Boas, and R. J. Fernandes, "An energy balance of the 200 m front crawl race," *European Journal of Applied Physiology*, vol. 111, no. 5, pp. 767–777, 2011.

[27] C. Capelli, D. R. Pendergast, and B. Termin, "Energetics of swimming at maximal speeds in humans," *European Journal of Applied Physiology and Occupational Physiology*, vol. 78, no. 5, pp. 385–393, 1998.

[28] P. Zamparo, C. Capelli, and D. Pendergast, "Energetics of swimming: a historical perspective," *European Journal of Applied Physiology*, vol. 111, no. 3, pp. 367–378, 2011.

[29] M. Alberty, M. Sidney, F. Huot-Marchand, J. M. Hespel, and P. Pelayo, "Intracyclic velocity variations and arm coordination during exhaustive exercise in front crawl stroke," *International Journal of Sports Medicine*, vol. 26, no. 6, pp. 471–475, 2005.

[30] M. Alberty, M. Sidney, P. Pelayo, and H. M. Toussaint, "Stroking characteristics during time to exhaustion tests," *Medicine and Science in Sports and Exercise*, vol. 41, no. 3, pp. 637–644, 2009.

[31] J. P. Clarys and J. Cabri, "Electromyography and the study of sports movements: a review," *Journal of Sports Sciences*, vol. 11, no. 5, pp. 379–448, 1993.

[32] I. Stirn, T. Jarm, V. Kapus, and V. Strojnik, "Evaluation of muscle fatigue during 100-m front crawl," *European Journal of Applied Physiology*, vol. 111, no. 1, pp. 101–113, 2011.

[33] H. J. Hermens, B. Freriks, C. Disselhorst-Klug, and G. Rau, "Development of recommendations for SEMG sensors and sensor placement procedures," *Journal of Electromyography and Kinesiology*, vol. 10, no. 5, pp. 361–374, 2000.

[34] K. de Jesus, K. de Jesus, P. Figueiredo et al., "Biomechanical analysis of backstroke swimming starts," *International Journal of Sports Medicine*, vol. 32, no. 7, pp. 546–551, 2011.

[35] A. H. Rouard, R. P. Billat, V. Deschodt, and J. P. Clarys, "Muscular activations during repetitions of sculling movements up to exhaustion in swimming," *Archives of Physiology and Biochemistry*, vol. 105, no. 7, pp. 655–662, 1997.

[36] V. Caty, Y. Aujouannet, F. Hintzy, M. Bonifazi, J. P. Clarys, and A. H. Rouard, "Wrist stabilisation and forearm muscle coactivation during freestyle swimming," *Journal of Electromyography and Kinesiology*, vol. 17, no. 3, pp. 285–291, 2007.

[37] G. V. Dimitrov, T. I. Arabadzhiev, K. N. Mileva, J. L. Bowtell, N. Crichton, and N. A. Dimitrova, "Muscle fatigue during dynamic contractions assessed by new spectral indices," *Medicine and Science in Sports and Exercise*, vol. 38, no. 11, pp. 1971–1979, 2006.

[38] M. González-Izal, A. Malanda, I. Navarro-Amézqueta et al., "EMG spectral indices and muscle power fatigue during dynamic contractions," *Journal of Electromyography and Kinesiology*, vol. 20, no. 2, pp. 233–240, 2010.

[39] J. Cohen, *Statistical Power Analysis for the Behavioral Sciences*, Lawrence Erlbaum, Hillsdale, NJ, USA, 2nd edition, 1988.

[40] S. G. Psycharakis, R. Naemi, C. Connaboy, C. McCabe, and R. H. Sanders, "Three-dimensional analysis of intracycle velocity fluctuations in frontcrawl swimming," *Scandinavian Journal of Medicine and Science in Sports*, vol. 20, no. 1, pp. 128–135, 2010.

[41] H. M. Toussaint, A. Carol, H. Kranenborg, and M. J. Truijens, "Effect of fatigue on stroking characteristics in an arms-only 100-m front-crawl race," *Medicine and Science in Sports and Exercise*, vol. 38, no. 9, pp. 1635–1642, 2006.

[42] R. J. Fernandes, V. L. Billat, A. C. Cruz, P. J. Colaço, C. S. Cardoso, and J. P. Vilas-Boas, "Does net energy cost of swimming affect time to exhaustion at the individual's maximal oxygen consumption velocity?" *Journal of Sports Medicine and Physical Fitness*, vol. 46, no. 3, pp. 373–380, 2006.

[43] Y. A. Aujouannet, M. Bonifazi, F. Hintzy, N. Vuillerme, and A. H. Rouard, "Effects of a high-intensity swim test on kinematic parameters in high-level athletes," *Applied Physiology, Nutrition and Metabolism*, vol. 31, no. 2, pp. 150–158, 2006.

[44] T. M. Barbosa, M. Costa, D. A. Marinho, J. Coelho, M. Moreira, and A. J. Silva, "Modeling the links between young swimmers' performance: energetic and biomechanic profiles," *Pediatric Exercise Science*, vol. 22, no. 3, pp. 379–391, 2010.

[45] P. Figueiredo, J. P. Vilas-Boas, L. Seifert, D. Chollet, and R. J. Fernandes, "Inter-limb coordinative structure in a 200 m front

crawl event," *Open Sports Sciences Journal*, vol. 3, pp. 25–27, 2010.

[46] C. Schnitzler, L. Seifert, M. Alberty, and D. Chollet, "Hip velocity and arm coordination in front crawl swimming," *International Journal of Sports Medicine*, vol. 31, no. 12, pp. 875–881, 2010.

[47] K. Wakayoshi, L. J. D'Acquisto, J. M. Cappaert, and J. P. Troup, "Relationship between oxygen uptake, stroke rate and swimming velocity in competitive swimming," *International Journal of Sports Medicine*, vol. 16, no. 1, pp. 19–23, 1995.

[48] K. Wakayoshi, J. Acquisto, J. M. Cappaert, and J. P. Troup, "Relationship between metabolic parameters and stroking technique characteristics in front crawl," in *Biomechanics and Medicine in Swimming VII*, J. P. Troup, A. P. Hollander, D. Strasse, S. W. Trappe, J. M. Cappaert, and T. A. Trappe, Eds., pp. 152–158, E & FN Spon, London, UK, 1996.

[49] P. Figueiredo, R. Sanders, T. Gorski, J. P. Vilas-Boas, and R. J. Fernandes, "Kinematic and electromyographic changes during 200 m front crawl at race pace," *International Journal of Sports Medicine*, vol. 34, no. 1, pp. 49–55, 2013.

[50] P. Zamparo, M. Bonifazi, M. Faina et al., "Energy cost of swimming of elite long-distance swimmers," *European Journal of Applied Physiology*, vol. 94, no. 5-6, pp. 697–704, 2005.

[51] V. Y. Caty, A. H. Rouard, F. Hintzy, Y. A. Aujouannet, F. Molinari, and M. Knaflitz, "Time-frequency parameters of wrist muscles EMG after an exhaustive freestyle test," *Portuguese Journal of Sport Sciences*, vol. 6, no. S2, pp. 28–30, 2006.

[52] K. N. Mileva, J. Morgan, and J. Bowtell, "Differentiation of power and endurance athletes based on their muscle fatigability assessed by new spectral electromyographic indices," *Journal of Sports Sciences*, vol. 27, no. 6, pp. 611–623, 2009.

[53] K. Watanabe, K. Katayama, K. Ishida, and H. Akima, "Electromyographic analysis of hip adductor muscles during incremental fatiguing pedaling exercise," *European Journal of Applied Physiology*, vol. 106, no. 6, pp. 815–825, 2009.

[54] K. M. Newell, "Constraints on the development of coordination," in *Motor Development in Children: Aspect of Coordination and Control*, M. G. Wade and H. T. A. Whiting, Eds., pp. 341–360, Nijhoff, Dordrecht, The Netherlands, 1986.

[55] P. S. Glazier, J. S. Wheat, D. L. Pease, and R. M. Bartlett, "Dynamic systems theory and the functional role of movement variability," in *Movement System Variability*, K. Davids, S. Bennett, and K. M. Newell, Eds., pp. 49–72, Human Kinetics, Champaign, Ill, USA, 2006.

[56] T. M. Barbosa, F. Lima, A. Portela et al., "Relationships between energy cost, swimming velocity and speed fluctuation in competitive swimming strokes," *Portuguese Journal of Sport Sciences*, vol. 6, no. S2, pp. 28–29, 2006.

[57] H. M. Toussaint, T. Janssen, and M. Kluft, "Effect of propelling surface size on the mechanics and energetics of front crawl swimming," *Journal of Biomechanics*, vol. 24, no. 3-4, pp. 205–211, 1991.

[58] M. Sidney, S. Paillette, J. Hespel, D. Chollet, and P. Pelayo, "Effect of swim paddles on the intra-cycle velocity variations and on the arm coordination of front crawl stroke," in *Proceedings of the 21st International Symposium on Biomechanics in Sports*, J. R. Blackwell and R. H. Sanders, Eds., pp. 39–42, University of San Francisco, San Francisco, Calif, USA, 2001.

[59] H. M. Toussaint and P. J. Beek, "Biomechanics of competitive front crawl swimming," *Sports Medicine*, vol. 13, no. 1, pp. 8–24, 1992.

[60] K. L. Keskinen and P. V. Komi, "Stroking characteristics of front crawl swimming during exercice," *Journal of Biomechanics*, vol. 9, pp. 219–226, 1993.

[61] L. Seifert, J. Komar, P. M. Leprêtre et al., "Swim specialty affects energy cost and motor organization," *International Journal of Sports Medicine*, vol. 31, no. 9, pp. 624–630, 2010.

[62] W. A. Sparrow and K. M. Newell, "Metabolic energy expenditure and the regulation of movement economy," *Psychonomic Bulletin and Review*, vol. 5, no. 2, pp. 173–196, 1998.

[63] D. L. Costill, M. G. Flynn, J. P. Kirwan et al., "Effects of repeated days of intensified training on muscle glycogen and swimming performance," *Medicine and Science in Sports and Exercise*, vol. 20, no. 3, pp. 249–254, 1988.

[64] S. P. McLean, D. Palmer, G. Ice, M. Truijens, and J. C. Smith, "Oxygen uptake response to stroke rate manipulation in freestyle swimming," *Medicine and Science in Sports and Exercise*, vol. 42, no. 10, pp. 1909–1913, 2010.

[65] M. R. Alberty, F. P. Potdevin, J. Dekerle, P. P. Pelayo, and M. C. Sidney, "Effect of stroke rate reduction on swimming technique during paced exercise," *Journal of Strength and Conditioning Research*, vol. 25, no. 2, pp. 392–397, 2011.

[66] D. G. Allen, G. D. Lamb, and H. Westerblad, "Skeletal muscle fatigue: cellular mechanisms," *Physiological Reviews*, vol. 88, no. 1, pp. 287–332, 2008.

[67] P. Pelayo, I. Mujika, M. Sidney, and J. C. Chatard, "Blood lactate recovery measurements, training, and performance during a 23-week period of competitive swimming," *European Journal of Applied Physiology and Occupational Physiology*, vol. 74, no. 1-2, pp. 107–113, 1996.

[68] K. L. Keskinen, L. J. Tilli, and P. V. Komi, "Maximum velocity swimming: interrelationships of stroking characteristics, force production and anthropometric variables," *Scandinavian Journal of Sports Sciences*, vol. 11, no. 2, pp. 87–92, 1989.

[69] J. M. H. Cabri, L. Annemans, J. P. Clarys, E. Bollens, and J. Publie, "The relation of stroke frequency, force, and EMG in front crawl tethered swimming," in *Swimming Science V*, pp. 183–189, 1988.

[70] S. C. Gandevia, "Spinal and supraspinal factors in human muscle fatigue," *Physiological Reviews*, vol. 81, no. 4, pp. 1725–1789, 2001.

# Kinematic Measures during a Clinical Diagnostic Technique for Human Neck Disorder: Inter- and Intraexaminer Comparisons

**Joseph Vorro,[1] Tamara R. Bush,[2] Brad Rutledge,[3] and Mingfei Li[4]**

[1] Department of Family Medicine, College of Osteopathic Medicine, Michigan State University, East Lansing, MI 48824, USA
[2] Department of Mechanical Engineering, College of Engineering, Michigan State University, East Lansing, MI 48824, USA
[3] Biomechanics Division, MEA Forensic Engineers & Scientists, Laguna Hills, CA 92653, USA
[4] Department of Mathematical Sciences and Center for Quantitative Analysis, Bentley University, Waltham, MA 02452, USA

Correspondence should be addressed to Joseph Vorro; vorro@msu.edu

Academic Editor: Miguel A. Rivero

Diagnoses of human musculoskeletal dysfunction of the cervical spine are indicated by palpable clues of a patient's structural compliance/noncompliance as this body segment responds to diagnostic motion demands applied by a clinician. This process includes assessments of motion range, motion performance, and changes in tissue responses. However, biomechanical quantification of these diagnostic actions and their reproducible components is lacking. As a result, this study sought to use objective kinematic measures to capture aspects of the diagnostic process to compare inter- and intraexaminer motion behaviors when performing a specific clinical diagnostic protocol. Pain-free volunteers and a group determined to be symptomatic based on a psychometric pain score were examined by two clinicians while three-dimensional kinematic data were collected. Intraexaminer diagnostic motion ranges of cervical lateral flexion and secondary rotations were consistent for each examiner and for each subject group. However, interexaminer comparisons for motion range, secondary rotations, and average velocities yielded consistently larger measures for one examiner for both subject groups ($P < 0.05$). This research demonstrates that fundamental aspects of the clinical diagnostic process for human neck disorders can be identified and measured using kinematic parameters. Further, these objective data have the potential to be linked to clinical decision making.

## 1. Introduction

Three-dimensional (3D) kinematic measures are commonly used in the diagnostic assessment of human gait [1]. These measures are collected using specialized motion capture systems with retroreflective targets on key anatomical landmarks. Data related to the locations of these targets in 3D space are used to compute joint angles, velocity measures, and other objective measures. Results from this type of research have also successfully related these kinematic parameters to general function, aging, disease, and dysfunction. However, similar use of this technology for other body regions to identify and relate dynamic 3D kinematic assessments to clinical diagnoses has been limited.

One area that has the potential to benefit from kinematic-based biomechanical parameters is the study of musculoskeletal disorders (MSDs) of the cervical spine. MSDs of the

cervical spine have a significant impact on society, frequently causing major disabilities and lasting functional limitations [2–4]. These disorders affect as many as two-thirds of the world's population and are second only to low back problems [5, 6]. A unique aspect associated with these neck disorders is that they are commonly diagnosed by palpation, which are clinician-directed techniques (passively induced motions) guided by physical clues produced by patient's regional physiologic responses [7]. Specifically, during these procedures, a clinician monitors a patient using light, nonintrusive touch, while continually assessing compliance/noncompliance of the anatomical structures as they respond to the motion demands of the clinical technique. The results can be deviations from normative ranges of motion (ROM) and changes in tissue resistances, particularly at the end of motion range.

However, many of these structural diagnostic techniques lack objective, scientific, evidence for their relevance and effectiveness. Attempts to develop reliable approaches to these clinical evaluations exist; however, none have yet captured essential components of the diagnostic procedures [8, 9]. As a result, a need exists to quantify clinician-directed diagnostic actions into components that are measurable and also meaningful to the diagnostic process, clinical decision making, and evaluation of treatment effectiveness.

Therefore, the purpose of this study was to use 3D kinematic measures to quantify and compare inter- and intraexaminer motion behaviors when performing a cervical diagnostic protocol on a group of symptom-free volunteers, and a group determined to be symptomatic based on a psychometric pain score. Specifically, three kinematic parameters were used to evaluate the diagnostic process: (1) the magnitude and variation of the primary diagnostic motion (lateral cervical flexion), (2) the magnitude and variation of secondary rotations that occurred at maximal diagnostic ROMs, and (3) the angular velocities (speeds) with which the examiners performed the passive diagnostic tests.

## 2. Methods

*2.1. Subjects and Subject Group Assignment.* Subjects were recruited from a university campus clinical center, and from the general university student, faculty, and staff populations. These volunteers provided written consent prior to participation. Everyone completed two questionnaires: a Visual-Analog Scale (VAS) [10] to determine the level of pain in the neck region and a Neck Pain and Disability Scale (NPDS) [11] to quantify the level of dysfunction.

As part of the protocol to establish subject groups, all volunteers also received an initial (reference) diagnostic assessment by a physician (Examiner 1) using the standard palpatory diagnostic test of right and left cervical lateral flexions (side-bending). Examiner 1 was blinded to the VAS and NPDS scores.

Based on this assessment by Examiner 1, and the results of the VAS questionnaire, two subject groups were established.

(1) *Control Group*: subjects who self-scored the VAS = 0, and who were symmetric for the right and left lateral flexion diagnostic motions as determined by Examiner 1.

(2) *Experimental Group*: subjects who self-scored the VAS ≥ 3 indicating cervical pain [10].

Evaluations were conducted on 131 total volunteers. Of these, 41 qualified for the study, including control group (*n* = 22; 16 males, 4 females and 2 who opted not to provide a response) and experimental group (*n* = 19; 14 males and 5 females). Volunteers not qualifying for the study were dismissed.

*2.2. Additional Diagnostic Evaluations by Examiners 2 and 3.* To be able to evaluate inter- and intraexaminer kinematic consistencies during the specific diagnostic test, control and experimental subjects experienced further diagnostic testing (cervical lateral flexion) from two additional physicians (Examiners 2 and 3). Both Examiners conducted the passive diagnostic motions in front of a six-camera motion capture system (Qualisys, Gothenburg, Sweden) while head and neck

motions relative to the thorax were recorded (detailed in Section 2.3).

Examiners 2 and 3 were blinded to each subject's VAS and NPDS scores, subject diagnostic categories, and each other's assessments. Two separate diagnostic trials (Trial 1 and Trial 2) were conducted by each examiner, with each trial consisting of three right and left repeated motions.

The cervical palpatory diagnostic technique used by all three examiners during all subject screening and testing is a commonly practiced, standard test of cervical lateral flexion [12, 13]. Each examiner was a practicing osteopathic physician specialized in manual medicine for over 10 years. The procedure was as follows:

(1) the examiner aligned himself/herself posterior to a seated subject. Each subject had to remain passive as the diagnostic motions were performed (Figures 1(a) and 1(b)).

(2) One of the examiner's hands (the moving hand) was placed gently on a subject's head, while the contralateral hand was placed lightly on the posterior thoracic midline.

(3) The examiner passively guided the subject's head in lateral flexion to the right, taking the right ear toward the ipsilateral shoulder until a palpable sense of end ROM was achieved. End ROM was defined as the point where a tissue texture change required a substantial increase in force to continue the diagnostic motion [13].

(4) At the conclusion of the initial motion, the subject's head was guided back to neutral, the examiner's hand placement was changed, and movement to the contralateral side was conducted.

(5) The subject's head was guided back to neutral again, and steps 2–5 were repeated so that one trial consisted of three right lateral flexions and three left lateral flexions.

Examiners then made their clinical evaluations based upon the following, previously established criteria [10, 12, 13]:

(1) visual and proprioceptive evaluations of the magnitude and symmetry of right and left cervical lateral flexions,

(2) palpatory assessments from the moving hand to determine quality of motion, specifically, smoothness and tissue resistance, and

(3) "end-feel" was considered as any specific resistance to the diagnostic movements.

*2.3. Kinematic Data Collection.* A six-camera Qualisys motion capture system, in conjunction with retroreflective markers, was used to capture motions of the head relative to the thorax (Figure 1(a)). Retroreflective markers were placed on each subject's temples, and one marker was centered on the forehead to capture head motions. Three additional markers, in the form of a rigid triad, were attached to the skin over the central sternum of each subject (Figure 1(b)). The duration

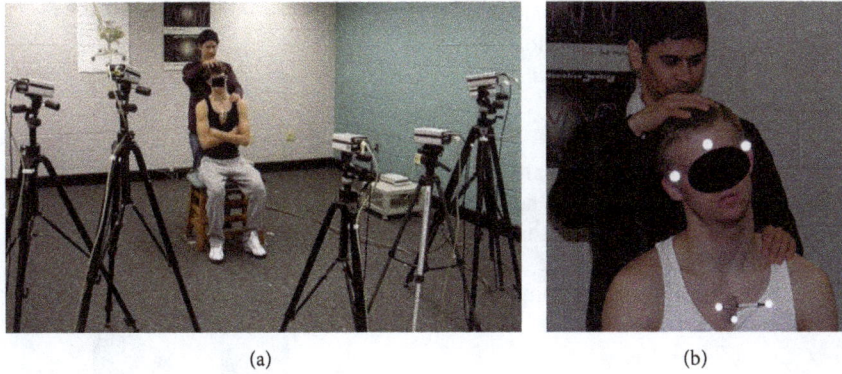

FIGURE 1: (a) Examiner and subject positions, with motion capture cameras in place. (b) Retroreflective marker positions during execution of the lateral flexion diagnostic test.

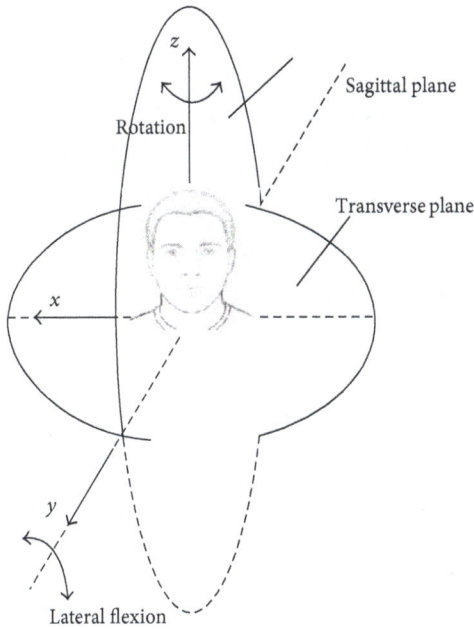

FIGURE 2: Lateral flexion and coupled axial rotations of the head during the diagnostic test.

of each trial was not controlled, so that examiners moved the subjects at their preferred motion rates for the diagnostic process. Test trial times ranged from 30 to 65 seconds, and data were collected at 20 Hz.

To establish subject-specific neutral head reference locations (0° angle), each trial began with a three-second period where baseline data were collected as subjects were instructed to remain still and face forward. After baseline data were collected, the examiner's hands were placed on the subject to begin the diagnostic movements.

The motion capture system was calibrated daily with the global reference system oriented such that the $x$-axis progressed horizontally from the equivalent of a subject's left to right, the $y$-axis from posterior to anterior, and the $z$-axis vertically from inferior to superior (Figure 2). The system error was less than ±2 mm and ±1°.

2.4. *Data Analysis.* Euler angles were used to compute motions of the head relative to the thorax [14]. Thus, any movement of the thorax was accounted for by the relative segment assessment. First, local Cartesian coordinate systems were established on the head and thorax in the form of unit vectors ($\hat{i}$, $\hat{j}$, and $\hat{k}$). These coordinate systems were generated from the coordinates of the markers on the head (forehead, left temple, and right temple) and the sternum (middle sternum, left sternum, and right sternum). These local coordinate systems were then aligned with respect to the global coordinate system. Based on the unit vectors for a local coordinate system at frame "$n$" and frame "$n + 1$," the rotation matrix between the two frames could be calculated, and, thus, the angles of rotation could be determined. The rotation matrix, based upon a rotation sequence of $yzx$, was determined from the summation of the rotation matrices for rotations around the $y$-axis, $z$-axis, and $x$-axis independently as follows:

$$R_y(\theta_1) = \begin{bmatrix} \cos\theta_1 & 0 & \sin\theta_1 \\ 0 & 1 & 0 \\ -\sin\theta_1 & 0 & \cos\theta_1 \end{bmatrix},$$

$$R_z(\theta_2) = \begin{bmatrix} \cos\theta_2 & -\sin\theta_2 & 0 \\ \sin\theta_2 & \cos\theta_2 & 0 \\ 0 & 0 & 1 \end{bmatrix},$$

$$R_x(\theta_3) = \begin{bmatrix} 1 & 0 & 0 \\ 0 & \cos\theta_3 & -\sin\theta_3 \\ 0 & \sin\theta_3 & \cos\theta_3 \end{bmatrix},$$

$$R(\theta_1, \theta_2, \theta_3) = R_y(\theta_1) R_z(\theta_2) R_x(\theta_3) = [R],$$

where $\theta_1$ was the angle of rotation about the $y$-axis (lateral flexion), $\theta_2$ was the angle of rotation around the $z$-axis (axial rotation), and $\theta_3$, was the angle of rotation around the $x$-axis (flexion and extension) [15]. The rotation matrix, $R(\theta_1, \theta_2, \theta_3)$, was then multiplied with the unit vectors of the local coordinate system at the original frame, $X_n$, to determine the location of the local coordinate system in the next frame, $X_{n+1}$, allowing the computation of the three angles throughout the entire diagnostic motion. Although three angles were computed, forward flexion and extension data were not analyzed since the focus of the palpatory diagnostic movement was on lateral flexion and rotation (also the largest two motions) (Figure 2). Three variables were analyzed.

*2.4.1. Cervical ROM.* Maximal diagnostic right and left motions were identified as angles greater than 10° that were greater than the previous 10 values and greater than the following 10 values. Additionally, total diagnostic ROMs (from maximum right to maximum left) were computed. All angles were based on the subject's self-selected neutral position (0° angle).

*2.4.2. Secondary Rotations.* Axial rotations at frames corresponding to maximum right and left diagnostic ROMs were identified for each subject and averaged for each examiner. Additionally, the total rotation (from maximum diagnostic right to maximum diagnostic left) was computed. Positive axial rotations were associated with ipsilateral lateral flexions such that right lateral flexion ROM was associated with axial rotation to the right, and left lateral flexion ROM was accompanied by axial rotation to the left. Negative axial rotations indicated contralateral rotation, or axial rotation in the opposite direction to the lateral flexion being performed (Figure 2).

*2.4.3. Diagnostic Motion Angular Velocities.* Angular velocities (degrees/second) for each passive diagnostic motion were identified as the slopes of linear regressions calculated from the start-to-peak excursions for each motion. This portion of the curve corresponded to the movements where the examiners conducted their passive diagnoses. Average angular velocities for diagnostic movements to the right and to the left were analyzed. Additionally, the total average angular velocities (for both right and left movements) were computed and analyzed.

*2.5. Statistical Analyses.* One-way repeated measures ANOVAs assessed intraexaminer data, while *t*-tests compared interexaminer data for three kinematic factors: (1) lateral flexion ROMs, (2) secondary rotations at maximum ROM, and (3) the rates (velocities) at which the diagnostic motions were conducted.

Data for trials one and two were evaluated separately for intraexaminer comparisons so that the diagnostic motion consistencies could be assessed. However, for between-examiner comparisons, the two trials for each examiner were averaged. The criterion alpha level was set to 0.05.

# 3. Results

*3.1. Subjects.* The average age of the control subjects was 19.9 years (±1.9) and experimental subjects 27.5 years (±13.1). A statistical analysis indicated no significant between-group age differences. The VAS for control subjects was zero (the requirement for inclusion in this subject group); experimental subjects produced an average VAS of 4.6 out of 10.0. The average NPDS score for control subjects was 2.5 (±4.1) and

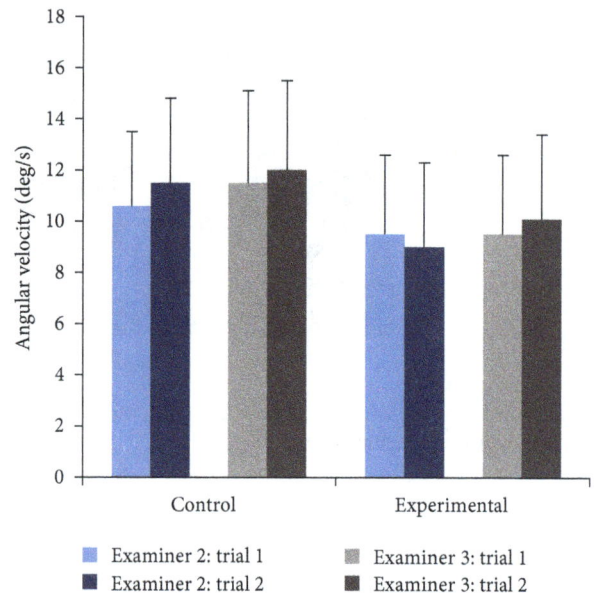

FIGURE 3: Angular velocities for diagnostic tests, subject group, trials and examiners. Data represent average velocities for each trial for all subjects. Bars represent one standard deviation.

46.9 (±21.0) for experimental subjects, with larger numbers indicating higher levels of head and neck disability.

*3.2. Intraexaminer ROMs.* When comparing the average total diagnostic ROM, Examiner 2 demonstrated between-trial differences of less than 1° for the control group and less than 3° for the experimental group; while Examiner 3's passive, diagnostic ROMs varied less than 1° for both subject groups (Tables 1 and 2). One-way ANOVAs indicated no within-examiner ROM differences for right, left, or total diagnostic ROM or subject group, and no differences for the repeated trials.

*3.3. Intraexaminer Secondary Rotations.* The total secondary rotations produced by each Examiner also demonstrated slight differences across repeat trials. Examiner 2 produced less than 1° of secondary rotation between trials for the control group and less than 3° for the experimental group. Examiner 3's passive between-trial secondary rotations were less than 1° for control subjects and slightly more than 1° for the experimental subjects (Tables 3 and 4). One-way ANOVAs indicated no within-examiner secondary rotations differences.

*3.4. Intraexaminer Rates of Motion.* Assessments of within-examiner diagnostic angular velocities indicated that on average both examiners produced between-trial differences of less than 1.0 degree/sec for both control and experimental subjects (Tables 5 and 6, Figure 3).

One-way repeated ANOVAs indicated differences in the velocity data; however, they were not consistent across examiner, trial, or group. Specifically, Examiner 2 produced significantly different between-trial velocities ($P = 0.045$)

TABLE 1: Examiner 2, comparisons of diagnostic motions for control and experimental subjects for right, left, and total passive (maximum right to left) diagnostic ROM.

| Subjects | Right diagnostic ROM | | Left diagnostic ROM | | Total diagnostic ROM | |
|---|---|---|---|---|---|---|
| | Trial 1 | Trial 2 | Trial 1 | Trial 2 | Trial 1 | Trial 2 |
| | Ave (SD) | | Ave (SD) | | Ave (SD) | |
| | (Degrees) | | (Degrees) | | (Degrees) | |
| Control | 35.6 | 36.2 | 35.5 | 35.5 | 71.1 | 71.7 |
| | (7.2) | (6.7) | (5.6) | (5.7) | (11.6) | (11.5) |
| Experimental | 33.0 | 31.8 | 34.6 | 33.0 | 67.6 | 64.8 |
| | (7.1) | (6.2) | (7.0) | (6.3) | (12.2) | (11.8) |

TABLE 2: Examiner 3, comparisons of diagnostic motions for control and experimental subjects for right, left, and total passive (maximum right to left) diagnostic ROM.

| Subjects | Right Diagnostic ROM | | Left Diagnostic ROM | | Total Diagnostic ROM | |
|---|---|---|---|---|---|---|
| | Trial 1 | Trial 2 | Trial 1 | Trial 2 | Trial 1 | Trial 2 |
| | Ave (SD) | | Ave (SD) | | Ave (SD) | |
| | (Degrees) | | (Degrees) | | (Degrees) | |
| Control | 32.6 | 32.7 | 33.1 | 33.6 | 65.7 | 66.3 |
| | (7.1) | (7.1) | (6.3) | (5.9) | (11.7) | (11.3) |
| Experimental | 28.5 | 28.3 | 29.7 | 29.4 | 58.2 | 57.7 |
| | (5.8) | (6.9) | (6.7) | (6.5) | (11.4) | (12.0) |

TABLE 3: Examiner 2, secondary rotations at maximum diagnostic right lateral flexion ROM, specifically, rotations at maximum right lateral flexion, at maximum left lateral flexion, and the total rotation (from maximum diagnostic right to maximum diagnostic left).

| Subjects | To the right | | To the left | | Total rotation | |
|---|---|---|---|---|---|---|
| | Trial 1 | Trial 2 | Trial 1 | Trial 2 | Trial 1 | Trial 2 |
| | Ave (SD) | | Ave (SD) | | Ave (SD) | |
| | (Degrees) | | (Degrees) | | (Degrees) | |
| Control | 8.8 | 9.3 | 10.8 | 10.4 | 19.7 | 19.6 |
| | (4.8) | (4.3) | (4.8) | (6.1) | (8.0) | (8.8) |
| Experimental | 10.9 | 12.1 | 10.7 | 11.6 | 21.6 | 23.7 |
| | (4.5) | (5.1) | (8.4) | (8.4) | (10.9) | (11.7) |

TABLE 4: Examiner 3, secondary rotations at maximum diagnostic right lateral flexion ROM, specifically, rotations at maximum right lateral flexion, at maximum left lateral flexion, and the total rotation (from maximum diagnostic right to maximum diagnostic left).

| Subjects | To the right | | To the left | | Total rotation | |
|---|---|---|---|---|---|---|
| | Trial 1 | Trial 2 | Trial 1 | Trial 2 | Trial 1 | Trial 2 |
| | Ave (SD) | | Ave (SD) | | Ave (SD) | |
| | (Degrees) | | (Degrees) | | (Degrees) | |
| Control | 8.7 | 8.5 | 6.7 | 7.0 | 15.4 | 15.5 |
| | (4.6) | (4.8) | (5.2) | (5.2) | (7.7) | (8.5) |
| Experimental | 11.2 | 9.2 | 6.3 | 7.2 | 17.5 | 16.4 |
| | (4.6) | (4.5) | (8.0) | (8.6) | (11.1) | (11.6) |

when moving control subjects to the right. Further, Examiner 2 produced significantly different between-trial velocities when moving experimental subjects to the right ($P = 0.030$ and to the left ($P = 0.044$). Examiner 3 produced significantly different between-trail velocities when moving control subjects only, to the right and left ($P = 0.006$ and $P = 0.035$, resp.).

3.5. Interexaminer ROM. Between-examiner comparisons for diagnostic ROM indicated that Examiner 2 consistently moved control and experimental subjects through greater ranges than Examiner 3 (average percent difference between examiners: for the control group = 7.4%; for the experimental group = 12.4%) (Tables 1 and 2, Figure 4). Additionally, both examiners produced greater total diagnostic ROM for control subjects.

Independent samples $t$-tests indicated significant differences in between-Examiner passive motions for experimental subjects only, right ROM ($P = 0.029$), left ROM ($P = 0.053$), and total ROM ($P = 0.040$).

3.6. Interexaminer Secondary Rotations. Secondary rotations for all trials, Examiners and subject groups are presented in Tables 3 and 4. Between-examiner comparisons indicated that Examiner 2 consistently produced greater average secondary rotations than Examiner 3 (Tables 3 and 4, Figure 5).

TABLE 5: Comparisons of Examiner 2 angular velocities for subject group and trials, including average angular velocities for movements to the right, left, and total average angular velocities for right and left movements.

| Subjects | Right velocity | | Left velocity | | Total average angular velocity | |
|---|---|---|---|---|---|---|
| | Trial 1 | Trial 2 | Trial 1 | Trial 2 | Trial 1 | Trial 2 |
| | Ave (SD) | | Ave (SD) | | Ave (SD) | |
| | (°/second) | | (°/second) | | (°/second) | |
| Control | 10.5 | 11.8 | 10.7 | 11.3 | 10.6 | 11.5 |
| | (2.9) | (3.4) | (2.9) | (3.3) | (2.9) | (3.3) |
| Experimental | 9.7 | 9.1 | 9.4 | 9.0 | 9.5 | 9.0 |
| | (3.1) | (3.1) | (3.0) | (3.4) | (3.0) | (3.3) |

TABLE 6: Comparisons of Examiner 3 angular velocities for subject group and trials, including average angular velocities for movements to the right, left, and total average angular velocities for right and left movements.

| Subjects | Right velocity | | Left velocity | | Total average angular velocity | |
|---|---|---|---|---|---|---|
| | Ex. 2 | Ex. 3 | Ex. 2 | Ex. 3 | Ex. 2 | Ex. 3 |
| | Ave (SD) | | Ave (SD) | | Ave (SD) | |
| | (°/second) | | (°/second) | | (°/second) | |
| Control | 11.1 | 11.6 | 11.8 | 12.3 | 11.5 | 12.0 |
| | (3.7) | (3.2) | (3.4) | (3.5) | (3.6) | (3.4) |
| Experimental | 9.7 | 10.0 | 9.4 | 10.1 | 9.6 | 10.0 |
| | (3.2) | (3.3) | (2.9) | (3.4) | (3.6) | (3.3) |

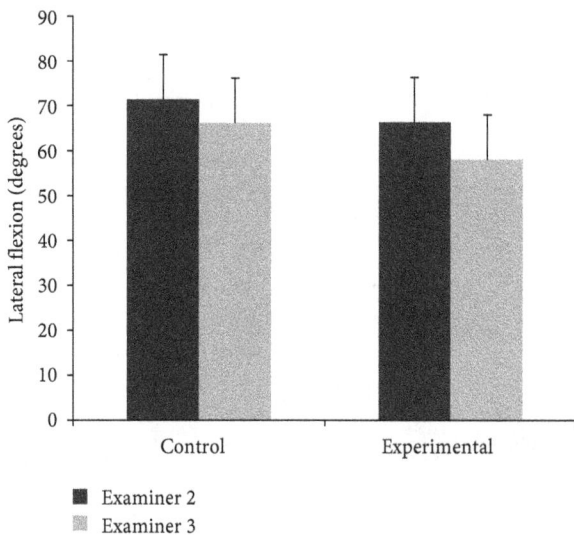

FIGURE 4: Comparisons between Examiner 2 (dark blue) and Examiner 3 (grey) for average total passive diagnostic ROM. Bars represent one standard deviation.

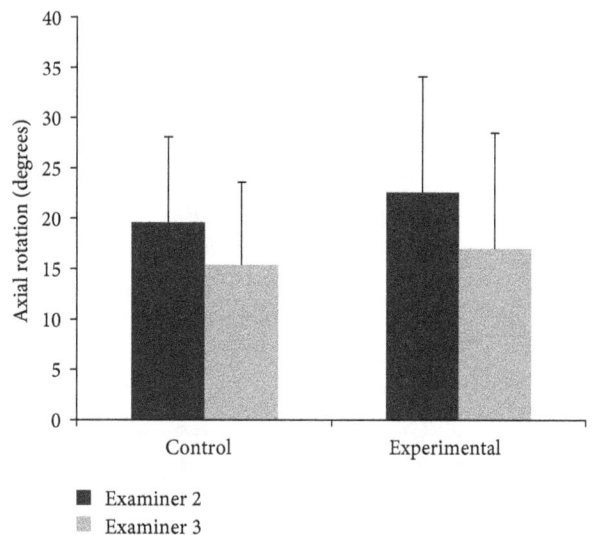

FIGURE 5: Comparisons between Examiner 2 (dark blue) and Examiner 3 (grey) for average secondary rotations. Bars represent one standard deviation.

The average percent difference between examiners for the control group was 21.3% and 25.1% for the experimental group. In addition, both Examiners produced greater secondary rotations when passively moving experimental subjects.

Independent samples $t$-tests indicated significant between-examiner differences for passive secondary rotations for control subjects only: for diagnostic motions to the right ($P = 0.036$), to the left ($P = 0.026$), and for total right and left motions ($P = 0.018$). No between-examiner differences for secondary rotations occurred for experimental subjects.

*3.7. Interexaminer Rates of Motion.* The right, left, and total average velocities for control and experimental subjects for each examiner are presented in (Tables 5 and 6). Examiner 3 consistently produced higher average diagnostic motion velocities than Examiner 2. Greater angular velocities were produced by each examiner during diagnostic motions for control subjects.

Between-examiner average angular velocities varied less than 6% for diagnostic motions and subject groups. Independent samples $t$-tests indicated a significant difference between Examiners for average velocities to the left for Control subjects ($P = 0.050$).

## 4. Discussion

Manual diagnostic assessments of human neck motions are a standard part of the clinical examination for patients suffering from neck disorders, neck pain, and discomfort. However, little biomechanical evidence exists that quantifies these clinician-directed palpatory diagnostic techniques and comparisons across subject groups and different examiners.

As a result, this study used kinematic measures to capture biomechanical aspects of a standard clinical screening test (cervical lateral flexion) so that inter- and intraexaminer diagnostic motions could be compared through objective data sets. Control subjects ($n = 22$) demonstrated diagnostic motion symmetry as determined by a trained, experienced examiner. These subjects also self-reported no pain based on results from a psychometric test. Experimental subjects ($n = 19$) consisted of volunteers with positive scores of $\geq 3$ from the same psychometric test indicating neck pain.

Two additional examiners (2 and 3) conducted passive diagnostic motions for each subject in front of a motion capture system. Three specific parameters were evaluated from the kinematic data: diagnostic ROM, secondary rotations around the primary diagnostic motion, and angular velocities.

Intraexaminer data for two kinematic variables, total diagnostic ROM, and secondary rotations were consistent within each examiner's data sets and yielded no significant differences. The statistical analysis did indicate several differences in velocity data; however, they were not consistent across examiner, trial, or subject group.

Interexaminer kinematic data varied. Examiner 2 consistently produced greater average passive diagnostic ROMs than Examiner 3, and Examiner 2 produced greater average secondary rotations. Between-examiner velocities were not significantly different except for one comparison, movement to the left for control subjects.

Modern medicine uses multiple standard, objective, evidence-based tests to determine and confirm the presence of disease, for example, microbiological cultures, blood samples and medical images. These tests are used to the extent needed to establish an accurate medical diagnosis. Once the diagnosis is made, appropriate treatments can be implemented, and prevention options can be sought.

While laboratory tests and clinical evaluations make up the bulk of medical diagnostic techniques for most of the body's organ systems, the health and integrity of the musculoskeletal system is typically evaluated via palpation. This structural (haptic) diagnosis uses passive gross motions introduced as tests to gain palpable information about the body's regional and segmental motor functions and integrity.

Review articles indicate that much of the related literature is associated with manual medicine treatment techniques, and the majority is qualitative in nature. However, while some excellent manual medicine diagnostic studies exist, review articles indicate that findings often vary between studies because of methodological differences, different clinician experiences, lack of examiner training and consensus before the study, lack of concomitant comparisons between symptomatic and asymptomatic subjects, and lack of objective markers [7, 16–19]. Further, many studies of inter- and intraexaminer comparisons do not use clinically relevant diagnostic tests (i.e., the guiding of a body part to a palpable (comfortable) sense of diagnostic motion end range) [13], but rather required the production of maximal anatomical joint excursions [20–23].

No within-examiner ROM differences for movements, subject group, or repeated trials were observed in this experiment. Within-examiner consistencies of this nature are well documented in the literature [7, 24–26]. However, a few within-examiner differences occurred for average angular velocities. Also, when evaluating the subject groups, Examiners consistently moved the experimental group slower than the control group. It is suggested that the reason for the observed average velocity differences in this paper relates specifically to the clinical intent of the palpatory process. The two examiners performed the diagnostic motions while simultaneously monitoring the motions for specific palpable cues, for example, tissue texture changes and restrictive barriers. As a result, the examiners moved the pain group at a slower rate, even though they were blinded to the group assignment and they did not have any verbal communication with the subjects. Other studies have looked at angular velocities; however, they differed from this paper in that the motions were active instead of passive, and, motions other than cervical lateral flexion were evaluated. For example, Sjölander et al. [27] and Prushansky et al. [28] studied active cervical motions for a group of acute whiplash patients and no-pain controls and found reduced peak velocities for the whiplash groups. Bahat et al. [29] used a virtual reality motion assessment system to evaluate pain and no-pain subjects and reported reductions in both peak and average velocities for motions produced by pain subjects. However, this group only studied cervical sagittal flexion/extension and rotation.

The current findings also indicated that between-examiner diagnostic ROM, secondary rotations and average motion velocities varied. Interexaminer differences of this nature are well reported [7, 16–19, 30]. However, these papers point out that the comparisons were typically made on qualitative assessments, and that reasons for the reported differences were due to the lack of objective markers by which to make evaluations.

A few studies have made objective measurements of clinical cervical motions with between-examiner comparisons. Lantz et al. [31] reported good between-examiner agreement for cervical axial rotations and lateral flexions; however, for normal subjects only, for maximal ROMs instead of diagnostic ROMs, and their protocol required the use of a thoracic harness during testing which is not a standard clinical protocol. Morphett et al. [32] also reported good between-examiner ROM agreement for both pain and nonpain subjects, although their protocol required maximal cervical motion ranges. Strimpakos et al. [26] examined maximal ROM for nonpain subjects only and found that one examiner yielded significantly lower neck ROM values in all assessments than the other examiner. Further, their subjects were fitted to a stabilization system that isolated the cervical region from the rest of the body, and there were no reports of secondary motions or associated velocities.

In terms of general kinematic differences between pain and no-pain subjects, two of the three variables in this experiment, diagnostic ROM and average motion velocities were reduced for the pain subjects as compared to the nonpain group and supported previous research [21, 29, 33–37]. The third kinematic variable, secondary rotations, were greater for pain subjects, directly supporting a previous investigation [33], while also supporting additional work that included different experimental intentions and protocols [20, 25, 26, 34, 38–41].

The results presented here indicate that diagnostic techniques for MSD have measurable anatomical and physiological characteristics. As a result, MSD yields objective measures that manifest as altered motion qualities detected through palpation, for example, differences in motion range and motion symmetry, modified proprioception, changes in muscle recruitment patterns, and altered kinematics.

The importance and necessity of this work are confirmed by numerous review articles that document contradicting results, while also indicating the need for objective measures to eliminate inconsistencies found from subjective measures [7, 16–19]. In addition, the future potential of this work can be seen in training new clinicians—specifically, using objective information to help learners understand diagnostic movements, developing consistent approaches to conducting these assessments, and relating specific changes in kinematic measures to levels of dysfunction to assist with the diagnosis, and to document treatment effects.

## Funding

This study was financed by research funds from the American Osteopathic Association Council on Research and the Osteopathic Heritage Foundation.

## Ethical Approval

This study was approved by the Institutional Review Board of Michigan State University (no. 06-464) and http://ClinicalTrials.gov/ (NCT01186718).

## Conflict of Interests

The authors declare no conflict of interests with institutes, organizations or companies relevant to this paper.

## Acknowledgments

The authors wish to thank the following individuals for their assistance: Jessica Buschman, Lisa DeStefano, D.O., Timothy Francisco, D.O., Sherman Gorbis, D.O., and Michael Seffinger, D.O.

## References

[1] J. Perry and J. M. Burnfield, *Gait Analysis: Normal and Pathological Function, Thorofare*, New Jersey, NJ, USA, 2010.

[2] A. D. Woolf and B. Pfleger, "Burden of major musculoskeletal conditions," *Bulletin of the World Health Organization*, vol. 81, no. 9, pp. 646–656, 2003.

[3] World Health Organization, "The burden of musculoskeletal conditions at the start of the new millenium," WHO Technical Report 919, Geneva, Switzerland, 2003.

[4] Bone and Joint Decade, 2002-USA-2011, "The Burden of Musculoskeletal Diseases in the United States, Executive Summary," *American Academy of Orthopaedic Surgeons*, pp. 1–9, 2008.

[5] R. Fejer, K. O. Kyvik, and J. Hartvigsen, "The prevalence of neck pain in the world population: a systematic critical review of the literature," *European Spine Journal*, vol. 15, no. 6, pp. 834–848, 2006.

[6] J. Hartvigsen, K. Christensen, and H. Frederiksen, "Back and neck pain exhibit many common features in old age: a population-based study of 4,486 danish twins 70–102 years of age," *Spine*, vol. 29, no. 5, pp. 576–580, 2004.

[7] M. A. Seffinger, W. I. Najm, S. I. Mishra et al., "Reliability of spinal palpation for diagnosis of back and neck pain: a systematic review of the literature," *Spine*, vol. 29, no. 19, pp. E413–E425, 2004.

[8] T. S. Carey, J. Garrett, A. Jackman, C. McLaughlin, J. Fryer, and D. R. Smucker, "The outcomes and costs of care for acute low back pain among patients seen by primary care practitioners, chiropractors, and orthopedic surgeons," *The New England Journal of Medicine*, vol. 333, no. 14, pp. 913–917, 1995.

[9] P. S. Khalsa, A. Eberhart, A. Cotler, and R. Nahin, "The 2005 conference on the biology of manual therapies," *Journal of Manipulative and Physiological Therapeutics*, vol. 29, no. 5, pp. 341–346, 2006.

[10] D. J. Magee, *Orthopedic Physical Assessment*, Saunders, Philadelphia, Pa, USA, 2002.

[11] A. H. Wheeler, P. Goolkasian, A. C. Baird, and B. V. Darden, "Development of the neck pain and disability scale: item analysis, face, and criterion-related validity," *Spine*, vol. 24, no. 13, pp. 1290–1294, 1999.

[12] W. Johnston, H. Friedman, and D. Eland, *Functional Methods: A Manual for Palpatory Skill Development in Osteropathic Examination & Manipulation of Motor Function*, American Academy of Osteopathy, Indianapolis, Ind, USA, 2005.

[13] L. DeStefano, *Greenman'S Principles of Manual Medicine*, Wolters Kluwer, Lippincott, Williams & Wilkins, Philadelphia, Pa, USA, 2011.

[14] C. Reinschmidt and A. J. van den Bogert, "KineMat: A MATLAB Toolbox for Three-Dimensional Kinematic Analyses International Society of Biomechanics," 1997, http://www.isbweb.org/software/movanal/kinemat.

[15] M. Whittle and J. Walker, "The three dimensional measurement of head movement," in *Proceedings of the 8th International Symposium on the 3-D Analysis of Human Movement University of South Florida*, Tampa, Fla, USA, 2004.

[16] M. A. Williams, C. J. McCarthy, A. Chorti, M. W. Cooke, and S. Gates, "A systematic review of reliability and validity studies of methods for measuring active and passive cervical range of motion," *Journal of Manipulative and Physiological Therapeutics*, vol. 33, no. 2, pp. 138–155, 2010.

[17] M. T. Haneline and M. Young, "A review of intraexaminer and interexaminer reliability of static spinal palpation: a literature synthesis," *Journal of Manipulative and Physiological Therapeutics*, vol. 32, no. 5, pp. 379–386, 2009.

[18] J. J. Deeks, "Systematic reviews in health care: systematic reviews of evaluations of diagnostic and screening tests," *British Medical Journal*, vol. 323, no. 7305, pp. 157–162, 2001.

[19] D. Hollerwöger, "Methodological quality and outcomes of studies addressing manual cervical spine examinations: a review," *Manual Therapy*, vol. 11, no. 2, pp. 93–98, 2006.

[20] J. M. W. Ngan, D. H. K. Chow, and A. D. Holmes, "The kinematics and intra- and inter-therapist consistencies of lower cervical rotational manipulation," *Medical Engineering and Physics*, vol. 27, no. 5, pp. 395–401, 2005.

[21] L. Vogt, C. Segieth, W. Banzer, and H. Himmelreich, "Movement behaviour in patients with chronic neck pain," *Physiotherapy Research International*, vol. 12, no. 4, pp. 206–212, 2007.

[22] B. Cagnie, A. Cools, V. De Loose, D. Cambier, and L. Danneels, "Reliability and normative database of the zebris cervical range-of-motion system in healthy controls with preliminary validation in a group of patients with neck pain," *Journal of Manipulative and Physiological Therapeutics*, vol. 30, no. 6, pp. 450–455, 2007.

[23] J. Y. Maigne, F. Chantelot, and G. Chatellier, "Interexaminer agreement of clinical examination of the neck in manual medicine," *Annals of Physical and Rehabilitation Medicine*, vol. 52, no. 1, pp. 41–48, 2009.

[24] J. J. Pool, J. L. Hoving, H. C. De Vet, H. Van Mameren, and L. M. Bouter, "The interexaminer reproducibility of physical examination of the cervical spine," *Journal of Manipulative and Physiological Therapeutics*, vol. 27, no. 2, pp. 84–90, 2004.

[25] N. Strimpakos, "The assessment of the cervical spine. Part 1: range of motion and proprioception," *Journal of Bodywork and Movement Therapies*, vol. 15, no. 1, pp. 114–124, 2011.

[26] N. Strimpakos, V. Sakellari, G. Gioftsos et al., "Cervical spine ROM measurements: optimizing the testing protocol by using a 3D ultrasound-based motion analysis system," *Cephalalgia*, vol. 25, no. 12, pp. 1133–1145, 2005.

[27] P. Sjölander, P. Michaelson, S. Jaric, and M. Djupsjöbacka, "Sensorimotor disturbances in chronic neck pain-Range of motion, peak velocity, smoothness of movement, and repositioning acuity," *Manual Therapy*, vol. 13, no. 2, pp. 122–131, 2008.

[28] T. Prushansky, E. Pevzner, C. Gordon, and Z. Dvir, "Performance of cervical motion in chronic whiplash patients and healthy subjects: the case of atypical patients," *Spine*, vol. 31, no. 1, pp. 37–43, 2006.

[29] H. S. Bahat, P. L. Weiss, and Y. Laufer, "The effect of neck pain on cervical kinematics, as assessed in a virtual environment," *Archives of Physical Medicine and Rehabilitation*, vol. 91, no. 12, pp. 1884–1890, 2010.

[30] W. I. Najm, M. A. Seffinger, S. I. Mishra et al., "Content validity of manual spinal palpatory exams—a systematic review," *BMC Complementary and Alternative Medicine*, vol. 3, no. 1, pp. 1–4, 2003.

[31] C. A. Lantz, J. Chen, and D. Buch, "Clinical validity and stability of active and passive cervical range of motion with regard to total and unilateral uniplanar motion," *Spine*, vol. 24, no. 11, pp. 1082–1089, 1999.

[32] A. L. Morphett, C. M. Crawford, and D. Lee, "The use of electromagnetic tracking technology for measurement of passive cervical range of motion: a pilot study," *Journal of Manipulative and Physiological Therapeutics*, vol. 26, no. 3, pp. 152–159, 2003.

[33] T. R. Bush and J. Vorro, "Kinematic measures to objectify head and neck motions in palpatory diagnosis: a pilot study," *Journal of the American Osteopathic Association*, vol. 108, no. 2, pp. 55–62, 2008.

[34] Z. Dvir, T. Prushansky, and C. Peretz, "Maximal versus feigned active cervical motion in healthy patients: the coefficient of variation as an indicator for sincerity of effort," *Spine*, vol. 26, no. 15, pp. 1680–1688, 2001.

[35] Z. Dvir, N. Gal-Eshel, B. Shamir, E. Pevzner, C. Peretz, and N. Knoller, "Simulated pain and cervical motion in patients with chronic disorders of the cervical spine," *Pain Research and Management*, vol. 9, no. 3, pp. 131–136, 2004.

[36] P. T. Dall'Alba, M. M. Sterling, J. M. Treleaven, S. L. Edwards, and G. A. Jull, "Cervical range of motion discriminates between asymptomatic persons and those with whiplash," *Spine*, vol. 26, no. 19, pp. 2090–2094, 2001.

[37] H. Grip, G. Sundelin, B. Gerdle, and J. S. Karlsson, "Variations in the axis of motion during head repositioning—a comparison of subjects with whiplash-associated disorders or non-specific neck pain and healthy controls," *Clinical Biomechanics*, vol. 22, no. 8, pp. 865–873, 2007.

[38] T. R. Bush, J. Vorro, G. Alderink, S. Gorbis, M. Li, and S. Leitkam, "Relating a manual medicine diagnostic test of cervical motion function to specific three-dimensional kinematic variables," *International Journal of Osteopathic Medicine*, vol. 13, no. 2, pp. 48–55, 2010.

[39] P. H. Trott, M. J. Pearcy, S. A. Ruston, I. Fulton, and C. Brien, "Three-dimensional analysis of active cervical motion: the effect of age and gender," *Clinical Biomechanics*, vol. 11, no. 4, pp. 201–206, 1996.

[40] V. Feipel, P. Salvia, H. Klein, and M. Rooze, "Head repositioning accuracy in patients with whiplash-associated disorders," *Spine*, vol. 31, no. 2, pp. E51–E58, 2006.

[41] M. Alund, S. E. Larsson, and T. Lewin, "Work-related persistent neck impairment: a study on former steelworks grinders," *Ergonomics*, vol. 37, no. 7, pp. 1253–1260, 1994.

# Exploratory Study on the Methodology of Fast Imaging of Unilateral Stroke Lesions by Electrical Impedance Asymmetry in Human Heads

**Jieshi Ma, Canhua Xu, Meng Dai, Fusheng You, Xuetao Shi, Xiuzhen Dong, and Feng Fu**

*Department of Biomedical Engineering, Fourth Military Medical University, Xi'an 710032, China*

Correspondence should be addressed to Xiuzhen Dong; dongxiuzhen@fmmu.edu.cn and Feng Fu; fufeng@fmmu.edu.cn

Academic Editor: Francisco Javier Carod-Artal

Stroke has a high mortality and disability rate and should be rapidly diagnosed to improve prognosis. Diagnosing stroke is not a problem for hospitals with CT, MRI, and other imaging devices but is difficult for community hospitals without these devices. Based on the mechanism that the electrical impedance of the two hemispheres of a normal human head is basically symmetrical and a stroke can alter this symmetry, a fast electrical impedance imaging method called symmetrical electrical impedance tomography (SEIT) is proposed. In this technique, electrical impedance tomography (EIT) data measured from the undamaged craniocerebral hemisphere (CCH) is regarded as reference data for the remaining EIT data measured from the other CCH for difference imaging to identify the differences in resistivity distribution between the two CCHs. The results of SEIT imaging based on simulation data from the 2D human head finite element model and that from the physical phantom of human head verified this method in detection of unilateral stroke.

## 1. Introduction

Stroke refers to the rapid loss of brain function because of a disturbance in blood supply to the brain. Stroke can be classified into two major categories: ischemic and hemorrhagic. Related studies show that the disease is the second leading cause of death in the world [1]. The three-year cumulative incidence could reach 16.8 per 1000 [2]. Thirty-day mortality could reach 20% [3], and the disability rate is higher than 65% [4]. The timely detection and treatment of stroke reduces the risk of death of patients [5] and is important in improving the prognosis [6].

The devices utilized to diagnose stroke include CT and MRI, which are noninvasive means of detection involving images of high spatial resolution. However, these devices are bulky and expensive and are available only in well-equipped hospitals. Community hospitals generally do not have such devices. Hence, a portable technique for the detection of stroke is needed to preliminarily diagnose sudden stroke cases in community hospitals.

Biological tissues are characterized by electrical properties (conductivity and permittivity) that allow electrical current to flow in the presence of an electric field [7]. Electrical properties depend on the constituent elements and structure of tissue; therefore, each type of human tissue and body fluids is with specific conductivity and permittivity. Thus, measurements of electrical impedance can be used to differentiate between different types of tissues or to assess the state of tissue [8]. Changes in the composition and structure of a tissue modify its electrical properties and consequently change its electrical impedance. The occurrence of stroke can make such changes and modify the electrical impedance distribution in the human head.

Based on the fact that different biological tissues have different electrical properties, electrical impedance tomography (EIT), also known as bioimpedance tomography [9], applies a safe alternating current to the body surface, measures body surface potential, and reconstructs images of resistivity distribution or its change within the body through a certain image reconstruction algorithm. Considering its portability,

Exploratory Study on the Methodology of Fast Imaging of Unilateral Stroke Lesions by Electrical Impedance Asymmetry in Human Heads

141

noninvasiveness, and the absence of ionizing radiation, EIT has been studied as a monitoring and assessment tool for medical applications, such as detecting breast cancer [10], assessing abdominal hemorrhage [11], imaging lung ventilation [12], and assessing intracranial impedance variation [13, 14]. Thus, EIT has potential for detection of stroke [15].

EIT imaging approaches include time difference EIT (tdEIT) or dynamic EIT, static EIT, and frequency difference EIT (fdEIT) [13]. Difference imaging is conducted between two frames of EIT data measured at two time points in tdEIT. The distribution of the absolute resistivity value in the measured objects with one frame of EIT data is reconstructed in static EIT and difference imaging between two frames of EIT data measured at two frequencies is conducted in fdEIT. In theory, static EIT satisfies the requirements of reconstructing an image of a stroke lesion at a single time. However, the method has been limited by fundamental ill-posedness as well as technical difficulties due to a limited amount of measurements, unknown boundary geometry, and uncertainty in electrode positions, while systematic measurement artifacts and random noise are also the limitations of the method [16]. In principle, fdEIT also satisfies the requirements of reconstructing an image of a stroke lesion [17]. However, fdEIT requires high-performance hardware systems and imaging algorithms [16–19]. fdEIT is in the stage of system optimization and algorithm improvement and can only image objects, such as carrots and bananas, in physical phantoms without skull.

Through extensive research, our research group established a dynamic EIT platform for monitoring cerebral impedance and utilized it to dynamically monitor the conditionally controllable progress of stroke in animal models [20, 21]. Given that dynamic EIT requires time-referenced data, it is inapplicable to cases where a single image in time is required or when time-referenced data are unavailable. Hence, dynamic EIT cannot be utilized to conduct rapid single-time imaging of presented stroke lesions. A rapid electrical impedance imaging method with no high demands on imaging systems should be investigated to rapidly diagnose stroke patients. Accordingly, the objective of this study was to establish a new EIT method to determine whether there was a lesion in a subject's head and estimate approximately the location and the size of the lesion.

The median sagittal plane of the brain separates the brain into two basically symmetrical hemispheres. Published reports have demonstrated that a stroke lesion is usually located in one cerebral hemisphere [22]. No significant difference in impedance exists between the two cerebral hemispheres of normal humans, but stroke significantly increases the difference [23–25]. The bilateral impedance of normal human heads is almost symmetrical; however, a stroke can alter such symmetry. Thus, a stroke may be detectable by measuring the electrical impedance asymmetry in the head of a stroke patient.

Given that unilateral stroke lesions often generate significant symptoms on the other parts of the body [26], this can be utilized to determine which craniocerebral hemisphere (CCH) with the lesion. Effective measurements to determine which CCH was damaged by unilateral stroke lesion were also studied [27, 28]. Thus, the damaged and undamaged CCHs can be determined in practice.

An EIT method called symmetrical electrical impedance tomography (SEIT) was then proposed to image the asymmetry of the bilateral impedance of human heads to detect unilateral cerebral lesions. In this technique, electrical impedance tomography (EIT) data measured from the undamaged craniocerebral hemisphere (CCH) is regarded as reference data for the remaining EIT data measured from the other CCH for difference imaging to identify the differences in resistivity distribution between the two CCHs. The technical validation of this approach in the rapid detection of unilateral cerebral lesions was verified with finite element simulation and physical phantom experiments.

## 2. Materials and Methods

### 2.1. Evaluation of Electrical Impedance Asymmetry in the Human Head with EIT Data

*2.1.1. The Symmetrical Relationship of EIT Data Measured from the Two Hemispheres of the Human Head.* Craniocerebral EIT data refer to the boundary voltages (BVs) obtained by injecting the driving current to the human head by EIT. Differences in EIT data from the two CCHs can be utilized to assess the impedance asymmetry of the two CCHs caused by a unilateral lesion.

Considering that the EIT data-measuring pattern involves multiple measurements based on multiple polar drives [29], the BVs in one frame of EIT data acquired by a 16-electrode EIT system were marked as $U_{i,j}$, where the subscript $i$ is the drive number ranging from 1 to 16 and corresponding to the polar-drive electrode pairs (1, 9), (2, 10), (3, 11), ..., (16, 8); the subscript $j$ is the measurement number ranging from 1 to 16 and corresponding to the adjacent-measurement electrode pairs (1, 2), (2, 3), (3, 4), ..., (16, 1). The positions and corresponding numbers of the 16 electrodes are shown in Figure 1. When the driving and data-measuring electrodes have a common electrode, $U_{i,j}$ (e.g., $U_{1,1}$) is considered as invalid data for EIT imaging. Sixty-four $U_{i,j}$ were deemed invalid, and the remaining 192 $U_{i,j}$ were considered valid EIT data to be used in subsequent EIT image reconstruction.

The EIT data from the two hemispheres utilized to evaluate craniocerebral impedance asymmetry were transformed into symmetrical boundary voltage pairs (SBVPs) as follows. For one frame of EIT data, two BVs measured on two pairs of data-measuring electrodes symmetrical to the median sagittal plane of the brain were defined as SBVP. SBVP was generated by anterior-posterior drive and symmetrical drives. An anterior-posterior drive refers to injecting the driving current through electrode pair (1, 9) or (9, 1). When electrode pair (1, 9) carried the current, six groups of SBVP were generated (Figure 1(a)); thus, the two anterior-posterior drives can generate 12 groups of SBVP. In addition, a pair of symmetrical drives refers to two drives, where the line joining one pair of driving electrodes and the line joining the other are symmetrical to the line joining electrodes 1

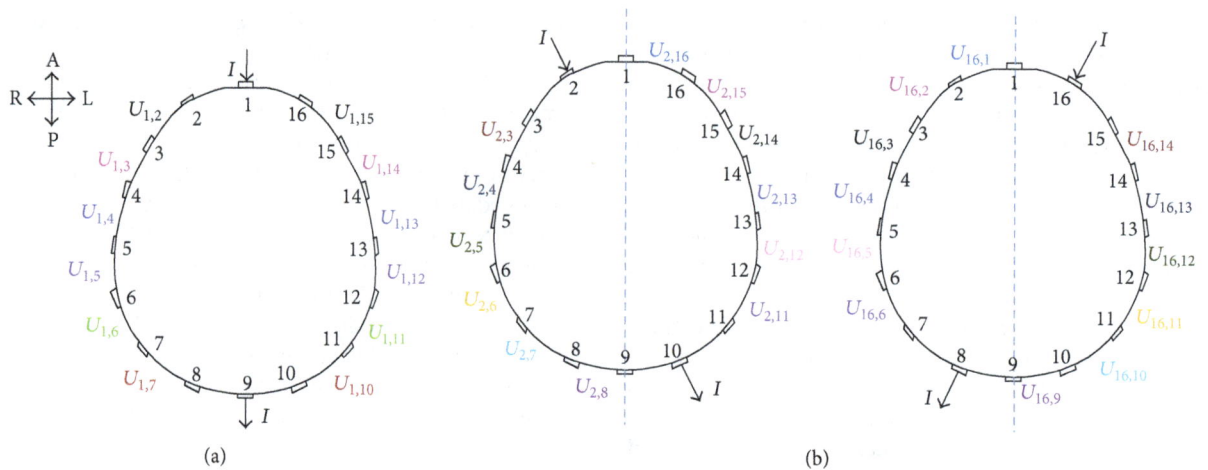

(a)                                                                                                          (b)

FIGURE 1: Illustrations of SBVP formation. (a) When the current was injected through electrode pair (1, 9), six groups of SBVP were formed by six BVs measured from the electrodes on the right hemisphere of the head and six other BVs from the left hemisphere. (b) The drives on electrode pair (2, 10) and (16, 8) constructed a pair of symmetrical drives. Six groups of SBVP were formed by six BVs measured from the right hemisphere during the drive on electrode pair (2, 10) and six BVs from the left hemisphere during the drive on electrode pair (16, 8). Similarly, another six groups of SBVP were formed by six BVs from the right hemisphere during the drive on electrode pair (16, 8) and six BVs from the left hemisphere during the drive on electrode pair (2, 10). This pair of symmetrical drives generated 12 groups of SBVP. Two BVs that formed one group of SBVP are marked with the same color (A: anterior; P: posterior; L: left; and R: right).

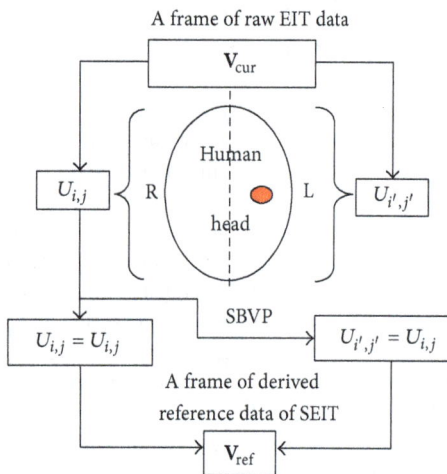

FIGURE 2: Construction of SEIT reference data. During construction, the raw EIT data measured from the undamaged CCH (e.g., the right CCH) was copied to the contralateral side according to the relationship of SBVP.

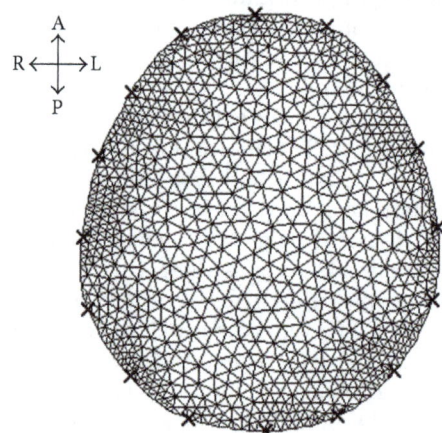

FIGURE 3: Image reconstruction model. The image reconstruction model consisted of 1804 triangular elements, 984 nodes, and 16 electrodes. The model was used for image reconstructions in all imaging experiments, including simulation and physical phantom experiments (A: anterior; P: posterior; L: left; and R: right).

and 9. Examples include the drives on electrode pair (2, 10) and (16, 8), drives on electrode pair (3, 11) and (15, 7), ..., drives on electrode pair (8, 16) and (10, 2). Seven pairs of symmetrical drives were obtained. In one pair of symmetrical drives, such as the drives on electrode pair (2, 10) and (16, 8), 12 groups of SBVP were generated (Figure 1(b)). Thus, seven pairs of symmetrical drives generated 84 groups of SBVP. Overall, valid data in one frame of EIT data can be divided into 96 SBVP groups, containing 192 BVs. Among them, 96 BVs $U_{i,j}$ ($j = 1, \ldots, 8$; $i = 1, \ldots, 16$) measured from the right hemisphere of the head (Tables 2, 3, 4, 5, 6, 7, and 9,

first row) had a relationship with the remaining 96 BVs $U_{i',j'}$ ($j' = 16, 15, 14, 13, 12, 11, 10, 9$; $i' = 1, 16, 15, 14, 13, 12, 11, 10, 9, 8, 7, 6, 5, 4, 3, 2$) measured from the left hemisphere of the head (Tables 2–9, second row). The numbers of SBVPs are shown in the third row of Tables 2–9.

2.1.2. *Evaluation of the Electrical Impedance Asymmetry.* As mentioned above, 192 BVs in one frame of EIT data were divided into 96 groups of SBVP. Each group contained two BVs measured from two symmetrical data-measuring electrode pairs. Their numerical difference

Exploratory Study on the Methodology of Fast Imaging of Unilateral Stroke Lesions by Electrical Impedance Asymmetry in Human Heads

143

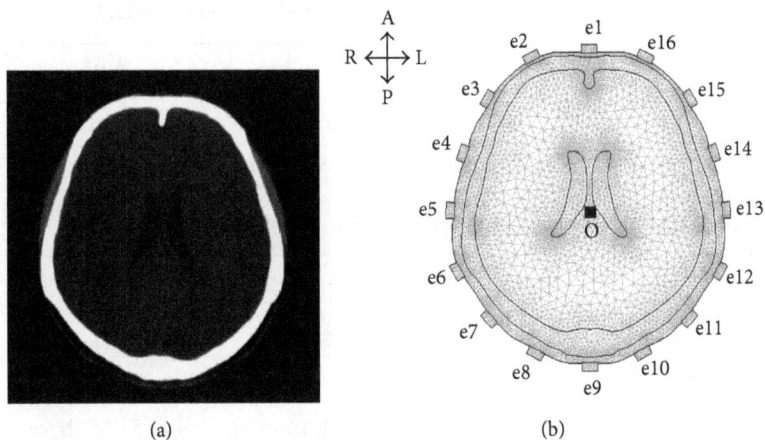

FIGURE 4: CT image and 2D finite element model of human head used for simulation experiments. (a) A head CT image of a healthy volunteer was used to construct a finite element model. (b) A finite element model (FEM) with an ideally symmetrical structure was constructed according to the right boundary of each layer of head tissues in the head CT image. The 2D head model consisted of 17659 triangular elements, 9200 nodes, and 16 electrodes (A: anterior; P: posterior; L: left; and R: right).

TABLE 1: RGB color mapping for the pixel $(x, y)$ based on the mapping index $g(x, y)$.

| $g(x, y)$ | R | G | B |
|---|---|---|---|
| $0.75 < g(x, y) < 1$ | 0 | $1022 - 1020 * g(x, y)$ | 255 |
| $0.5 < g(x, y) < 0.75$ | 0 | 255 | $1020 * g(x, y) - 510$ |
| $0.25 < g(x, y) < 0.5$ | $510 - 1020 * g(x, y)$ | 255 | 0 |
| $0 < g(x, y) < 0.25$ | 255 | $1020 * g(x, y)$ | 0 |

TABLE 2: SBVPs from electrode pair $(1, 2)$ and electrode pair $(1, 16)$.

| $U_{3,1}$ | $U_{4,1}$ | $U_{5,1}$ | $U_{6,1}$ | $U_{7,1}$ | $U_{8,1}$ | $U_{11,1}$ | $U_{12,1}$ | $U_{13,1}$ | $U_{14,1}$ | $U_{15,1}$ | $U_{16,1}$ |
|---|---|---|---|---|---|---|---|---|---|---|---|
| $U_{15,16}$ | $U_{14,16}$ | $U_{13,16}$ | $U_{12,16}$ | $U_{11,16}$ | $U_{10,16}$ | $U_{7,16}$ | $U_{6,16}$ | $U_{5,16}$ | $U_{4,6}$ | $U_{3,16}$ | $U_{2,16}$ |
| 1 | 2 | 3 | 4 | 5 | 6 | 7 | 8 | 9 | 10 | 11 | 12 |

TABLE 3: SBVPs from electrode pair $(2, 3)$ and electrode pair $(16, 15)$.

| $U_{1,2}$ | $U_{4,2}$ | $U_{5,2}$ | $U_{6,2}$ | $U_{7,2}$ | $U_{8,2}$ | $U_{9,2}$ | $U_{12,2}$ | $U_{13,2}$ | $U_{14,2}$ | $U_{15,2}$ | $U_{16,2}$ |
|---|---|---|---|---|---|---|---|---|---|---|---|
| $U_{1,15}$ | $U_{14,15}$ | $U_{13,15}$ | $U_{12,15}$ | $U_{11,15}$ | $U_{10,15}$ | $U_{9,15}$ | $U_{6,15}$ | $U_{5,15}$ | $U_{4,15}$ | $U_{3,15}$ | $U_{2,15}$ |
| 13 | 14 | 15 | 16 | 17 | 18 | 19 | 20 | 21 | 22 | 23 | 24 |

TABLE 4: SBVPs from electrode pair $(3, 4)$ and electrode pair $(15, 14)$.

| $U_{1,3}$ | $U_{2,3}$ | $U_{5,3}$ | $U_{6,3}$ | $U_{7,3}$ | $U_{8,3}$ | $U_{9,3}$ | $U_{10,3}$ | $U_{13,3}$ | $U_{14,3}$ | $U_{15,3}$ | $U_{16,3}$ |
|---|---|---|---|---|---|---|---|---|---|---|---|
| $U_{1,14}$ | $U_{16,14}$ | $U_{13,14}$ | $U_{12,14}$ | $U_{11,14}$ | $U_{10,14}$ | $U_{9,14}$ | $U_{8,14}$ | $U_{5,14}$ | $U_{4,14}$ | $U_{3,14}$ | $U_{2,14}$ |
| 25 | 26 | 27 | 28 | 29 | 30 | 31 | 32 | 33 | 34 | 35 | 36 |

TABLE 5: SBVPs from electrode pair $(4, 5)$ and electrode pair $(14, 13)$.

| $U_{1,4}$ | $U_{2,4}$ | $U_{3,4}$ | $U_{6,4}$ | $U_{7,4}$ | $U_{8,4}$ | $U_{9,4}$ | $U_{10,4}$ | $U_{11,4}$ | $U_{14,4}$ | $U_{15,4}$ | $U_{16,4}$ |
|---|---|---|---|---|---|---|---|---|---|---|---|
| $U_{1,13}$ | $U_{16,13}$ | $U_{15,13}$ | $U_{12,13}$ | $U_{11,13}$ | $U_{10,13}$ | $U_{9,13}$ | $U_{8,13}$ | $U_{7,13}$ | $U_{4,13}$ | $U_{3,13}$ | $U_{2,13}$ |
| 37 | 38 | 39 | 40 | 41 | 42 | 43 | 44 | 45 | 46 | 47 | 48 |

TABLE 6: SBVPs from electrode pair $(5, 6)$ and electrode pair $(13, 12)$.

| $U_{1,5}$ | $U_{2,5}$ | $U_{3,5}$ | $U_{4,5}$ | $U_{7,5}$ | $U_{8,5}$ | $U_{9,5}$ | $U_{10,5}$ | $U_{11,5}$ | $U_{12,5}$ | $U_{15,5}$ | $U_{16,5}$ |
|---|---|---|---|---|---|---|---|---|---|---|---|
| $U_{1,12}$ | $U_{16,12}$ | $U_{15,12}$ | $U_{14,12}$ | $U_{11,12}$ | $U_{10,12}$ | $U_{9,12}$ | $U_{8,12}$ | $U_{7,12}$ | $U_{6,12}$ | $U_{3,12}$ | $U_{2,12}$ |
| 49 | 50 | 51 | 52 | 53 | 54 | 55 | 56 | 57 | 58 | 59 | 60 |

TABLE 7: SBVPs from electrode pair (6, 7) and electrode pair (12, 11).

| $U_{1,6}$ | $U_{2,6}$ | $U_{3,6}$ | $U_{4,6}$ | $U_{5,6}$ | $U_{8,6}$ | $U_{9,6}$ | $U_{10,6}$ | $U_{11,6}$ | $U_{12,6}$ | $U_{13,6}$ | $U_{16,6}$ |
|---|---|---|---|---|---|---|---|---|---|---|---|
| $U_{1,11}$ | $U_{16,11}$ | $U_{15,11}$ | $U_{14,11}$ | $U_{13,11}$ | $U_{10,11}$ | $U_{9,11}$ | $U_{8,11}$ | $U_{7,11}$ | $U_{6,11}$ | $U_{5,11}$ | $U_{2,11}$ |
| 61 | 62 | 63 | 64 | 65 | 66 | 67 | 68 | 69 | 70 | 71 | 72 |

TABLE 8: SBVPs from electrode pair (7, 8) and electrode pair (11, 10).

| $U_{1,7}$ | $U_{2,7}$ | $U_{3,7}$ | $U_{4,7}$ | $U_{5,7}$ | $U_{6,7}$ | $U_{9,7}$ | $U_{10,7}$ | $U_{11,7}$ | $U_{12,7}$ | $U_{13,7}$ | $U_{14,7}$ |
|---|---|---|---|---|---|---|---|---|---|---|---|
| $U_{1,10}$ | $U_{16,10}$ | $U_{15,10}$ | $U_{14,10}$ | $U_{13,10}$ | $U_{12,10}$ | $U_{9,10}$ | $U_{8,10}$ | $U_{7,10}$ | $U_{6,10}$ | $U_{5,10}$ | $U_{4,10}$ |
| 73 | 74 | 75 | 76 | 77 | 78 | 79 | 80 | 81 | 82 | 83 | 84 |

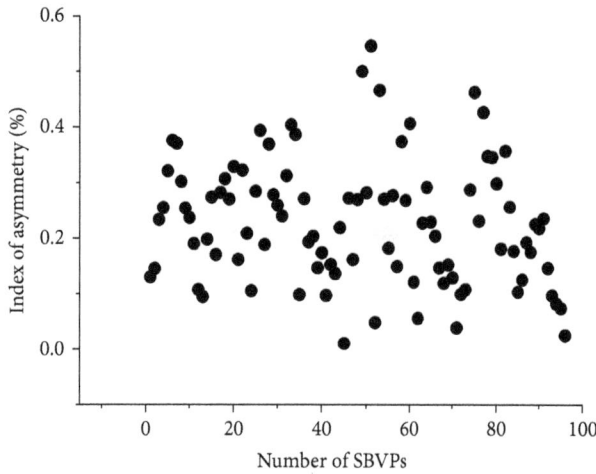

FIGURE 5: Index of asymmetry (IA) of the 2D FEM of human head.

reflected the impedance asymmetrical degree of symmetrical regions in the two CCHs. The difference was defined as the index of asymmetry (IA) and was calculated by

$$\text{IA} = 100 \times \frac{|L - R|}{(L + R)/2}\%. \tag{1}$$

In (1), $L$ and $R$ are the two BVs in one group of SBVP and are measured from the left and right CCHs, respectively.

When a small difference exists in the two BVs in one group of SBVP, IA is close to 0 and the asymmetry level of these two impedance data is relatively low and vice versa. The maximum value of IA ($\text{IA}_{\text{max}}$) corresponds to the highest level of impedance asymmetry of the two hemispheres of the head and decides the upper limit of the range of IA. Therefore, $\text{IA}_{\text{max}}$ was utilized as the index to evaluate the asymmetry of EIT impedance data measured from the two hemispheres of the head.

2.2. *Symmetrical EIT (SEIT)*. SEIT is a difference EIT method, within which one frame of EIT reference data $\mathbf{V}_{\text{ref}}$ was first constructed based on one frame of EIT raw data $\mathbf{V}_{\text{cur}}$ and changes of $\mathbf{V}_{\text{cur}}$ with respect to $\mathbf{V}_{\text{ref}}$ were then identified. Therefore, SEIT mainly involves two main procedures: constructing SEIT reference data and reconstructing the SEIT image.

*2.2.1. Construction of SEIT Reference Data.* A frame of EIT raw data can be divided into two parts measured from the left and right CCHs. Half of a frame of EIT raw data measured from the undamaged CCH can be utilized to construct SEIT reference data. The undamaged CCH is referred to as the "reference hemisphere."

In this research, the CCH without anomaly was regarded as the reference hemisphere. SEIT reference data were constructed as follows. Firstly, 192 valid BVs of a frame of EIT raw data $\mathbf{V}_{\text{cur}}$ were divided into two parts: 96 BVs $U_{i,j}$ ($j = 1, \ldots, 8$; $i = 1, \ldots, 16$) were measured from the right (undamaged) CCH and the other 96 BVs $U_{i',j'}$ ($j' = 16, 15, 14, 13, 12, 11, 10, 9$; $i' = 1, 16, 15, 14, 13, 12, 11, 10, 9, 8, 7, 6, 5, 4, 3, 2$) were measured from the left (damaged) CCH. This procedure revealed the relationship of SBVP between $U_{i',j'}$ and $U_{i,j}$. Secondly, a frame of SEIT reference data $\mathbf{V}_{\text{ref}}$ with all zero values was constructed, and the 96 BVs $U_{i,j}$ measured from the undamaged CCH were duplicated and stored into the same positions of $\mathbf{V}_{\text{ref}}$. These BVs were also stored into the positions of the 96 BVs $U_{i',j'}$ of $\mathbf{V}_{\text{ref}}$ according to the relationship of SBVP (Tables 2–9). A frame of SEIT reference data $\mathbf{V}_{\text{ref}}$ containing 192 valid BVs was then constructed. The construction of SEIT reference data is shown in Figure 2.

With the undamaged CCH as the reference hemisphere, the object of SEIT image reconstruction is to solve the resistivity distribution change in the damaged CCH with respect to that in the reference hemisphere. Thus, a stroke lesion was expected to be observed in the damaged side.

*2.2.2. SEIT Image Reconstruction.* SEIT image reconstruction is similar to general difference EIT imaging. When a frame of SEIT reference data $\mathbf{V}_{\text{ref}}$ was constructed, the damped least squares image reconstruction algorithm [30] was utilized for difference EIT image reconstruction. According to the changes in EIT boundary voltages ($\Delta\mathbf{V}$), difference EIT image reconstruction solved the changes in the internal resistivity of the measured subject with

$$\Delta\boldsymbol{\rho} = \mathbf{B}\Delta\mathbf{V}. \tag{2}$$

In (2), matrix $\mathbf{B}$ is the image reconstruction matrix calculated from sensitivity coefficient matrix $\mathbf{S}$ [30]. Matrix $\mathbf{S}$ is the linearized sensitivity matrix, and its elements reflect the relationship between the resistivity changes in the finite elements of the imaging region and the changes in the EIT

Exploratory Study on the Methodology of Fast Imaging of Unilateral Stroke Lesions by Electrical Impedance Asymmetry in Human Heads

145

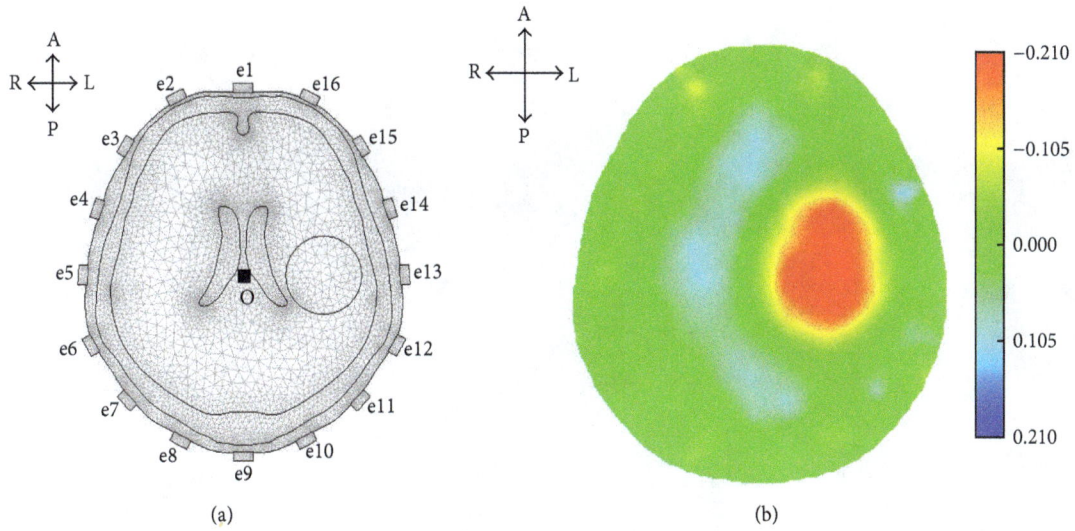

FIGURE 6: SEIT imaging based on simulated EIT data. (a) A simulated stroke lesion was set on the left side of the model. (b) The reconstructed SEIT image reflected the simulated lesion (A: anterior; P: posterior; L: left; and R: right).

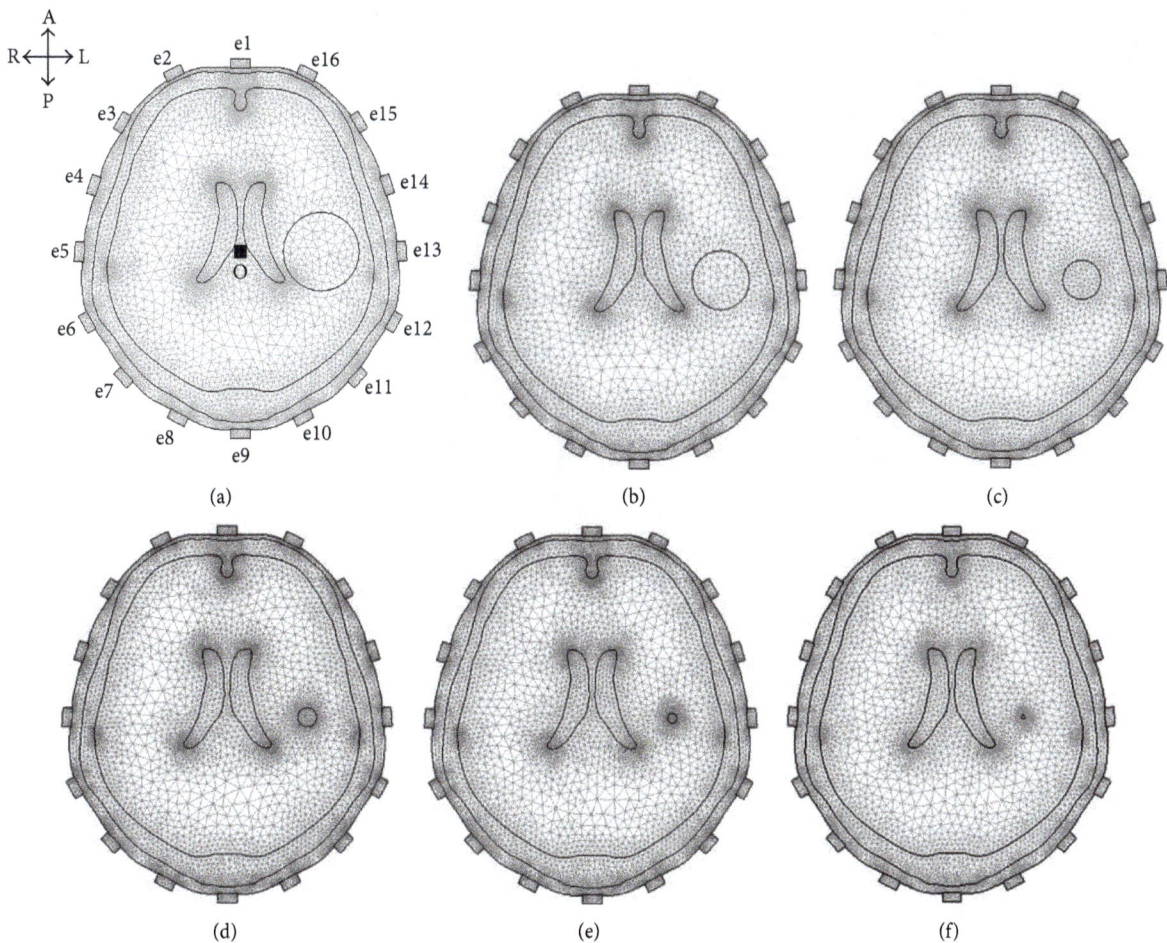

FIGURE 7: Simulated stroke lesions of different sizes.The radius of the object was set to 2.00, 1.50, 1.00, 0.50, 0.25, and 0.10 cm (from (a) to (f)) (A: anterior; P: posterior; L: left; and R: right). The conductivity of the object was set to 0.65 S/m in the case of simulated hemorrhagic stroke and 0.13 S/m in the case of simulated ischemic stroke.

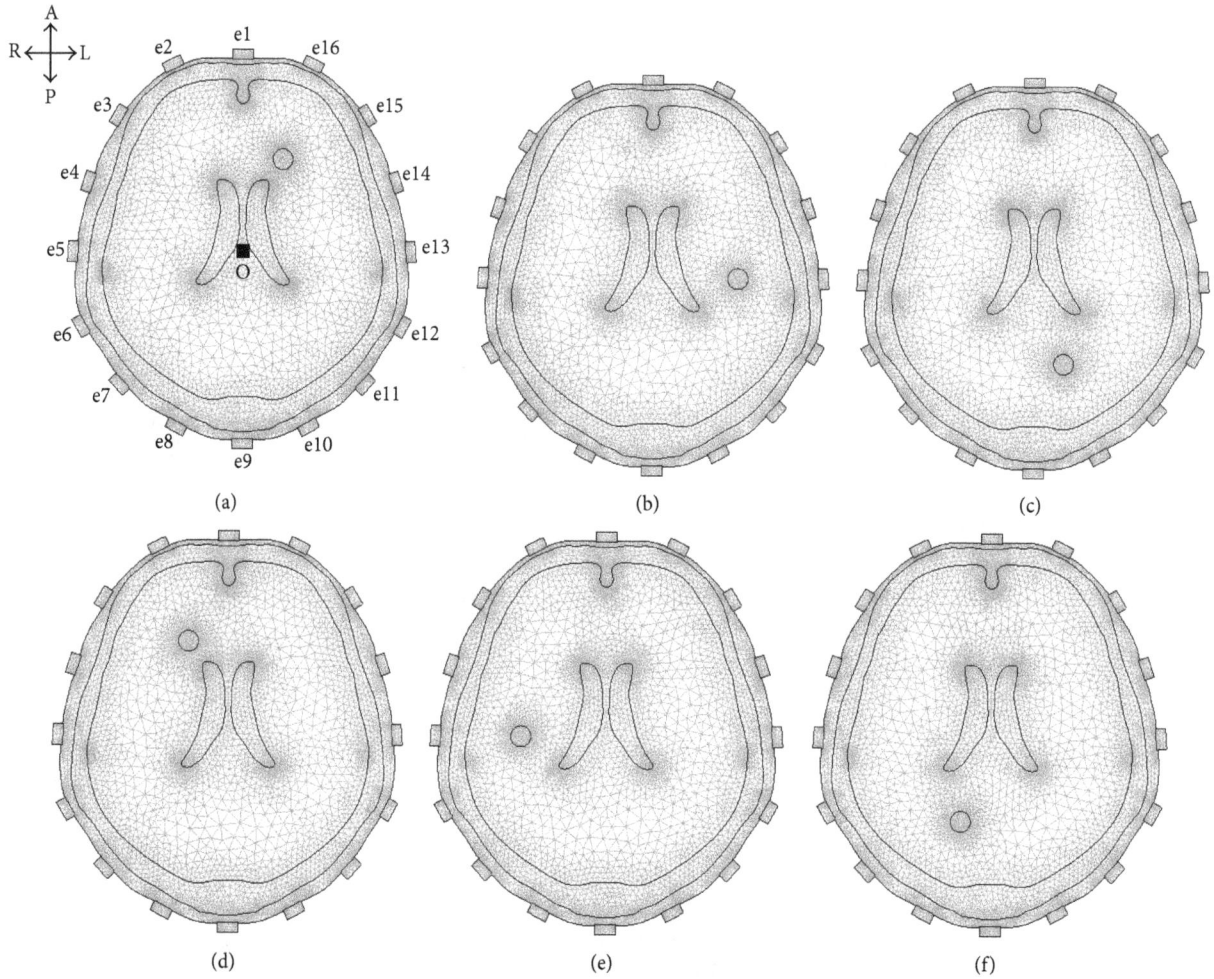

FIGURE 8: Different locations of simulated stroke lesion set in the model. The radius of the simulated lesion was 0.50 cm. The conductivity of the object was set to 0.65 S/m in the case of simulated hemorrhagic stroke and 0.13 S/m in the case of simulated ischemic stroke (A: anterior; P: posterior; L: left; and R: right).

TABLE 9: SBVPs from electrode pair (8, 9) and electrode pair (10, 9).

| $U_{2,8}$ | $U_{3,8}$ | $U_{4,8}$ | $U_{5,8}$ | $U_{6,8}$ | $U_{7,8}$ | $U_{10,8}$ | $U_{11,8}$ | $U_{12,8}$ | $U_{13,8}$ | $U_{14,8}$ | $U_{15,8}$ |
|-----------|-----------|-----------|-----------|-----------|-----------|------------|------------|------------|------------|------------|------------|
| $U_{16,9}$ | $U_{15,9}$ | $U_{14,9}$ | $U_{13,9}$ | $U_{12,9}$ | $U_{11,9}$ | $U_{8,9}$ | $U_{7,9}$ | $U_{6,9}$ | $U_{5,9}$ | $U_{4,9}$ | $U_{3,9}$ |
| 85 | 86 | 87 | 88 | 89 | 90 | 91 | 92 | 93 | 94 | 95 | 96 |

boundary voltages. Matrix **S** was acquired by solving EIT forward problems [30] with an image reconstruction model (Figure 3). The changes in EIT boundary voltages were calculated according to

$$\Delta \mathbf{V} = \frac{\mathbf{V}_{cur} - \mathbf{V}_{ref}}{\mathbf{V}_{ref}}. \tag{3}$$

During image reconstruction, the image reconstruction software directly employed the reconstruction model and image reconstruction matrix **B** to calculate (2) and to visualize the reconstructed image.

Each pixel in the reconstructed image was with a single value of relative resistivity change ($\Delta \rho$) in arbitrary units (AU), and the values were mapped to a colorbar showing the change of resistivity distribution. EIT image was displayed using a false color mapping, and the mapping index $g(x, y)$ of the pixel $(x, y)$ was calculated according to the following [31]:

$$g(x, y) = \frac{\Delta \rho(x, y) + \Delta \rho_{max}}{2 \Delta \rho_{max}}, \tag{4}$$

in (4) $\Delta \rho(x, y)$ is the resistivity change at the pixel coordinates $(x, y)$ in the SEIT image. $\Delta \rho_{max}$ is the maximum absolute value of $\Delta \rho(x, y)$. After the mapping index $g(x, y)$ was calculated, the color of the pixel $(x, y)$ was determined according to Table 1 [31].

In the color map, red indicated a decrease in resistivity distribution while blue indicated the opposite change.

Exploratory Study on the Methodology of Fast Imaging of Unilateral Stroke Lesions by Electrical Impedance Asymmetry in Human Heads

147

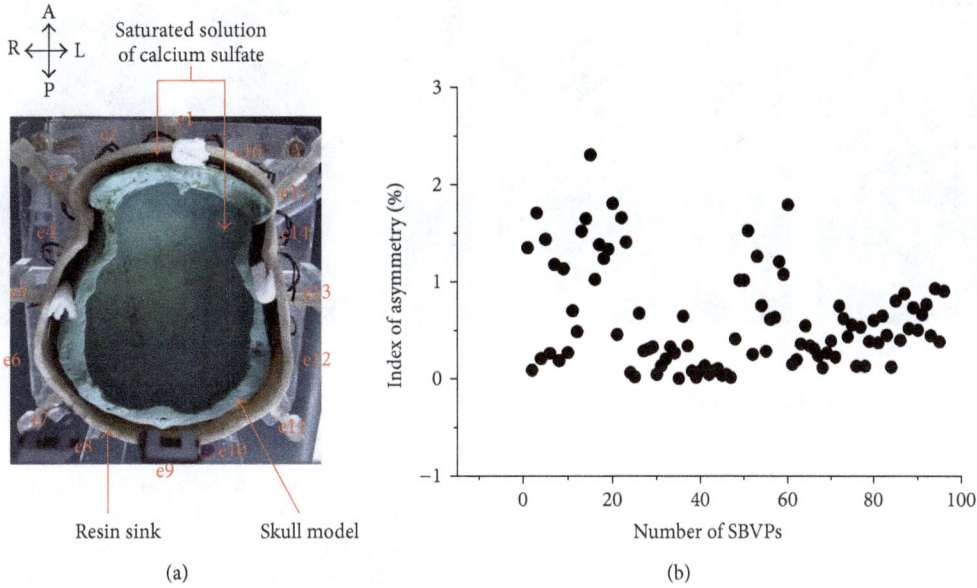

FIGURE 9: Physical phantom experiments: (a) physical phantom of realistic human head (A: anterior; P: posterior; L: left; and R: right); (b) index of asymmetry (IA) of the physical phantom.

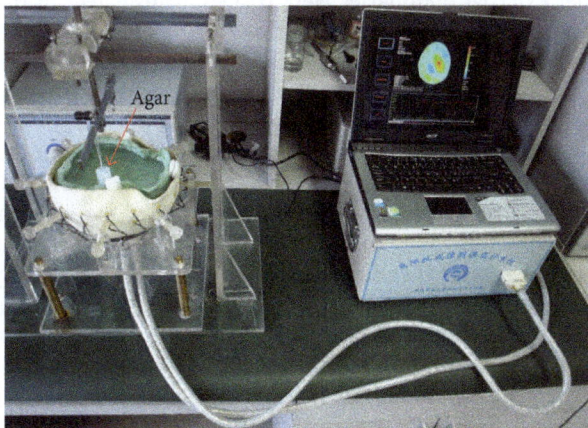

FIGURE 10: EIT data collection on the physical phantom with an agar cylinder.

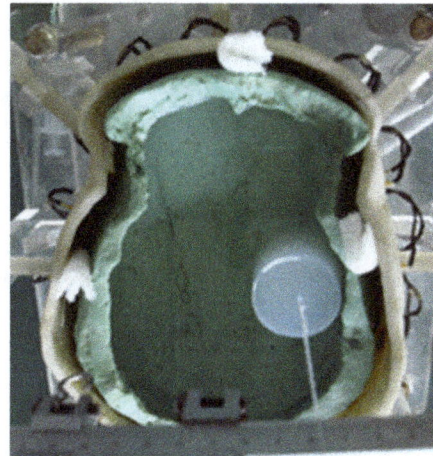

FIGURE 11: Location of the agar cylinders in the phantom: The initial radius of those cylinders was 2 cm.

*2.2.3. Analysis of Reconstructed SEIT Images.* To quantitatively analyze the pixel intensities in the abnormal resistivity distribution region in the reconstructed image, the mean abnormal resistivity value (MARV), which is the absolute value of the average of abnormal resistivity for all of the pixels in the region of interest (ROI), is calculated as

$$\text{MARV} = \left| \frac{1}{N} \sum \Delta\rho\left(x, y\right) \right| \quad xy \in \text{ROI}, \tag{5}$$

where $\Delta\rho(x, y)$ is the resistivity change at the pixel coordinates $(x, y)$ in the SEIT image and $N$ is the total number of the pixels within ROI, which is determined by a preset threshold defined by

$$\frac{\Delta\rho_{\max} - \left|\Delta\rho\left(x, y\right)\right|}{\Delta\rho_{\max}} \leq t. \tag{6}$$

In the above formula, $\Delta\rho_{\max}$ is the maximum absolute value of $\Delta\rho(x, y)$ and $t$ is the threshold parameter. Previous studies have indicated that 0.2 is an appropriate value for $t$ [20].

### 2.3. SEIT Imaging Based on Simulation Data

*2.3.1. Finite Element Modeling and Simulation of EIT Electric Field.* The 2D EIT electrodes are lying on an axial plane approximately 3 cm above inion of the human head; therefore, a head CT image (Figure 4(a)) of a healthy volunteer was utilized to construct a 2D head model. We duplicated the right boundary of each layer of head tissues in the head CT image and mirrored it to the left side to construct a finite element model for the purposes of studying

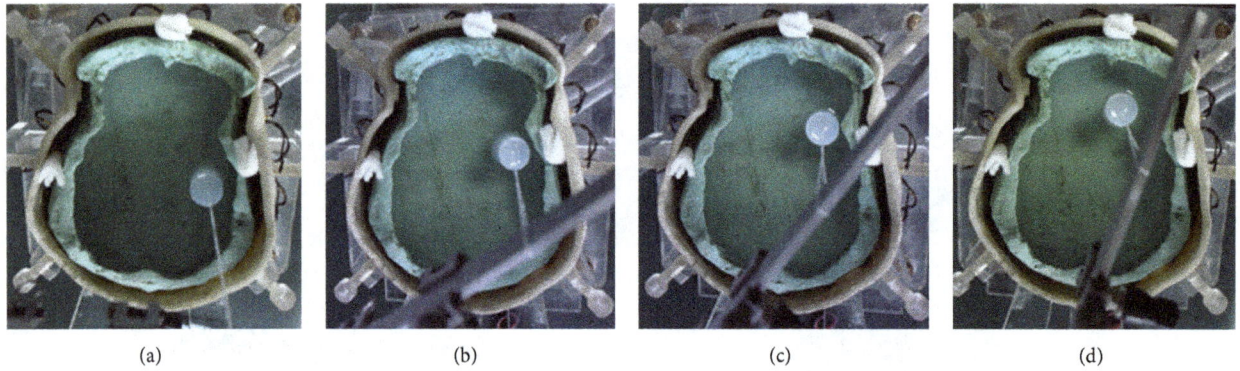

(a)  (b)  (c)  (d)

FIGURE 12: Four locations of the agar cylinder at the anterior and left part of the phantom.

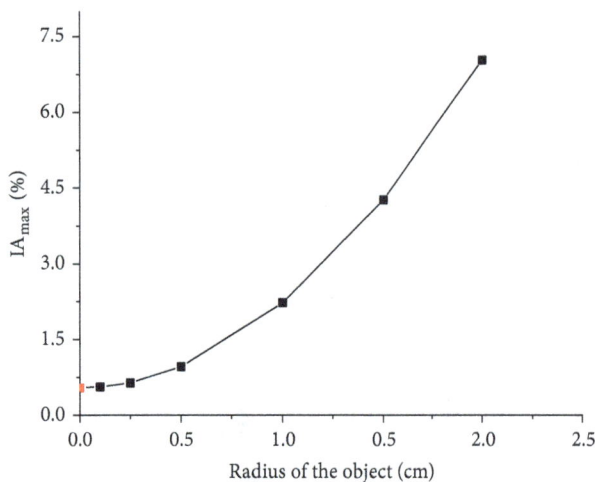

FIGURE 13: $IA_{max}$ of EIT data from the two hemispheres of the model with and without a hemorrhagic lesion. Black points represent $IA_{max}$ with a simulated hemorrhagic stroke lesion, while the red point represents $IA_{max}$ without a simulated lesion.

the feasibility of reconstructing the stroke lesion by SEIT. According to finite element modeling [32], a 2D human head finite element model (FEM) with ideally bilateral symmetry was established with COMSOL Multiphysics 3.5a (COMSOL, Inc., Stockholm, Sweden) (Figure 4(b)). The model contained the layers of scalp, skull, brain tissue, and cerebrospinal fluid from the outside to the inside. The conductivity of the layers was set as follows: 0.4400 S/m for the scalp [33], 0.0126 S/m for the skull (the conductivity of standard trilayer skull in the literature [34]), 0.2500 S/m for brain tissues [35], and 1.2500 S/m for cerebrospinal fluid (CSF) [36]. Sixteen rectangles with a width of 1 cm and height of 0.5 cm were set on the outermost layer of the scalp. The material property was set as brass, representing 16 EIT electrodes (Electrodes 1 to 16). The centers of these electrodes were set as e1 to e16. Electrodes 1 and 9 were at frontal and occipital poles of the head, respectively. The midpoint of Lines e1–e9 (Point O) was considered the center of the model. Electrodes 5 and 13 were on the horizontal line through Point O. Then, the distance between electrodes 1 and 5 on the scalp was measured, and the connective line between the two electrodes

was equally divided into four parts to determine the locations of electrodes 2, 3, and 4. Similarly, the locations of the remaining electrodes were determined.

In one polar drive, the used boundary conditions were as follows: (1) inward current flow (current density = 15.92 A/m$^2$) was applied to the outer boundary of electrode 1 to simulate the EIT driving current of 1250 $\mu$A, (2) ground was applied to the outer boundary of electrode 9, and (3) all other external boundaries were treated as insulated. In this way, one polar drive simulation was completed. Thus, the electric field formed by the current applied to the head by EIT was simulated. The potential on each electrode was then calculated by solving the Laplace equation, $\nabla \cdot (\sigma \nabla V) = 0$ ($\sigma$, conductivity; $V$, potential), with the stationary solver of a direct solving method, UMFPACK, in the COMSOL AC/DC Module. The absolute potential differences of the adjacent electrodes were EIT boundary voltages. The remaining 15 polar drives were then simulated according to the EIT driving order. A frame of simulated EIT raw data was obtained after 16 drives. A frame of simulated EIT data was calculated when there was no object in the model, which was utilized to calculate the index of asymmetry (IA) to evaluate the electrical impedance asymmetry of the 2D FEM of human head (Figure 5). The maximum value of IA ($IA_{max}$) was 0.55%.

*2.3.2. The Method of SEIT Imaging Based on Simulation Data.* A simulated stroke lesion was set in the model. For example, a circular region was set at the midpoint of line O-e13 of the model, with a conductivity of 0.65 S/m [36] and a radius of 2 cm, to simulate a hemorrhagic stroke lesion (Figure 6(a)); one frame of EIT raw data was then calculated. With half of the frame of EIT raw data measured from the undamaged CCH (the right CCH), one frame of SEIT reference data was constructed. Finally, difference imaging was conducted with EIT raw data and SEIT reference data.

Given the simulated lesion, the maximum absolute value of $\Delta \rho$ in the image was about 0.210. Thus, $\Delta \rho_{max}$ in (4) was set as 0.210 to obtain an SEIT image as shown in Figure 6(b). Abnormal resistivity distribution was reconstructed in the SEIT image.

Simulated stroke lesions of different sizes were firstly set in the middle of one hemisphere of the FEM model and the minimum area of the stroke lesion detectable by SEIT was

Exploratory Study on the Methodology of Fast Imaging of Unilateral Stroke Lesions by Electrical Impedance Asymmetry in Human Heads

149

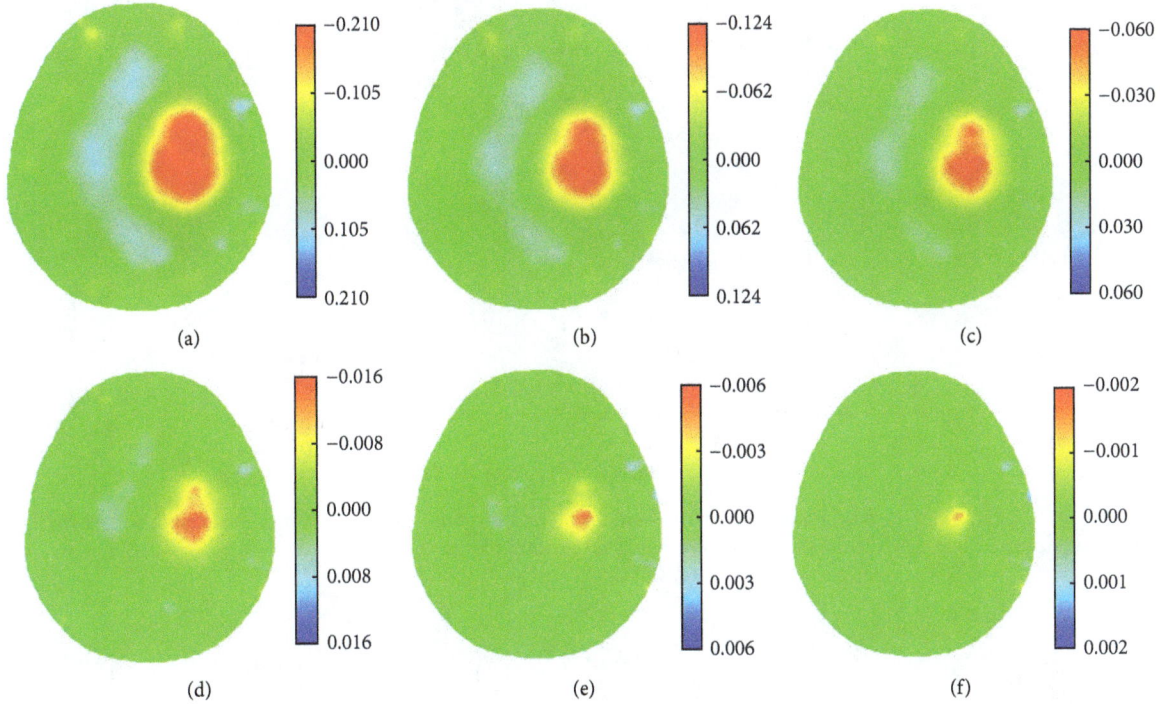

FIGURE 14: SEIT reconstructions of the simulated hemorrhagic stroke lesions of different sizes. Images (a) to (f) are SEIT reconstructions corresponding to the simulated hemorrhagic stroke lesion with a radius of 2.00, 1.50, 1.00, 0.50, 0.25, and 0.10 cm, respectively.

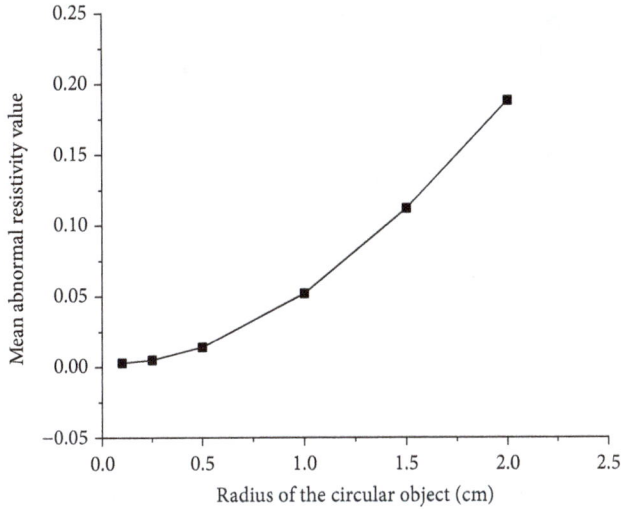

FIGURE 15: MARV of ROI in the SEIT reconstructions of the simulated hemorrhagic stroke lesions of different sizes: 2.00, 1.50, 1.00, 0.50, 0.25, and 0.10 cm.

identified. We then verified whether SEIT could reconstruct the stroke lesion in the different lobes of the head model with the identified minimum area. The conductivity of the circular region was set as 0.65 S/m in hemorrhagic stroke lesion modeling; the conductivity of ischemic stroke lesion was set as 0.13 S/m [37] in ischemic stroke lesion modeling.

*2.3.3. Simulation-Based SEIT Imaging of Stroke Lesions of Different Sizes.* We successively set a circular region at the mid-

point of line O-e13 on the left of FEM, with the radii of 2.00, 1.50, 1.00, 0.50, 0.25, and 0.10 cm, to simulate a hemorrhagic or ischemic stroke lesion of different sizes. The locations and sizes of the simulated stroke lesions are shown in Figure 7. One frame of EIT raw data was obtained in every lesion setting. The corresponding $IA_{max}$ of each frame of EIT data was calculated, and SEIT imaging was conducted. The degree of abnormal resistivity distribution resulted from the object was quantitatively evaluated by calculating MARV of ROI in SEIT images.

*2.3.4. Simulation-Based SEIT Imaging of the Stroke Lesion at Different Locations.* After the smallest area of stroke lesions detectable by SEIT was identified, we successively set a circular region with the minimal area at the midpoint of lines O-e16, O-e13, and O-e10 on the left of FEM to simulate a hemorrhagic or ischemic stroke lesion in the left frontal lobe, left temporal lobe, and left occipital lobe (Figures 8(a)–8(c)). We also set the same-sized circular region at the midpoint of lines O-e2, O-e5, and O-e8 on the right of FEM to simulate a hemorrhagic or ischemic stroke lesion in the right frontal lobe, right temporal lobe, and right occipital lobe (Figures 8(d)–8(f)). A frame of EIT raw data was obtained after each lesion was set. We also calculated the corresponding $IA_{max}$ of each frame of EIT data and conducted SEIT imaging.

*2.4. Experiments on SEIT Imaging of the Physical Phantom*

*2.4.1. Experiment Setups.* An EIT system named FMEIT-5 [38] was used to measure EIT data. The working frequency

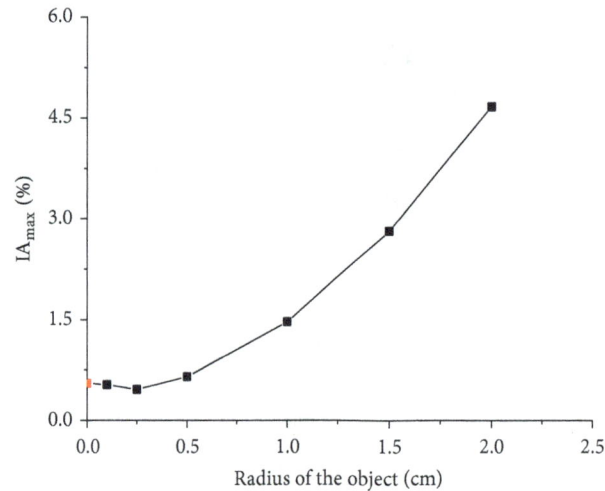

FIGURE 16: $IA_{max}$ of EIT data from the two hemispheres of the model with and without an ischemic lesion. Black points represent $IA_{max}$ with a simulated ischemic stroke lesion, while the red point represents $IA_{max}$ without a simulated lesion.

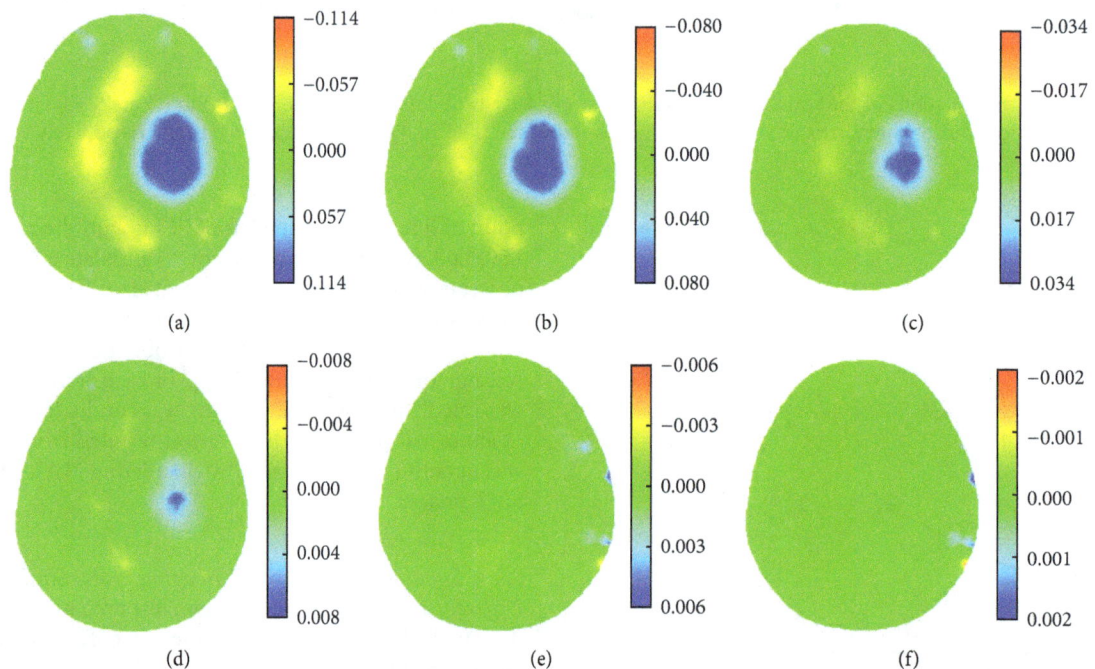

FIGURE 17: SEIT reconstructions of the simulated ischemic stroke lesions of different sizes. Images (a) to (f) are SEIT reconstructions corresponding to the simulated ischemic stroke lesion with a radius of 2.00, 1.50, 1.00, 0.50, 0.25, and 0.10 cm, respectively.

of this system is from 1 kHz to 190 kHz with measurement accuracy of ±0.01% and common mode rejection ratio over 80 dB. The system is serial in data acquisition and there are driven screens for the leads connecting the electrodes. The drive current was 1250 $\mu$A at 50 kHz. The experiments were conducted with a physical phantom of realistic human head developed by our research group [39], as shown in Figure 9(a). There were three parts, which were a white resin sink, a skull model made of plaster with conductivity of 0.013 S/m, and saturated solution of calcium sulfate with conductivity of 0.2 S/m. The sink and the skull model were

made in strict accordance with 3D reconstruction of human head CT, which should guarantee the maximum conformity to real human head in terms of structure and resistivity distribution. Before the experiments, the skull was put on the base of the sink and saturated solution of calcium sulfate was injected into the sink as well as the skull model to simulate scalp and brain tissues. 16 Ag-AgCl electrodes (Electrode 1 to 16) with diameter of 1 cm on the inner wall of the sink were used to collect EIT data. The centers of these electrodes were set as e1 to e16. The locations of the electrodes (Figure 9(a)) are similar to those in the 2D head model (Figure 4(b)).

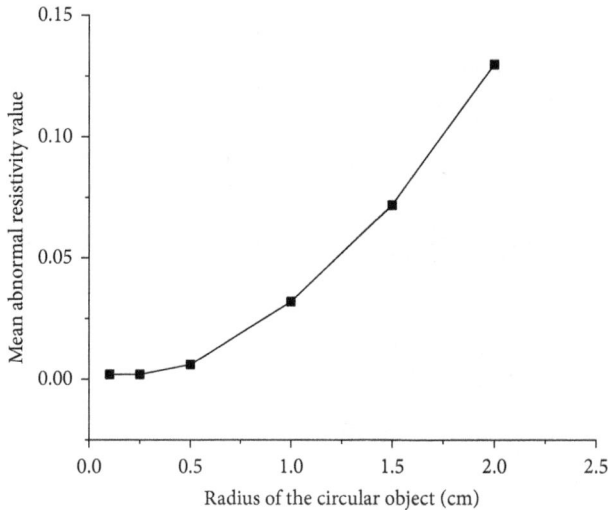

FIGURE 18: MARV of ROI in the SEIT reconstructions of the simulated ischemic stroke lesions of different sizes: 2.00, 1.50, 1.00, 0.50, 0.25, and 0.10 cm.

The surface of the solution must be higher than the top of the measurement electrodes by 1 cm. Agar cylinders of 6 cm in height were considered to simulate stroke lesions, of which the radii were 2, 1.5, 1, 0.5, and 0.25 cm, respectively. The conductivity of the agar was 0.65 S/m in the case of simulated hemorrhagic stroke and 0.13 S/m in the case of simulated ischemic stroke. The immersion depth of the agar cylinders into the solution was 5 cm. During the experiments, the room temperature was controlled at $20 \pm 1°C$.

We first collected data on the physical phantom without agar using EIT for 1 h. After that one frame of EIT data was utilized to calculate the index of asymmetry (IA) to evaluate the electrical impedance asymmetry of the physical phantom (Figure 9(b)). The maximum value of IA ($IA_{max}$) was 2.31%.

Next we put agar cylinders into the physical phantom (the location and size of each agar are shown in the next section) and measured EIT data, as shown in Figure 10. Then SEIT reference data was constructed for each frame of measured EIT data by using the EIT data measured from the hemisphere of the phantom without the agar. SEIT images were acquired by difference imaging method using the measured EIT data and their reference data. The degree of abnormal resistivity distribution in each SEIT image was quantitatively evaluated by calculating MARV of ROI in SEIT images when simulated stroke lesions of different sizes were placed consequently in the phantom.

### 2.4.2. Placement of Agars in the Physical Phantom

*(1) Simulated Stroke Lesions of Different Sizes in the Phantom.* The crossing point of line e1–e9 and line e5–e13 (Point O) was considered the center of the model. Agar cylinders with radii of 2.00, 1.50, 1.00 cm, 0.50 cm, and 0.25 cm were placed at the midpoint of line O-e13 to simulate stroke lesions of different sizes (Figure 11). The conductivity of the agar was 0.65 S/m in

the case of simulated hemorrhagic stroke and 0.13 S/m in the case of simulated ischemic stroke. Then the minimum size of simulated stroke lesions detectable by SEIT was determined.

*(2) Simulated Stroke Lesions at Different Locations of the Phantom.* Figure 9(b) indicated that the electrical impedance asymmetry of the anterior part of the phantom is higher than that of the posterior part; therefore, the anterior part of the phantom was used to place agar cylinders to test whether a simulated stroke lesion could be detectable when the head was not well symmetrical. An agar cylinder with a conductivity of 0.65 S/m and a radius of 1 cm (the detected minimum size) was placed at midpoints of the Lines O-e13, O-e14, O-e15, and O-e16 consequently to simulate a hematoma at different locations (Figures 12(a)–12(d)). Then the agar cylinder was placed at symmetrical locations of those four locations (i.e., at midpoints of the lines O-e5, O-e4, O-e3, and O-e2 consequently). Similarly, an agar cylinder with a conductivity of 0.13 S/m was placed at those eight locations to simulate an ischemic lesion at different locations.

## 3. Results and Discussion

### 3.1. Results of Simulation

*3.1.1. SEIT Imaging of Stroke Lesions of Different Sizes.* When the simulated cerebral hemorrhage lesion gradually decreased (Figures 7(a)–7(f)), the $IA_{max}$ of EIT data from the two hemispheres of the model was gradually reduced (Figure 13); the area of reconstructed object in the SEIT image was also gradually reduced (Figures 14(a)–14(f)). The simulated cerebral hemorrhage lesion on the left of the model exhibited a decrease in resistivity distribution (as the red region) on the left side of the SEIT image. The mean abnormal resistivity value (MARV) also decreased with the decrease in the radius of the simulated hematoma (Figure 15). When the radius of the hemorrhagic lesion was smaller than 0.50 cm, $IA_{max}$ was close to the value with the lesion not set. In these cases, although the SEIT image presented abnormal resistivity distribution, resistivity changed only slightly (<0.006, which can be considered normal).

When the simulated cerebral ischemia lesion gradually decreased (Figures 7(a)–7(f)), the $IA_{max}$ of EIT data from the two hemispheres of the model gradually reduced (Figure 16); the area of the reconstructed object in the SEIT image also gradually reduced (Figures 17(a)–17(f)). The simulated cerebral ischemic lesion on the left of the model exhibited an increase in resistivity distribution (as the blue region) on the left side of the SEIT image. The MARV also decreased with the decrease in the radius of the simulated hematoma (Figure 18). When the radius of the ischemic lesion was smaller than 0.50 cm, the $IA_{max}$ was close to the value with no lesion set, and the SEIT image did not present an abnormal resistivity distribution.

Hence, according to the simulation imaging of stroke lesions with different sizes, SEIT clearly demonstrated that the minimal radius of the circular simulated stroke lesion was about 0.5 cm. The corresponding SEIT images are as shown in Figures 14(d) and 17(d).

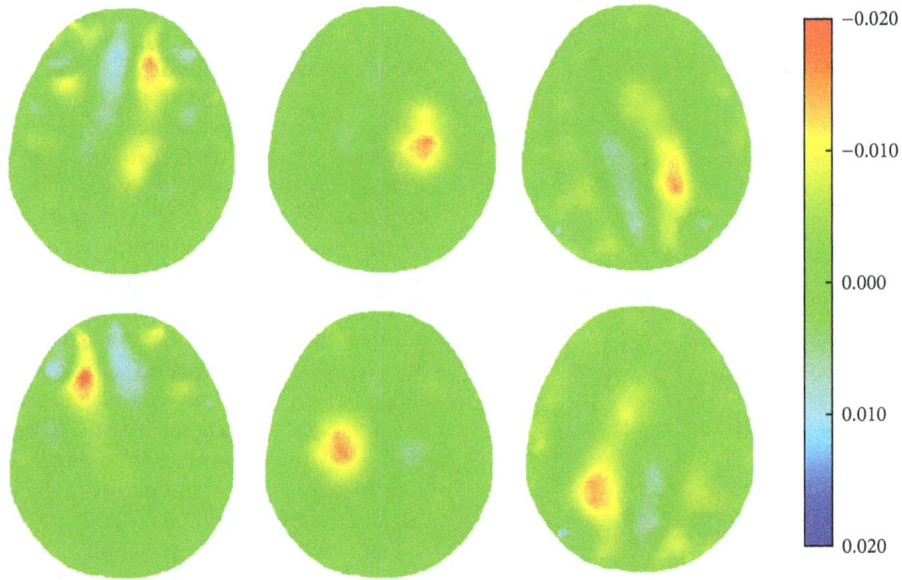

FIGURE 19: SEIT reconstructions of the simulated hemorrhagic stroke lesion at different locations of the 2D head model.

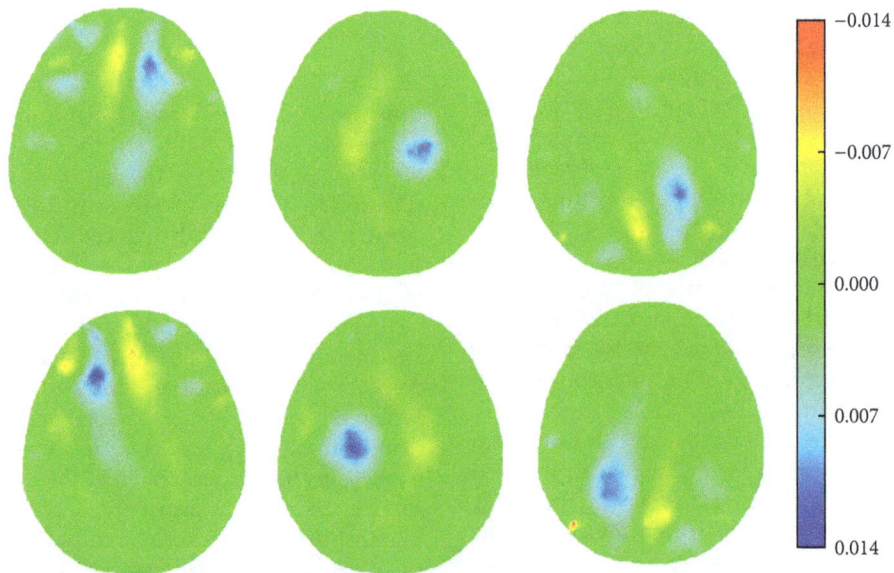

FIGURE 20: SEIT reconstructions of the simulated ischemic stroke lesion at different locations of the 2D head model.

*3.1.2. SEIT Imaging of the Stroke Lesion at Different Locations.* When a simulated hemorrhagic or ischemic stroke lesion with a radius of 0.50 cm was set at six different locations of the model (Figures 8(a)–8(f)), the image of the hemorrhagic or ischemic stroke lesion was reconstructed by SEIT, as shown in Figures 19 and 20, respectively.

### 3.2. Results of Experiments on the Physical Phantom

*3.2.1. Imaging Agar of Different Sizes in the Physical Phantom.* When the radius of the agar cylinder for simulating a hemorrhagic or ischemic stroke lesion decreased from 2 cm to 0.25 cm (Figure 11), the area of the reconstructed object also

reduced (Figures 21 and 22). The mean abnormal resistivity value (MARV) also decreased with the decrease of the size of agar cylinders (Figure 23). The minimum radius of the agar cylinders detected by SEIT was 1 cm. Those agar cylinders smaller than the minimum size were not reconstructed by SEIT (as shown in the last two images of Figures 21 and 22) and the corresponding maximum values of index of asymmetry ($IA_{max}$) were close to the value with the simulated lesion not set in the phantom (red point in Figure 24).

*3.2.2. Imaging Agar at Different Locations of the Phantom.* When an agar simulating a hematoma or an ischemic lesion (radius: 1 cm) at four different locations of the phantom

Exploratory Study on the Methodology of Fast Imaging of Unilateral Stroke Lesions by Electrical Impedance Asymmetry in Human Heads

153

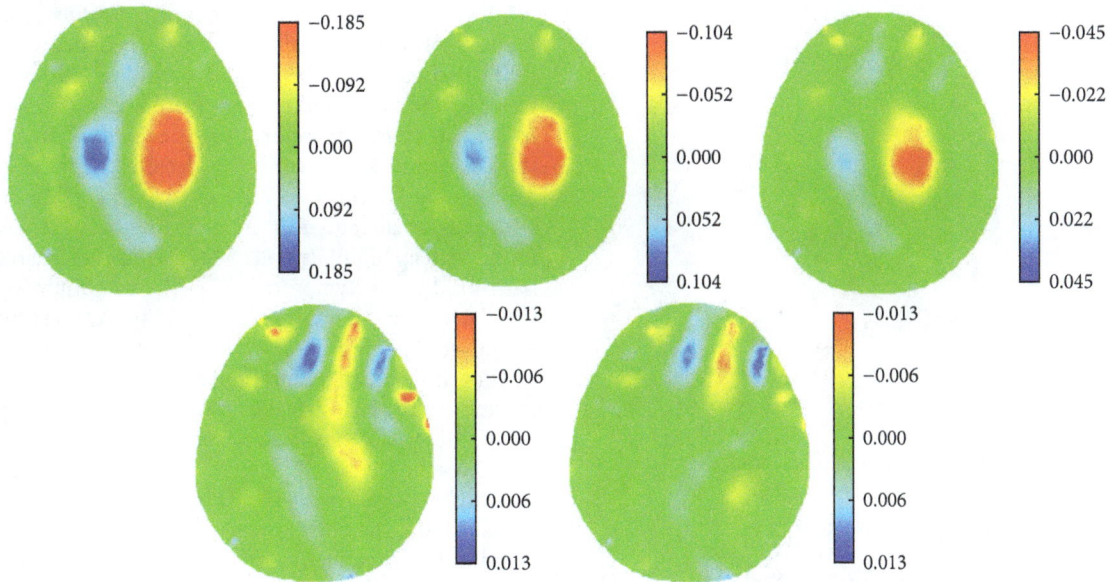

FIGURE 21: SEIT reconstructions of agar cylinders with radii of 2, 1.5, 1, 0.5, and 0.25 cm to simulate hemorrhagic stroke lesions of different sizes.

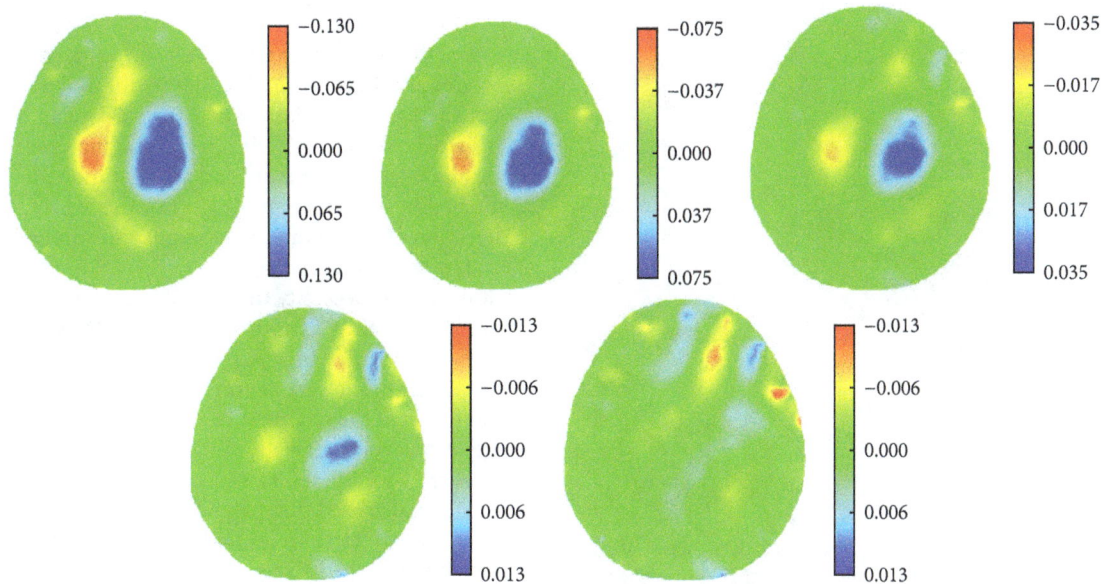

FIGURE 22: SEIT reconstructions of agar cylinders with radii of 2, 1.5, 1, 0.5, and 0.25 cm to simulate ischemic stroke lesions of different sizes.

(Figure 12), the reconstructed SEIT images are in row 1 of Figures 25 and 26, respectively; while the agar cylinder was placed at the symmetrical locations of those in Figure 12, the reconstructed SEIT images are in row 2 of Figures 25 and 26, respectively.

### 3.3. Discussion

*3.3.1. Summary of Results.* In this work, an SEIT method was proposed to image the difference in resistivity distribution between the two craniocerebral hemispheres (CCHs) of the human head. The significance of this study is that it proposes a novel EIT approach that reconstructs the image of a unilateral cerebral lesion based on one frame of EIT data to provide information on the lesion. Therefore, long-time impedance monitoring adopted by dynamic EIT becomes unnecessary when the method of SEIT is utilized to detection of stroke. Patients with large stroke lesions face high risk of mortality [40] and lesion-led injuries vary with the anatomical location of the lesion [41]. Thus, SEIT should evaluate lesion information on both aspects. The imaging experiments on the 2D head model and those on the physical phantom confirmed that SEIT can use abnormal resistivity

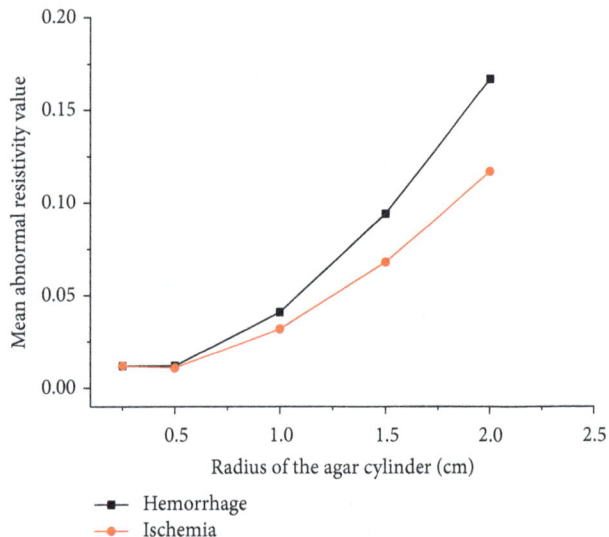

FIGURE 23: Mean abnormal resistivity value of ROI in each SEIT image of Figures 21 and 22 (hemorrhage for Figure 21 and ischemia for Figure 22).

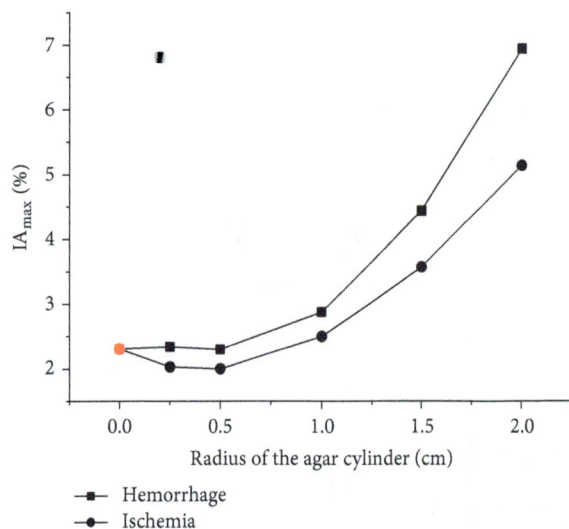

FIGURE 24: $IA_{max}$ of EIT data from the two hemispheres of the phantom with and without a simulated hemorrhagic or ischemic stroke lesion. Black points represent $IA_{max}$ with a simulated stroke lesion, while the red point represents $IA_{max}$ without a simulated lesion.

distribution to reflect stroke lesions at different locations and with different sizes when the lesion can significantly alter the electrical impedance asymmetry of the head. The electrical impedance asymmetry of the physical phantom was higher than that of the 2D head model; therefore, the detected minimum radius of the objects in the 2D head model was 0.5 cm, which was 1 cm in the physical model. Moreover, the image quality of the physical phantom experiments was lower than that of simulation experiments. Improving the quality of SEIT images is an important issue that needs further study when the imaged object is not well symmetrical.

*3.3.2. Possible Limitations of Clinical Application of SEIT.* The method of symmetrical electrical impedance tomography (SEIT) reconstructs the images of stroke lesions primarily based on the stroke-induced electrical impedance asymmetry in human heads; therefore, the greater the asymmetry is, the easier the lesion could be detected. This is a notable feature of SEIT. The lesion size of lacunar cerebral infarction is usually smaller than 2 cm in diameter, which is usually found in deep brain. Patients with this kind of infarction may have mild or no symptoms [42]. The minimum size of the simulated cerebral infarcts that could be detected by SEIT in the physical phantom was 2 cm in diameter. Due to the small sizes of the lacunar cerebral infarcts and their deep locations in the brain, the electrical impedance asymmetry in human heads induced by the infarction would be insignificant, which may be ignored by SEIT in clinical application. Imaging experiments on stroke patients by SEIT were not conducted in this study; therefore it was not confirmed whether SEIT could detect lacunar cerebral infarcts in clinical practice. Lacunar cerebral infarcts could be diagnosed by brain CT or MRI currently. They may be detected by SEIT in the future by optimizing the performances of the EIT imaging algorithm and the hardware system.

*3.3.3. Further Considerations of This Study.* This study is a methodology study to validate the method of SEIT in technique; therefore, simulation and physical phantom experiments were conducted. Preliminary imaging experiments on the heads of stroke patients and normal individuals will be conducted in the future studies to test the feasibility of SEIT in detection of stroke in practice and some important factors should be considered.

It is of great necessity to correlate the neuroimaging data obtained by brain CT and/or MRI with those obtained by SEIT in the further imaging experiments on the heads of stroke patients. The correlation analysis should include location and area of the reconstructed objects by the two kinds of image reconstruction means. In location correlation analysis the central coordinates of the two kinds of reconstructed objects are calculated first [43, 44] and then linear regression analysis is conducted. In area correlation analysis the area of the two kinds of the reconstructed objects are calculated [43, 44] followed by the linear regression analysis. The feasibility of clinical application of SEIT in detection of stroke would be verified by the above analyses.

The symmetrical distribution of EIT electrodes at the two sides of the imaged head by SEIT should be ensured. The asymmetrical placement of electrodes may lead to notable artifacts in reconstructed images. This was not a problem in this study, since the electrodes were precisely located in the 2D head model and the physical phantom. To ensure the symmetrical distribution of electrodes on the head in the further experiments, the location of each electrode should be precisely measured on human subjects before the placement of EIT electrodes. The data collection maybe time-consuming. Some researchers employed the Polhemus FASTRAK digitizer to accurately position electrodes [45], and several other researchers proposed a hydrogel elasticated

Exploratory Study on the Methodology of Fast Imaging of Unilateral Stroke Lesions by Electrical Impedance Asymmetry in Human Heads

155

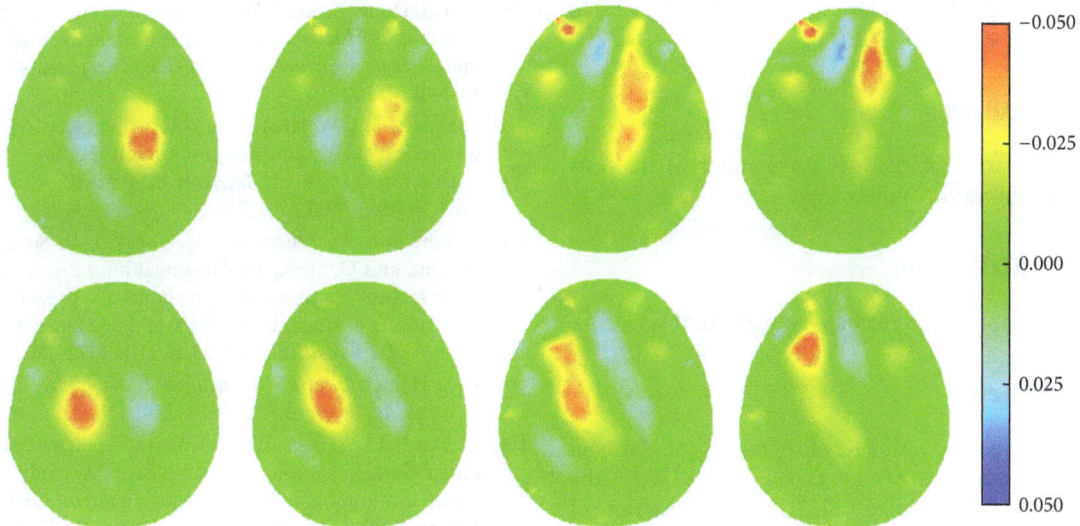

FIGURE 25: SEIT reconstructions of a simulated hematoma at different locations of the phantom.

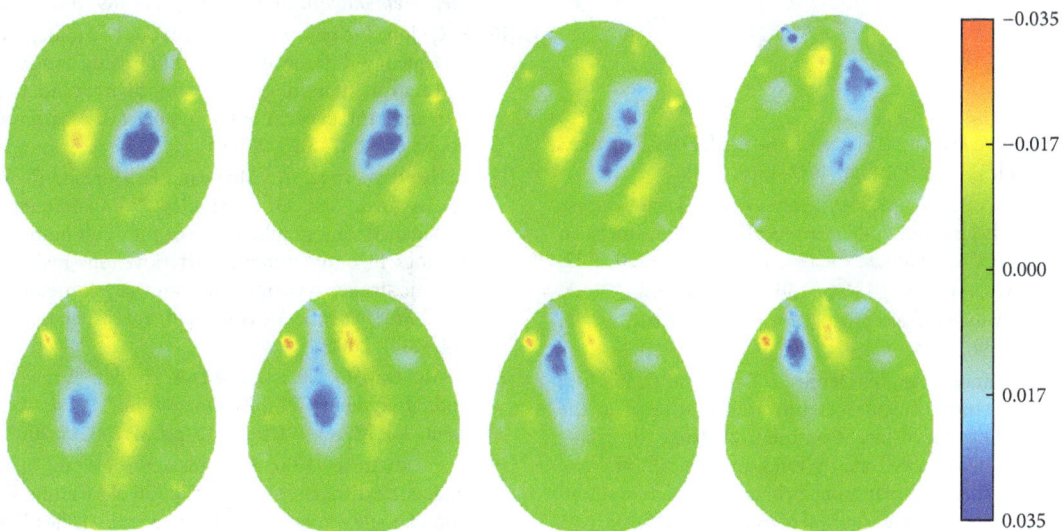

FIGURE 26: SEIT reconstructions of a simulated ischemic lesion at different locations of the phantom.

headnet suitable for EIT data collection by comparing several EIT electrode headnets [46]. Therefore, we will develop a headnet suitable for SEIT imaging to quickly and accurately place EIT electrodes on human heads.

*3.3.4. Potential in Determination of Stroke Type by SEIT.* Since the symptoms of a hemorrhagic stroke and an ischemic stroke are similar, it is necessary to identify different types of stroke. For instance, thrombolytic therapy must not be given to patients with hemorrhagic stroke, as the hemorrhage may extend. Therefore, identification of stroke type is essential to permit appropriate treatment to stroke patients.

The resistivity of a hemorrhagic lesion is lower than that of normal brain tissue in the contralateral hemisphere, whereas the resistivity of the lesion is higher when cerebral infarction occurs. This was demonstrated by the reconstructions of simulated hemorrhagic and ischemic stroke lesions in the current study. The SEIT images of the two types of simulated stroke lesions were clearly different. Thus, SEIT may characterize the two types of stroke with opposite resistivity changes.

## 4. Conclusion

A new EIT method (SEIT) was proposed to reconstruct the impedance image of unilateral stroke lesions. The preliminary imaging results of the 2D head model and the physical phantom verified the method in rapid detection of unilateral stroke lesions. In our future study, we will perfect this method and conduct experiments on suspected stroke patients in community hospitals to validate the feasibility of SEIT in rapid detection of stroke.

## Conflict of Interests

The authors declare that there is no conflict of interests regarding the publication of this paper.

## Authors' Contribution

Jieshi Ma and Canhua Xu contributed equally to this work.

## Acknowledgments

This work was supported by the Key Program of the National Nature Science Foundation of China (NSFC) (50937005), the General Program of the NSFC (61271101), the program of the NSFC (51207161), the program of the Natural Science Foundation of Shaanxi Province (2012JQ4012), the National Key Technology R&D Program of China (2011BAI08B13 and 2012BAI20B02), and the military foundation of China (CWS12J102).

## References

[1] G. A. Donnan, M. Fisher, M. Macleod, and S. M. Davis, "Stroke," *The Lancet*, vol. 371, no. 9624, pp. 1612–1623, 2008.

[2] V. L. Roger, A. S. Go, D. M. Lloyd-Jones et al., "Heart disease and stroke statistics–2012 update: a report from the American heart association," *Circulation*, vol. 125, no. 1, pp. e2–e220, 2012.

[3] F. L. Silver, J. W. Norris, A. J. Lewis, and V. C. Hachinski, "Early mortality following stroke: a prospective review," *Stroke*, vol. 15, no. 3, pp. 492–496, 1984.

[4] J. Bamford, P. Sandercock, M. Dennis, J. Burn, and C. Warlow, "A prospective study of acute cerebrovascular disease in the community: the Oxfordshire Community Stroke Project—1981-86. 2. Incidence, case fatality rates and overall outcome at one year of cerebral infarction, primary intracerebral and subarachnoid haemorrhage," *Journal of Neurology Neurosurgery and Psychiatry*, vol. 53, no. 1, pp. 16–22, 1990.

[5] K. S. Wong, "Risk factors for early death in acute ischemic stroke and intracerebral hemorrhage: a prospective hospital-based study in Asia," *Stroke*, vol. 30, no. 11, pp. 2326–2330, 1999.

[6] H. P. Adams Jr., G. del Zoppo, M. J. Alberts et al., "Guidelines for the early management of adults with ischemic stroke: a guideline from the American Heart Association/American Stroke Association Stroke Council, Clinical Cardiology Council, Cardiovascular Radiology and Intervention Council, and the Atherosclerotic Peripheral Vascular Disease and Quality of Care Outcomes in Research Interdisciplinary Working Groups: the American Academy of Neurology," *Circulation*, vol. 115, no. 20, pp. e478–e534, 2007.

[7] K. S. Cole, *Membranes, Ions, and Impulses*, University of California Press, Berkeley, Calif, USA, 1968.

[8] F. Seoane and K. Lindecrantz, "Electrical bioimpedance cerebral monitoring," in *Encyclopedia of Healthcare Information Systems*, pp. 480–486, 2008.

[9] R. H. Bayford, "Bioimpedance tomography (electrical impedance tomography)," *Annual Review of Biomedical Engineering*, vol. 8, pp. 63–91, 2006.

[10] Z. Ji, X. Dong, X. Shi, F. You, F. Fu, and R. Liu, "Novel electrode-skin interface for breast electrical impedance scanning," *Medical and Biological Engineering and Computing*, vol. 47, no. 10, pp. 1045–1052, 2009.

[11] F. You, X. Shi, W. Shuai et al., "Applying electrical impedance tomography to dynamically monitor retroperitoneal bleeding in a renal trauma patient," *Intensive Care Medicine*, vol. 39, no. 6, pp. 1159–1160, 2013.

[12] S. Lindgren, H. Odenstedt, C. Olegård, S. Söndergaard, S. Lundin, and O. Stenqvist, "Regional lung derecruitment after endotracheal suction during volume- or pressure-controlled ventilation: a study using electric impedance tomography," *Intensive Care Medicine*, vol. 33, no. 1, pp. 172–180, 2007.

[13] D. S. Holder, *Electrical Impedance Tomography: Methods, History and Applications*, IOP Publishing Ltd., London, UK, 2005.

[14] M. Dai, B. Li, S. Hu et al., "In vivo imaging of twist drill drainage for subdural hematoma: a clinical feasibility study on electrical impedance tomography for measuring intracranial bleeding in humans," *PLoS ONE*, vol. 8, no. 1, Article ID e55020, 2013.

[15] R. J. Yerworth, R. H. Bayford, B. Brown, P. Milnes, M. Conway, and D. S. Holder, "Electrical impedance tomography spectroscopy (EITS) for human head imaging," *Physiological Measurement*, vol. 24, no. 2, pp. 477–489, 2003.

[16] S. C. Jun, J. Kuen, J. Lee, E. J. Woo, D. Holder, and J. K. Seo, "Frequency-difference EIT (fdEIT) using weighted difference and equivalent homogeneous admittivity: validation by simulation and tank experiment," *Physiological Measurement*, vol. 30, no. 10, pp. 1087–1099, 2009.

[17] A. Romsauerova, A. McEwan, L. Horesh, R. Yerworth, R. H. Bayford, and D. S. Holder, "Multi-frequency electrical impedance tomography (EIT) of the adult human head: initial findings in brain tumours, arteriovenous malformations and chronic stroke, development of an analysis method and calibration," *Physiological Measurement*, vol. 27, no. 5, article S13, pp. S147–S161, 2006.

[18] A. McEwan, G. Cusick, and D. S. Holder, "A review of errors in multi-frequency EIT instrumentation," *Physiological Measurement*, vol. 28, no. 7, article S15, pp. S197–S215, 2007.

[19] B. Packham, H. Koo, A. Romsauerova et al., "Comparison of frequency difference reconstruction algorithms for the detection of acute stroke using EIT in a realistic head-shaped tank," *Physiological Measurement*, vol. 33, no. 5, pp. 767–786, 2012.

[20] C. H. Xu, L. Wang, X. T. Shi et al., "Real-time imaging and detection of intracranial haemorrhage by electrical impedance tomography in a piglet model," *Journal of International Medical Research*, vol. 38, no. 5, pp. 1596–1604, 2010.

[21] M. Dai, L. Wang, C. Xu, L. Li, G. Gao, and X. Dong, "Real-time imaging of subarachnoid hemorrhage in piglets with electrical impedance tomography," *Physiological Measurement*, vol. 31, no. 9, pp. 1229–1239, 2010.

[22] A. E. Baird, K. O. Lövblad, G. Schlaug, R. R. Edelman, and S. Warach, "Multiple acute stroke syndrome: marker of embolic disease?" *Neurology*, vol. 54, no. 3, pp. 674–678, 2000.

[23] L. X. Liu, W. W. Dong, J. Wang, Q. Wu, W. He, and Y. J. Jia, "The role of noninvasive monitoring of cerebral electrical impedance in stroke," *Acta Neurochirurgica*, no. 95, pp. S137–S140, 2005.

[24] L. Liu, W. Dong, X. Ji et al., "A new method of noninvasive brain-edema monitoring in stroke: cerebral electrical impedance measurement," *Neurological Research*, vol. 28, no. 1, pp. 31–37, 2006.

[25] G. Bonmassar, S. Iwaki, G. Goldmakher, L. M. Angelone, J. W. Belliveau, and M. H. Lev, "On the measurement of

electrical impedance spectroscopy (EIS) of the human head," *International Journal of Bioelectromagnetism*, vol. 12, no. 1, pp. 32–46, 2010.

[26] R. Kothari, L. Sauerbeck, E. Jauch et al., "Patients' awareness of stroke signs, symptoms, and risk factors," *Stroke*, vol. 28, no. 10, pp. 1871–1875, 1997.

[27] S. R. Atefi, F. Seoane, and K. Lindecrantz, "Electrical Bioimpedance cerebral monitoring. Preliminary results from measurements on stroke patients," in *Proceedings of the 2012 Annual International Conference of the IEEE on Engineering in Medicine and Biology Society (EMBC '12)*, pp. 126–129, 2012.

[28] S. R. Atefi, F. Seoane, T. Thorlin, and K. Lindecrantz, "Stroke damage detection using classification trees on electrical bioimpedance cerebral spectroscopy measurements," *Sensors*, vol. 13, no. 8, pp. 10074–10086, 2013.

[29] X. Shi, X. Dong, W. Shuai, F. You, F. Fu, and R. Liu, "Pseudo-polar drive patterns for brain electrical impedance tomography," *Physiological Measurement*, vol. 27, no. 11, article 002, pp. 1071–1080, 2006.

[30] C. Xu, M. Dai, F. You et al., "An optimized strategy for real-time hemorrhage monitoring with electrical impedance tomography," *Physiological Measurement*, vol. 32, no. 5, pp. 585–598, 2011.

[31] C. H. Xu, *Research on reconstruction algorithm and experiments of EIT for bedside dynamic image monitoring [Ph.D. thesis]*, Fourth Military Medical University, 2010.

[32] U. Topaloglu, Y. L. Yan, P. Novak, P. Spring, J. Suen, and G. Shafirstein, "Virtual thermal ablation in the head and neck using Comsol MultiPhysics," in *Proceedings of the COMSOL Conference*, pp. 1–7, Boston, Mass, USA, October 2008.

[33] H. C. Burger and J. B. Milaan, "Measurements of the specific resistance of the human body to direct current," *Acta Medica Scandinavica*, vol. 114, no. 6, pp. 584–607, 1943.

[34] C. Tang, F. You, G. Cheng, D. Gao, F. Fu, and X. Dong, "Modeling the frequency dependence of the electrical properties of the live human skull," *Physiological Measurement*, vol. 30, no. 12, pp. 1293–1301, 2009.

[35] J. Latikka, T. Kuurne, and H. Eskola, "Conductivity of living intracranial tissues," *Physics in Medicine and Biology*, vol. 46, no. 6, pp. 1611–1616, 2001.

[36] M. J. Peters, J. G. Stinstra, and M. Hendriks, "Estimation of the electrical conductivity of human tissue," *Electromagnetics*, vol. 21, no. 7-8, pp. 545–557, 2001.

[37] L. Horesh, *Some novel approaches in modelling and image reconstruction for multi-frequency electrical impedance tomography of the human brain [Ph.D. thesis]*, University College London, 2006.

[38] X. Shi, X. Dong, F. You, F. Fu, and R. Liu, "High precision multifrequency electrical impedance tomography system and preliminary imaging results on saline tank," in *Proceedings of the 2005 27th Annual International Conference of the Engineering in Medicine and Biology Society (IEEE-EMBS '05)*, vol. 2, pp. 1492–1495, September 2005.

[39] J. B. Li, X. Z. Dong, C. Tang et al., "Study on realistic skull model in electrical impedance tomography," *Chinese Medical Equipment Journal*, vol. 32, pp. 1–3, 2011.

[40] J. Kalita, U. K. Misra, A. Vajpeyee, R. V. Phadke, A. Handique, and V. Salwani, "Brain herniations in patients with intracerebral hemorrhage," *Acta Neurologica Scandinavica*, vol. 119, no. 4, pp. 254–260, 2009.

[41] J. Broderick, S. Connolly, E. Feldmann et al., "Guidelines for the management of spontaneous intracerebral hemorrhage in adults: 2007 update: a guideline from the American Heart Association/American Stroke Association Stroke Council, High Blood Pressure Research Council, and the Quality of Care and Outcomes in Research Interdisciplinary Working Group," *Circulation*, vol. 116, no. 16, pp. e391–e413, 2007.

[42] A. Arboix, C. García-Plata, L. García-Eroles et al., "Clinical study of 99 patients with pure sensory stroke," *Journal of Neurology*, vol. 252, no. 2, pp. 156–162, 2005.

[43] A. Padma and R. Sukanesh, "A wavelet based automatic segmentation of brain tumor in CT images using optimal statistical texture features," *International Journal of Image Processing*, vol. 5, no. 5, pp. 552–563, 2011.

[44] T. Tang, S. Oh, and R. J. Sadleir, "A robust current pattern for the detection of intraventricular hemorrhage in neonates using electrical impedance tomography," *Annals of Biomedical Engineering*, vol. 38, no. 8, pp. 2733–2747, 2010.

[45] G. Bonmassar and S. Iwaki, "The shape of electrical impedance spectroscopy (EIS) is altered in stroke patients," in *Proceedings of the 26th Annual International Conference of the IEEE Engineering in Medicine and Biology Society (EMBC '04)*, vol. 5, pp. 3443–3446, September 2004.

[46] A. T. Tidswell, A. P. Bagshaw, D. S. Holder et al., "A comparison of headnet electrode arrays for electrical impedance tomography of the human head," *Physiological Measurement*, vol. 24, no. 2, pp. 527–544, 2003.

# Membrane Properties Involved in Calcium-Stimulated Microparticle Release from the Plasma Membranes of S49 Lymphoma Cells

**Lauryl E. Campbell, Jennifer Nelson, Elizabeth Gibbons, Allan M. Judd, and John D. Bell**

*Department of Physiology and Developmental Biology, Brigham Young University, Provo, UT 84601, USA*

Correspondence should be addressed to John D. Bell; john_bell@byu.edu

Academic Editors: U. S. Gaipl, D. C. Rau, and H. Tuncel

This study answered the question of whether biophysical mechanisms for microparticle shedding discovered in platelets and erythrocytes also apply to nucleated cells: cytoskeletal disruption, potassium efflux, transbilayer phospholipid migration, and membrane disordering. The calcium ionophore, ionomycin, disrupted the actin cytoskeleton of S49 lymphoma cells and produced rapid release of microparticles. This release was significantly inhibited by interventions that impaired calcium-activated potassium current. Microparticle release was also greatly reduced in a lymphocyte cell line deficient in the expression of scramblase, the enzyme responsible for calcium-stimulated dismantling of the normal phospholipid transbilayer asymmetry. Rescue of the scrambling function at high ionophore concentration also resulted in enhanced particle shedding. The effect of membrane physical properties was addressed by varying the experimental temperature (32–42°C). A significant positive trend in the rate of microparticle release as a function of temperature was observed. Fluorescence experiments with trimethylammonium diphenylhexatriene and Patman revealed significant decrease in the level of apparent membrane order along that temperature range. These results demonstrated that biophysical mechanisms involved in microparticle release from platelets and erythrocytes apply also to lymphocytes.

## 1. Introduction

Microparticles are small vesicular structures (0.1–1 $\mu$m diameter) produced and released by exocytic blebbing of the cell plasma membrane from a variety of cell types including platelets, erythrocytes, leucocytes, endothelial cells, fibroblasts, epithelial cells, and tumor cells [1–4]. Microparticles are detectable basally in the blood of healthy individuals [5], and additional amounts may be shed from cells as a result of activation signals and/or during apoptosis [3, 6]. Microparticles appear to function as mediators of intercellular communication. For example, they may cause cellular activation or apoptosis depending on the target [7–10]. In addition, they are involved in regulation of inflammation, coagulation, and antigen presentation [3, 4, 6]. Hence, they may play a role in the pathogenesis of autoimmune diseases and inflammatory disorders. Moreover, elevated microparticle levels are typically seen in the blood of patients in a variety of disease states

such as various cardiovascular disorders including atherosclerosis, diabetes, certain infectious diseases such as HIV, Ebola, and cerebral malaria, and in several cancers [2, 3, 6, 9, 11, 12].

Although small microparticles (less than 100 nm) appear to have an endosomal origin, the majority are larger (100–1000 nm) and are shed through the process of "reverse budding" [13]. In this latter case, release is initiated by a sustained rise in intracellular calcium [1–3, 6]. The mechanism of calcium-stimulated microparticle release has been explored most extensively in platelets and erythrocytes where release requires reorganization of the cytoskeleton, translocation of phosphatidylserine (PS) and other phospholipids to the outer face of the cell membrane, and enhanced permeability to potassium with associated osmotic effects [14–28]. In addition, recent work with erythrocytes has demonstrated that vesicle shedding also depends on physical characteristic of the cell membrane, which can be detected with fluorescent

Membrane Properties Involved in Calcium-Stimulated Microparticle Release from the Plasma Membranes of S49 Lymphoma Cells

159

membrane probes sensitive to phospholipid order and organization in the bilayer [29].

The extent to which these various mechanisms apply to nucleated cells has not yet been adequately addressed [3]. Some evidence exists to suggest that the cytoskeletal changes are required in all cells that release microparticles [24–27], and it is reasonable to assume that cytoskeletal attachments would have to be broken for pieces of the membrane to be shed. Whether the other mechanisms (exposure of PS, transmembrane potassium flux, and favorable biophysical properties) are also necessary remains unknown. This study was designed to address that deficiency using S49 lymphoma cells as an experimental model.

## 2. Materials and Methods

*2.1. Reagents.* Ionomycin, 1-(trimethylammoniumphenyl)-6-phenyl-1,3,5-hexatriene p-toluenesulfonate (TMA-DPH), Alexa Fluor 488 Phalloidin Conjugate, and 6-hexadecanoyl-2-(((2-(trimethylammonium)ethyl)methyl)amino)naphthalene chloride (patman) were obtained from Life Technologies (Grand Island, NY, USA). Ionomycin and MC540 were dissolved in dimethyl sulfoxide (DMSO) as stock solutions, while TMA-DPH was suspended in dimethylformamide. Quinine was purchased from Sigma (St. Louis, MO, USA).

*2.2. Cell Preparation.* S49 mouse lymphoma cells were cultured in DMEM (10% horse serum) at 37°C in humidified air (10% $CO_2$). Raji human Burkitt's lymphoma cells were grown at 5% $CO_2$ in RPMI (10% fetal bovine serum and L-glutamine). Prior to experiments, unless otherwise stated, cells were isolated through centrifugation then washed and suspended in MBSS (134 mM NaCl, 6.2 mM KCl, 1.6 mM $CaCl_2$, 18.0 mM Hepes, 13.6 mM glucose, and pH 7.4 at 37°C) at a density of $0.4$–$3.0 \times 10^6$ cells/mL. Unless stated otherwise, experiments were conducted at 37°C.

*2.3. Fluorescence Spectroscopy and Light Scatter.* Washed cell samples (2 mL) were equilibrated 5 min in quartz fluorometer sample cells in either a Fluoromax 3 (Horiba, Edison, NJ, USA) or PC-1 (ISS, Champaign, IL, USA, anisotropy measurements) spectrofluorometer prior to data acquisition. Sample homogeneity was maintained by magnetic stirring, and temperature was regulated with circulating water baths.

Microparticle release was assayed by light scatter at 500 nm [15]. After data acquisition was initiated, ionomycin (300 nM) was added, and the rate of microparticle release was determined as the slope of the rise in light scatter intensity. That this procedure assesses microparticles released from cells was established previously for S49 cells by differential centrifugation and lipid analysis [30]. Note that typically many of the microparticles released are larger than 500 nm and therefore produce an elevation in the light scatter intensity. However, in rare cases microparticle size is uniformly smaller than 500 nm, and in those instances the shedding of particles produces a negative deflection in the light scatter due to shrinkage of the cells following release.

The fluorescence emission of patman was observed at 435 and 500 nm (250 nM final, excitation = 350 nm) by rapid (3 s resolution) sluing of the emission monochromator mirror. The probe was added to cell samples after measuring background intensity for 100 s, and the fluorescence intensity was then monitored for several hundred seconds until steady state was reached. The polarity of patman's environment was assessed by calculating the generalized polarization (GP) as follows [31]:

$$GP = \frac{I_{435} - I_{500}}{I_{435} + I_{500}}, \tag{1}$$

where $I_{435}$ and $I_{500}$ are the emission intensities at 435 and 500 nm. The intensity data were smoothed by nonlinear regression to an arbitrary function (sum of two exponentials) prior to calculation of GP.

The steady-state anisotropy of TMA-DPH (250 nM final, excitation = 350, emission = 452) was assessed using Glan-Thompson polarizers. Probe was equilibrated with cell samples for 10 min prior to acquisition of data with excitation and emission polarizers alternatively oriented parallel and then perpendicular to each other. Anisotropy was calculated as described previously, and at least 20 points were averaged in determining values for the figures and statistical analyses [32].

*2.4. Fluorescence Imaging of Actin Cytoskeleton.* Cells were washed and treated with ionomycin as in other experiments. The treated cells were then simultaneously fixed, permeabilized, and stained with fluorescent phalloidin (Alexa Fluor© 488) according to the manufacturer's protocol (Life Technologies). After being mounted onto microscopy slides, cells were stained with a solution containing 165 nM phalloidin, 3.7% formaldehyde, 1% bovine serum albumin, and 0.1 mg/mL lyso-PC (1-palmitoyl-2-hydroxy-sn-glycero-3-phosphocholine) for 20 min at 4°C. Slides were then washed with buffer before coverslips were mounted. Images were collected on an Olympus FluoView FV 300 confocal laser scanning microscope using a 60x oil immersion objective lens. A 3x digital zoom was also applied. The excitation light source was a 488 nm argon laser, and a 505–525 nm bandpass filter was used on the emission detector.

## 3. Results

Figure 1 displays fluorescence images of the actin cytoskeleton of S49 lymphoma cells before (Figure 1(a)) and after (Figure 1(b)) incubation with a calcium ionophore (ionomycin). As expected based on observations with other cell types [21–27], the cytoskeleton was disrupted rapidly (within 10 min) as indicated by a reduction of extended parallel fibers and an increase of scattered brightly staining actin aggregates. As shown in Figure 2, this disruptive effect of ionomycin was accompanied by a rise in the intensity of light scattered by the sample. This elevation in intensity began within 50 s after addition of ionomycin and continued for another 100 s. Previous work has demonstrated that the light scatter change reflects the release of membrane particles [15, 30]. Figure 3 demonstrates that the difference in light scatter intensity

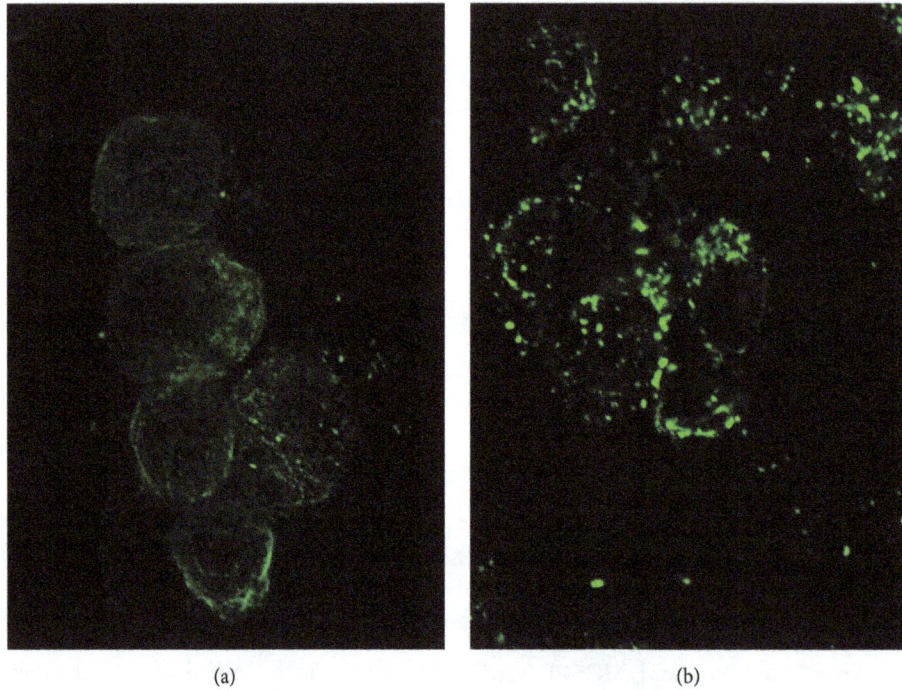

FIGURE 1: Confocal photographs of actin cytoskeletal without (a) or with (b) ionomycin treatment at 37°C. The actin cytoskeleton of S49 cells was stained with phalloidin.

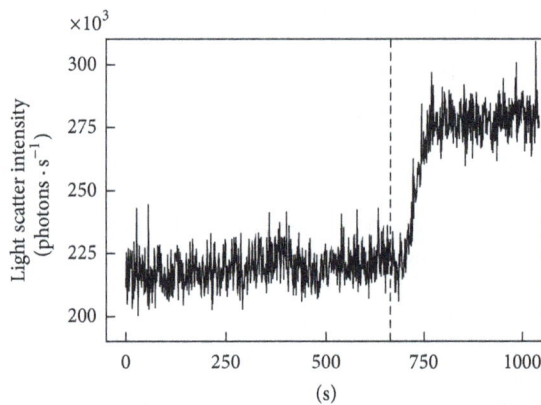

FIGURE 2: Ionomycin-stimulated microparticle release assayed by light scatter at 37°C. Ionomycin was added at the dotted line. An increase in scatter intensity indicates particle release [30].

before and after treatment with ionomycin was reproducible and statistically significant ($P < 0.0001$, one-sample $t$-test with $H_0 = 0$). Moreover, repetition of the experiment in the presence of a calcium chelator (EGTA) inhibited the response to ionomycin and therefore demonstrated that microparticle release was dependent on calcium and therefore not an artifact of the ionophore itself (Figure 3).

*3.1. Role of Potassium Channels.* To determine whether S49 cells require calcium-activated potassium flux for microparticle release similar to platelets and erythrocytes, the experiment of Figure 2 was repeated in the presence of a reduced

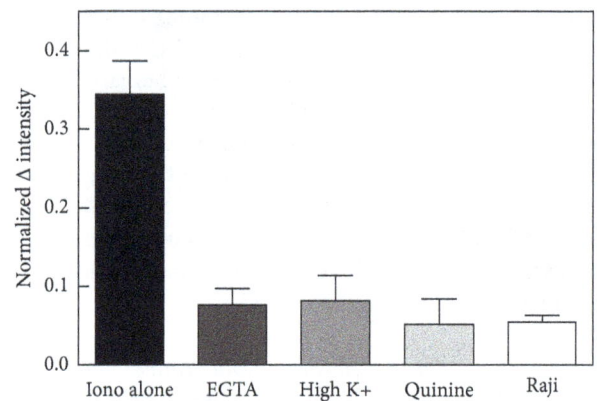

FIGURE 3: Ionomycin-stimulated microparticle release requires calcium-activated potassium current. Cells were washed and suspended in normal MBSS ("Iono alone" or "Raji") or in MBSS that contained EGTA (2 mM) instead of calcium, high potassium (83 mM KCl with equivalent reduction in NaCl), or quinine (1 mM) at 37°C. The normalized light scatter intensity was calculated by subtracting the average initial intensity immediately prior to ionomycin addition (20 points) from the average intensity at the plateau after ionomycin (about 350 s later; see Figure 2). This difference was then divided by the average initial intensity to standardize among trials. Differences in the normalized intensity among groups were significant by one-way analysis of variance ($P = 0.0004$, $n = 2$–9 per group). A posttest (Dunnett's) revealed that the group of S49 cells treated with normal MBSS was distinguishable from each of the other four ($P < 0.05$).

potassium gradient. In addition, the experiment was repeated with normal potassium concentrations in the presence of

Membrane Properties Involved in Calcium-Stimulated Microparticle Release from the Plasma Membranes of S49
Lymphoma Cells

161

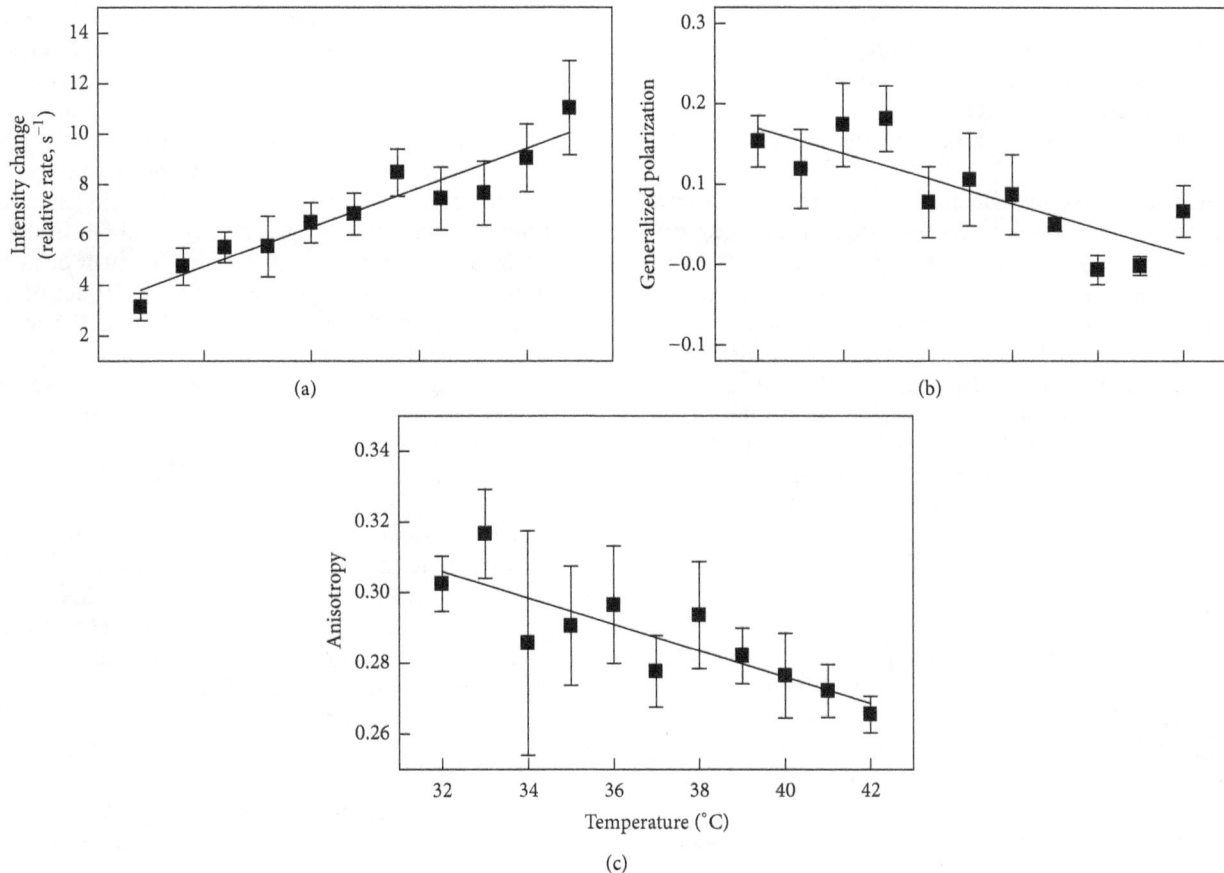

FIGURE 4: Relationship between the rate of microparticle release and membrane order as a function of temperature. (a) The relative rate of particle release upon addition of ionomycin was calculated from experiments such as that shown in Figure 2. Cells were equilibrated at 37°C and then adjusted to the indicated temperature and equilibrated for 10 min prior to adding ionomycin. The relative rate of release was determined from the maximum slope of the time profile following ionomycin addition and normalized to initial light scatter intensity as in Figure 3. Based on linear regression, the positive trend was significant ($P < 0.0001$, $r^2 = 0.28$, $n = 5$–15 per temperature, 83 total values). ((b)-(c)) The experiments of (a) were repeated with cells labeled with patman (b) or TMA-DPH (c). Nonlinear regression to an arbitrary function (sum of two exponentials) was used to smooth the Patman data prior to calculating the value of GP or TMA-DPH anisotropy was averaged from 7 points prior to addition of ionomycin. The negative trends in both cases were significant by linear regression ($P \leq 0.006$, $r^2 > 0.21$, and $n = 3$ per temperature).

the calcium-activated potassium channel blocker quinine [14–16]. Figure 3 shows that both interventions produced about 80% reduction in the amount of microparticles shed from the cells. These results indicated that calcium-activated potassium current was a necessary component of particle release. Presumably, this current is necessary in order to reduce the cell volume osmotically to accommodate the reduced membrane surface area associated with microparticle shedding [33, 34].

*3.2. Loss of Membrane Lipid Asymmetry.* Previous genetic data and experiments with a pharmacological inhibitor of scramblase (R5421, [35]) demonstrated in erythrocytes and platelets that the activity of that enzyme is important for microparticle shedding [19, 29, 36]. Since a scramblase inhibitor is no longer available, the requirement for migration of PS from the inner to the outer leaflet of the cell membrane was tested here for lymphoma cells by using the Raji lymphoma line, which is deficient in scramblase activity [37–40].

The white bar of Figure 3 shows the response to ionomycin in Raji cells. The ionomycin-stimulated change in light scatter intensity in Raji cells was only 16% of that observed in S49 cells. Furthermore, incubation of Raji cells with a higher dose of ionomycin sufficient to restore some of the ability of the cells to translocate PS [40] generated a twofold enhancement of apparent rate of microparticle release ($2.1 \pm 0.4$, $P = 0.01$, $n = 10$). These data argued that PS exposure is also important for microparticle release in nucleated cells.

*3.3. Membrane Lipid Order.* Recent studies suggested a role for membrane lipid order and fluidity in determining the ability of erythrocytes to release microvesicles in response to calcium [29]. To explore that possibility for S49 cells, experiments were conducted at various temperatures between 32 and 42°C. As shown in Figure 4(a), the rate of microparticle release increased monotonically with temperature. Linear regression analysis demonstrated that this trend was statistically significant (see legend). Control experiments assessing other known effects of ionomycin (cytoskeletal disruption, phospholipase sensitivity) demonstrated that the drug was equally effective throughout this temperature range (not shown). Therefore, the effects of temperature were presumably on the release process rather than on the efficacy of the drug.

The relative level of membrane order was also assessed at these temperatures by fluorescence spectroscopy using patman GP and TMA-DPH anisotropy. The data with both probes (Figures 4(b) and 4(c)) showed a significant decrease in apparent membrane order. Accordingly, a strong correlation between the level of order detected by these probes and the rate of vesicle release was observed ($P < 0.025$, $r^2 > 0.44$, $n = 11$ temperatures, by linear regression, values apply to both patman GP and TMA-DPH anisotropy). Overall, Figure 4 supports the hypothesis that membrane physical properties play a role in microparticle release for both nucleated and nonnucleated cells.

## 4. Discussion

It appears that the basic mechanisms governing microparticle shedding in platelets and erythrocytes may apply broadly to all cell types. This finding implies that microparticle release may simply be the consequence of a reduction in the stability of the cell membrane. Hence, factors that regulate the release process are biophysical in nature and represent withdrawal of elements that normally maintain stability. Based on this study and those focused on platelets and erythrocytes, these critical elements include the cellular osmotic balance, cytoskeletal attachments, the asymmetric distribution of phospholipids across the two faces of the membrane, and the elastic properties of the bilayer [14–27, 29, 41, 42].

The importance of osmotic balance is obvious because of associated effects on cell volume. A cell volume reduction would be required to compensate for the loss of plasma membrane surface area as microparticles are released. Moreover, the volume change would presumably promote membrane budding because of the stress imposed by surface area mismatch. In the case of calcium-induced particle release, the volume reduction appears to involve potassium efflux.

As shown in Figure 1, alterations to the actin cytoskeleton accompanied calcium influx as expected [21–27]. Attempts to verify the importance of this event through pharmacological inhibition of the cytoskeletal alterations (e.g., calpain inhibition) were unsuccessful, probably because of redundant mechanisms for the effect of calcium [3]. In erythrocytes and platelets, where regulatory mechanisms are simpler, inhibition of calpain is sufficient to impair microparticle release [15, 22, 23, 28]. To the extent that these findings apply broadly across cell types [24–27], they imply that one important role of the membrane-associated cytoskeleton is to maintain stability and prevent particle release.

The role of transbilayer asymmetry of phospholipid species in maintaining membrane stability seems less obvious. Loss of that asymmetry during internal calcium accumulation results in exposure of PS on the outer surface. Much attention has been paid to that exposure because of the roles of PS as a signal mechanism in hemostasis and recognition of

apoptotic cells by macrophages [43]. Why external exposure of that lipid would be permissive for vesicle release is unclear. In fact, it may be that the critical issue is actually the reduction of PS on the interior of the membrane. Since some proteins with C2-like domains are involved with cytoskeleton-membrane anchoring [44], the reduction of PS on the intracellular membrane face may lead to a loss of critical protein interactions resulting in a diminution of membrane stability. This explanation could rationalize why movement of such a minor component would have such a large impact on the membrane. Moreover, it is possible that PS is not the critical or sole participant in permitting microparticle release since calcium loading will also result in external exposure of phosphatidylethanolamine and intracellular accumulation of phosphatidylcholine through inhibition of aminophospholipid translocase and activation of scramblase [3].

The role of membrane order differed from that of potassium current, cytoskeleton breakdown, and scramblase activation. Each of the latter three appeared required, though not individually sufficient, for microparticle release. In contrast, membrane order, at least at the level accessible to experimental manipulation in living cells, functioned only as a modulator of the rate of release. Only a minimal trend in the apparent amount of membrane particles shed was observed across the temperature range, suggesting that there is a limit on how much is or can be released ($P = 0.05$, $r^2 = 0.047$ by linear regression, $n = 5$–15 per temperature, 83 total points).

It seems likely that the contribution of membrane order relates to the elasticity of the membrane. Presumably, a decrement in elasticity as temperature is raised creates flexibility in the membrane allowing for deformation sufficient for microparticles to bud and be released [41, 42]. The apparent upper boundary to the amount of particles discharged may be determined by the magnitude of the osmotically induced volume change and/or the extent of cytoskeletal alteration. The limit would be a key for cell survival since not all events resulting in microparticle shedding are associated with cell death [3].

## 5. Conclusions

The results of this study demonstrate that the biophysical mechanisms involved in microparticle release in erythrocytes and platelets probably apply broadly to all cell types. These include potassium ion efflux (presumably with concomitant osmotic effects), loss of membrane phospholipid asymmetry, and cytoskeletal disruption. Moreover, the level of membrane lipid order appears to modulate the release process with greater rates of release occurring from a more fluid membrane. This apparent effect of membrane order suggests that conditions in which these physical properties are altered may promote enhanced or inappropriate microparticle shedding. The obvious example is during apoptosis when the cell membrane becomes disordered and more fluid prior to membrane blebbing and microparticle shedding [45]. As a second example, recent investigations of blood microparticle levels of deep-sea divers have indicated an elevation of the particles after decompressing from the dive [46]. This reduction in pressure would surely have impact on the physical properties of the cell membrane similar to the elevation of temperature

Membrane Properties Involved in Calcium-Stimulated Microparticle Release from the Plasma Membranes of S49
Lymphoma Cells

163

in Figure 4. In addition, several reports have indicated that the plasma membrane of some tumor cells is more disordered than that of the corresponding nontransformed cells [47–50]. This raises the possibility that particles shed in greater numbers from these cells could spread signals from the tumor that may impact the pathogenesis of the disease.

## Conflict of Interests

There is no conflict of interests regarding the publication of this paper.

## References

[1] J. H. W. Distler, L. C. Huber, S. Gay, O. Distler, and D. S. Pisetsky, "Microparticles as mediators of cellular cross-talk in inflammatory disease," *Autoimmunity*, vol. 39, no. 8, pp. 683–690, 2006.

[2] L. Burnier, P. Fontana, B. R. Kwak, and A.-S. Anne, "Cell-derived microparticles in haemostasis and vascular medicine," *Thrombosis and Haemostasis*, vol. 101, no. 3, pp. 439–451, 2009.

[3] S. Montoro-García, E. Shantsila, F. Marín, A. Blann, and G. Y. H. Lip, "Circulating microparticles: new insights into the biochemical basis of microparticle release and activity," *Basic Research in Cardiology*, vol. 106, no. 6, pp. 911–923, 2011.

[4] A. P. Owens III and N. MacKman, "Microparticles in hemostasis and thrombosis," *Circulation Research*, vol. 108, no. 10, pp. 1284–1297, 2011.

[5] R. J. Berckmans, R. Nieuwland, A. N. Böing, F. P. H. T. M. Romijn, C. E. Hack, and A. Sturk, "Cell-derived microparticles circulate in healthy humans and support low grade thrombin generation," *Thrombosis and Haemostasis*, vol. 85, no. 4, pp. 639–646, 2001.

[6] J. M. Herring, M. A. McMichael, and S. A. Smith, "Microparticles in health and disease," *Journal of Veterinary Internal Medicine*, vol. 27, no. 5, pp. 1020–1033, 2013.

[7] K. N. Couper, T. Barnes, J. C. R. Hafalla et al., "Parasite-derived plasma microparticles contribute significantly to malaria infection-induced inflammation through potent macrophage stimulation," *PLoS Pathogens*, vol. 6, no. 1, Article ID e1000744, 2010.

[8] J. H. W. Distler, A. Akhmetshina, C. Dees et al., "Induction of apoptosis in circulating angiogenic cells by microparticles," *Arthritis and Rheumatism*, vol. 63, no. 7, pp. 2067–2077, 2011.

[9] J. E. Geddings and N. Mackman, "Tumor-derived tissue factor-positive microparticles and venous thrombosis in cancer patients," *Blood*, vol. 122, no. 11, pp. 1873–1880, 2013.

[10] S. B. Walters, J. Kieckbusch, G. Nagalingam et al., "Microparticles from mycobacteria-infected macrophages promote inflammation and cellular migration," *The Journal of Immunology*, vol. 190, no. 2, pp. 669–677, 2013.

[11] X. Delabranche, A. Berger, J. Boisramé-Helms, and F. Meziani, "Microparticles and infectious diseases," *Médecine et Maladies Infectieuses*, vol. 42, no. 8, pp. 335–343, 2012.

[12] M. T. Sartori, A. D. Puppa, A. Ballin et al., "Circulating microparticles of glial origin and tissue factor bearing in high-grade glioma: a potential prothrombotic role," *Thrombosis and Haemostasis*, vol. 110, no. 2, pp. 378–385, 2013.

[13] H. F. G. Heijnen, A. E. Schiel, R. Fijnheer, H. J. Geuze, and J. J. Sixma, "Activated platelets release two types of membrane vesicles: microvesicles by surface shedding and exosomes derived from exocytosis of multivesicular bodies and α-granules," *Blood*, vol. 94, no. 11, pp. 3791–3799, 1999.

[14] D. Allan and P. Thomas, "$Ca^{2+}$-induced biochemical changes in human erythrocytes and their relation to microvesiculation," *Biochemical Journal*, vol. 198, no. 3, pp. 433–440, 1981.

[15] S. K. Smith, A. R. Farnbach, F. M. Harris et al., "Mechanisms by which intracellular calcium induces susceptibility to secretory phospholipase $A_2$ in human erythrocytes," *The Journal of Biological Chemistry*, vol. 276, no. 25, pp. 22732–22741, 2001.

[16] E. Reichstein and A. Rothstein, "Effects of quinine on $Ca^{++}$-induced $K^+$ efflux from human red blood cells," *Journal of Membrane Biology*, vol. 59, no. 1, pp. 57–63, 1981.

[17] F. Bassé, J. G. Stout, P. J. Sims, and T. Wiedmer, "Isolation of an erythrocyte membrane protein that mediates $Ca^{2+}$-dependent transbilayer movement of phospholipid," *The Journal of Biological Chemistry*, vol. 271, no. 29, pp. 17205–17210, 1996.

[18] Q. Zhou, J. Zhao, J. G. Stout, R. A. Luhm, T. Wiedmer, and P. J. Sims, "Molecular cloning of human plasma membrane phospholipid scramblase. A protein mediating transbilayer movement of plasma membrane phospholipids," *The Journal of Biological Chemistry*, vol. 272, no. 29, pp. 18240–18244, 1997.

[19] E. M. Bevers, T. Wiedmer, P. Comfurius et al., "Defective $Ca^{2+}$-induced microvesiculation and deficient expression of procoagulant activity in erythrocytes from a patient with a bleeding disorder: a study of the red blood cells of Scott syndrome," *Blood*, vol. 79, no. 2, pp. 380–388, 1992.

[20] P. Comfurius, J. M. G. Senden, R. H. J. Tilly, A. J. Schroit, E. M. Bevers, and R. F. A. Zwaal, "Loss of membrane phospholipid asymmetry in platelets and red cells may be associated with calcium-induced shedding of plasma membrane and inhibition of aminophospholipid translocase," *Biochimica et Biophysica Acta*, vol. 1026, no. 2, pp. 153–160, 1990.

[21] S. Cauwenberghs, M. A. H. Feijge, A. G. S. Harper, S. O. Sage, J. Curvers, and J. W. M. Heemskerk, "Shedding of procoagulant microparticles from unstimulated platelets by integrin-mediated destabilization of actin cytoskeleton," *FEBS Letters*, vol. 580, no. 22, pp. 5313–5320, 2006.

[22] Y. Yano, E. Shiba, J.-I. Kambayashi et al., "The effects of calpeptin (a calpain specific inhibitor) on agonist induced microparticle formation from the platelet plasma membrane," *Thrombosis Research*, vol. 71, no. 5, pp. 385–396, 1993.

[23] J.-M. Pasquet, F. Toti, A. T. Nurden, and J. Dachary-Prigent, "Procoagulant activity and active calpain in platelet-derived microparticles," *Thrombosis Research*, vol. 82, no. 6, pp. 509–522, 1996.

[24] A. M. Curtis, P. F. Wilkinson, M. Gui, T. L. Gales, E. Hu, and J. M. Edelberg, "p38 mitogen-activated protein kinase targets the production of proinflammatory endothelial microparticles," *Journal of Thrombosis and Haemostasis*, vol. 7, no. 4, pp. 701–709, 2009.

[25] M. L. Coleman, E. A. Sahai, M. Yeo, M. Bosch, A. Dewar, and M. F. Olson, "Membrane blebbing during apoptosis results from caspase-mediated activation of ROCK I," *Nature Cell Biology*, vol. 3, no. 4, pp. 339–345, 2001.

[26] H. Miyoshi, K. Umeshita, M. Sakon et al., "Calpain activation in plasma membrane bleb formation during tert-butyl hydroperoxide-induced rat hepatocyte injury," *Gastroenterology*, vol. 110, no. 6, pp. 1897–1904, 1996.

[27] J. C. Mills, N. L. Stone, J. Erhardt, and R. N. Pittman, "Apoptotic membrane blebbing is regulated by myosin light chain phosphorylation," *The Journal of Cell Biology*, vol. 140, no. 3, pp. 627–636, 1998.

[28] F. Basse, P. Gaffet, and A. Bienvenue, "Correlation between inhibition of cytoskeleton proteolysis and anti-vesiculation effect of calpeptin during A23187-induced activation of human platelets: are vesicles shed by filopod fragmentation?" *Biochimica et Biophysica Acta*, vol. 1190, no. 2, pp. 217–224, 1994.

[29] L. J. Gonzalez, E. Gibbons, R. W. Bailey et al., "The influence of membrane physical properties on microvesicle release in human erythrocytes," *PMC Biophysics*, vol. 2, no. 1, article 7, 2009.

[30] H. A. Wilson, J. B. Waldrip, K. H. Nielson et al., "Mechanisms by which elevated intracellular calcium induces S49 cell membranes to become susceptible to the action of secretory phospholipase $A_2$," *The Journal of Biological Chemistry*, vol. 274, no. 17, pp. 11494–11504, 1999.

[31] T. Parasassi, G. de Stasio, G. Ravagnan, R. M. Rusch, and E. Gratton, "Quantitation of lipid phases in phospholipid vesicles by the generalized polarization of laurdan fluorescence," *Biophysical Journal*, vol. 60, no. 1, pp. 179–189, 1991.

[32] H. Franchino, E. Stevens, J. Nelson, T. A. Bell, and J. D. Bell, "Wavelength dependence of patman equilibration dynamics in phosphatidylcholine bilayers," *Biochimica et Biophysica Acta*, vol. 1828, no. 2, pp. 877–886, 2012.

[33] R. S. P. Benson, S. Heer, C. Dive, and A. J. M. Watson, "Characterization of cell volume loss in CEM-C7A cells during dexamethasone-induced apoptosis," *American Journal of Physiology—Cell Physiology*, vol. 270, no. 4, part 1, pp. C1190–C1203, 1996.

[34] E. Maeno, Y. Ishizaki, T. Kanaseki, A. Hazama, and Y. Okada, "Normotonic cell shrinkage because of disordered volume regulation is an early prerequisite to apoptosis," *Proceedings of the National Academy of Sciences of the United States of America*, vol. 97, no. 17, pp. 9487–9492, 2000.

[35] D. W. C. Dekkers, P. Comfurius, W. M. J. Vuist et al., "Impaired $Ca^{2+}$-induced tyrosine phosphorylation and defective lipid scrambling in erythrocytes from a patient with Scott syndrome: a study using an inhibitor for scramblase that mimics the defect in Scott syndrome," *Blood*, vol. 91, no. 6, pp. 2133–2138, 1998.

[36] J. Dachary-Prigent, J.-M. Pasquet, E. Fressinaud, F. Toti, J.-M. Freyssinet, and A. T. Nurden, "Aminophospholipid exposure, microvesiculation and abnormal protein tyrosine phosphorylation in the platelets of a patient with Scott syndrome: a study using physiologic agonists and local anaesthetics," *British Journal of Haematology*, vol. 99, no. 4, pp. 959–967, 1997.

[37] J. Zhao, Q. Zhou, T. Wiedmer, and P. J. Sims, "Level of expression of phospholipid scramblase regulates induced movement of phosphatidylserine to the cell surface," *The Journal of Biological Chemistry*, vol. 273, no. 12, pp. 6603–6606, 1998.

[38] A. D. Tepper, P. Ruurs, T. Wiedmer, P. J. Sims, J. Borst, and W. J. van Blitterswijk, "Sphingomyelin hydrolysis to ceramide during the execution phase of apoptosis results from phospholipid scrambling and alters cell-surface morphology," *The Journal of Cell Biology*, vol. 150, no. 1, pp. 155–164, 2000.

[39] V. E. Kagan, B. Gleiss, Y. Y. Tyurina et al., "A role for oxidative stress in apoptosis: oxidation and externalization of phosphatidylserine is required for macrophage clearance of cells undergoing Fas-mediated apoptosis," *Journal of Immunology*, vol. 169, no. 1, pp. 487–499, 2002.

[40] J. Nelson, L. L. Francom, L. Anderson et al., "Investigation into the role of phosphatidylserine in modifying the susceptibility of human lymphocytes to secretory phospholipase $A_2$ using cells deficient in the expression of scramblase," *Biochimica et Biophysica Acta*, vol. 1818, no. 5, pp. 1196–1204, 2012.

[41] T. Ruiz-Herrero, E. Velasco, and M. F. Hagan, "Mechanisms of budding of nanoscale particles through lipid bilayers," *The Journal of Physical Chemistry B*, vol. 116, no. 32, pp. 9595–9603, 2012.

[42] Y. Kozlovsky and M. M. Kozlov, "Membrane fission: model for intermediate structures," *Biophysical Journal*, vol. 85, no. 1, pp. 85–96, 2003.

[43] E. M. Bevers and P. L. Williamson, "Phospholipid scramblase: an update," *FEBS Letters*, vol. 584, no. 13, pp. 2724–2730, 2010.

[44] D. Zhang and L. Aravind, "Identification of novel families and classification of the C2 domain superfamily elucidate the origin and evolution of membrane targeting activities in eukaryotes," *Gene*, vol. 469, no. 1-2, pp. 18–30, 2010.

[45] R. W. Bailey, T. Nguyen, L. Robertson et al., "Sequence of physical changes to the cell membrane during glucocorticoid-induced apoptosis in S49 lymphoma cells," *Biophysical Journal*, vol. 96, no. 7, pp. 2709–2718, 2009.

[46] S. R. Thom, T. N. Milovanova, M. Bogush et al., "Microparticle production, neutrophil activation, and intravascular bubbles following open-water SCUBA diving," *Journal of Applied Physiology*, vol. 112, no. 8, pp. 1268–1278, 2012.

[47] J. Nelson, K. Barlow, D. O. Beck et al., "Synergistic effects of secretory phospholipase $A_2$ from the venom of *Agkistrodon piscivorus piscivorus* with cancer chemotherapeutic agents," *BioMed Research International*, vol. 2013, Article ID 565287, 5 pages, 2013.

[48] M. Sok, M. Šentjurc, and M. Schara, "Membrane fluidity characteristics of human lung cancer," *Cancer Letters*, vol. 139, no. 2, pp. 215–220, 1999.

[49] G. Taraboletti, L. Perin, B. Bottazzi, A. Mantovani, R. Giavazzi, and M. Salmona, "Membrane fluidity affects tumor-cell motility, invasion and lung-colonizing potential," *International Journal of Cancer*, vol. 44, no. 4, pp. 707–713, 1989.

[50] R. Zeisig, T. Koklič, B. Wiesner, I. Fichtner, and M. Sentjurč, "Increase in fluidity in the membrane of MT3 breast cancer cells correlates with enhanced cell adhesion *in vitro* and increased lung metastasis in NOD/SCID mice," *Archives of Biochemistry and Biophysics*, vol. 459, no. 1, pp. 98–106, 2007.

# 17

# Slow Diffusion Underlies Alternation of Fast and Slow Growth Periods of Microtubule Assembly

**Ming Yang**

*Department of Botany, Oklahoma State University, 301 Physical Sciences, Stillwater, OK 74078, USA*

Correspondence should be addressed to Ming Yang; ming.yang@okstate.edu

Academic Editors: A. Arcovito and A. Carotenuto

*In vitro* microtubule assembly exhibits a rhythmic phenomenon, that is, fast growth periods alternating with slow growth periods. Mechanism underlying this phenomenon is unknown. Here a simple diffusion mechanism coupled with small diffusion coefficients is proposed to underlie this phenomenon. Calculations based on previously published results demonstrate that such a mechanism can explain the differences in the average duration of the interval encompassing a fast growth period and a slow growth period in *in vitro* microtubule assembly experiments in different conditions. Because no parameter unique to the microtubule assembly process is involved in the analysis, the proposed mechanism is expected to be generally applicable to heterogeneous chemical reactions. Also because biological systems are characterized by heterogeneous chemical reactions, the diffusion-based rhythmic characteristic of heterogeneous reactions is postulated to be a fundamental element in generating rhythmic behaviors in biological systems.

## 1. Introduction

A living system such as a cell consists of multiple heterogeneous catalytic reactions that occur at the solid-liquid interphase. Some of the most prominent examples of such reactions include microtubule assembly during spindle formation, protein synthesis on ribosomes, and DNA and RNA formation from existing DNA templates. It is generally accepted that diffusion rate can be a limiting factor for the product output of a heterogeneous reaction system [1, 2]. However, how diffusion affects a heterogeneous reaction involving a cellular structure has not been described with the support of experimental data.

The diffusion coefficient $D$ of a molecule species, which defines the diffusion rate of the molecule in a solution, is determined by the size and shape of the molecule and the temperature and viscosity of the solution [3]. Diffusion coefficients of small and large molecules in cellular compartments have been experimentally determined, with the $D$s of proteins in the cytoplasm of various cells ranging from 0.15 to 40 $\mu m^2$/s [4–7]. The $D$ of tubulin dimers in the cytoplasm of embryonic cells of sea urchin was found to be 4–10 $\mu m^2$/s [5] while the $D$ of a protein of similar molecular

weight (111 kilo Daltons) was 5.5 $\mu m^2$/s in the cytoplasm of *E. coli* [7]. From the small values of $D$s of large molecules in cellular compartments one may speculate that slow diffusion rates in heterogeneous catalytic reactions in living systems may generate characteristics that are typically associated with living systems. However, the significance of slow diffusion rates to living systems has not been demonstrated with an actual biological process.

It has been reported that microtubule assembly occurs in a stepwise fashion both *in vivo* [8, 9] and *in vitro* [10]. Kerssemakers et al. [10] hypothesized that the stepwise microtubule assembly resulted from a simultaneous addition of more than one tubulin dimer to the microtubule plus end in one assembly step but they did not address the nature of the pausing period between two successive steps. The periods in individual microtubule assembly steps and the pausing periods in [10] in general were less than one second and several seconds, respectively. Scheck et al. [11] conducted a study similar to that in [10] and found that microtubule growth occurred in steps smaller than those in [10]. Scheck et al. thus concluded that microtubule assembly occurs one tubulin dimer at a time. However, data in [11] still show that during certain periods of several-second long, microtubule growth still occurred in successive peaks with pausing periods between the successive peaks, even though the pausing periods were shorter than those in [10]. No explanation of the pausing phenomenon was given in [11]. Furthermore, the durations of the pausing periods in two different experimental conditions in [10] also appear to differ from each other. These quantitative differences may reflect a fundamental characteristic in heterogeneous catalysis that warrants further investigation.

In this study, it is shown that the differences in the duration of the rhythmic cycle (alternation of a fast growth period with a slow growth period in an absolutely positive microtubule assembly phase) of microtubule assembly described above can be explained by the differences in the critical concentration and the diffusion coefficient of tubulin molecules in the different experimental settings. These results support the hypothesis that slow diffusion rate of tubulin molecules relative to the assembly rate generates the phenomenon of alternating fast and slow growth periods in microtubule assembly. It is also suggested that such a mechanism generally operates in heterogeneous reactions *in vivo* to produce rhythmic reactions in biological systems.

## 2. Methods

The theoretical basis for this analysis is described in the following imaginary experiment in which an enzyme activity exists on a solid structure that catalyzes a heterogeneous reaction using a soluble protein substrate in the solution. If the reaction has occurred for a sufficient period of time, and the reaction rate is fast so that the diffusion rate of the soluble protein is a limiting factor for the reaction, a dynamic chemical concentration gradient of the soluble protein should be established with the lowest concentration at the reaction site and a slope of increasing concentration extending outward from the reaction site. It is predicted that the reaction will come to a temporary halt or a dramatic reduction when the soluble protein reaches or comes close to a critical concentration at the reaction site; the critical concentration, $C_c$, is defined as such that below it the reaction ceases. Thus, this reaction system is expected to alternate between a fast reaction period $t_f$ and a slow reaction period $t_s$ (or even pause), and the sum of the averages of the two periods should be conceptually equal to the average diffusion time, $t_d$, of the soluble protein along the concentration gradient. That is,

$$t_d = t_f + t_s. \tag{1}$$

In this system, assuming one-dimensional diffusion to simplify the analysis, the diffusion time can be estimated by

$$t_d \approx \frac{x^2}{2D}, \tag{2}$$

where $x$ is the mean distance traveled by the diffusing protein molecules in one direction to the reaction site after elapsed time $t_d$ and $D$ is the diffusion coefficient of the protein in the solution. It is also noted here that $x$ should be proportional to

the length of the chemical gradient $L$ when the system reaches $C_c$ at the reaction site.

To conduct the analysis, the average $t_f$s and average $t_s$s (Table 1) were extrapolated from the left halves of Figures 2(a) and 2(b) in [10] and the time periods of 2.8 seconds to 4.6 seconds, 6.8 seconds to 8.5 seconds, and 15 to 17.5 seconds for the experiment with GTP-tubulin in Figure 3(a) in [11]. These segments of the microtubule assembly curves were chosen because they consist of at least two peaks during one microtubule growth period so that one or more plateaus (corresponding to $t_s$s) exist. A $t_f$ period is from an early rising point of a peak to an early point at the peak, and a $t_s$ is from an early point at the peak to an early rising point of the following peak, as illustrated in Figure 1. Two sets of experiments were conducted in [10]: one without the microtubule-associated protein XMAP215 and the other with XMAP215.

## 3. Results

All three average $t_f$s in Table 1 are not statistically different from each other (t-test, $P > 0.14$) but the $t_s$s in [10] are longer than that in [11] (t-test, $P < 6 \times 10^{-4}$). Furthermore, in [10], the $t_s$ without XMAP215 is also longer than the $t_s$ with XMAP215 (t-test, $P = 0.03$). These results indicate that the assembly reactions in all conditions occurred at a similar rate, and the variation in total diffusion time primarily derives from the variation in the duration of the slow growth period.

The average microtubule growth lengths per fast growth period in [10] were estimated, based on Figure 3(e) in the publication, to be 24 nm and 48 nm when without and with XMAP215, respectively. The average microtubule growth length per fast growth period in a particular condition is not given in [11]. The lengths of microtubule growth in the fast growth periods in Figure 3(a) in [11] were thus measured from the baseline of a preceding slow growth period to the topline of the subsequent fast growth period (Figure 1), and the average growth length per fast growth period is calculated to be 25.4 ± 3.7 nm (mean ± standard error, $n = 15$).

Considering that a microtubule consists of 13 protofilaments and each $\alpha\beta$ tubulin dimer adds approximately 8 nm to the length of a protofilament, the growth of 24 nm and 48 nm of a microtubule consumes 39 ((24/8) × 13) and 78 ((48/8) × 13) tubulin dimers, respectively. Assuming that a uniform tubulin concentration gradient is established at the end of a fast growth period and microtubule assembly comes to a temporary halt due to reaching the critical tubulin concentration around the microtubule assembly point [12], the flux of tubulin during the fast growth period can be expressed as follows according to Fick's First Law:

$$J = -D\left(\frac{\partial C}{\partial X}\right) = -D\left[\frac{(C_0 - C_c)}{L}\right], \tag{3}$$

where $D$ is the diffusion coefficient, $C_0$ is the tubulin concentration outside the tubulin gradient, and $L$ is the length of the tubulin concentration gradient. The flux can also be calculated by dividing the amount of tubulin consumed in a fast growth period by the area of the microtubule assembly

TABLE 1: Average durations (in second) of fast growth periods ($t_f$) and slow growth periods ($t_s$) in *in vitro* microtubule assembly and tubulin diffusion times ($t_d = t_f + t_s$).

| Mean $t_f$ ± standard error | Mean $t_s$ ± standard error | Mean $t_d$ | Seed for microtubule assembly | Reference |
|---|---|---|---|---|
| 0.55 ± 0.09 ($n = 5$) | 3.85 ± 0.57 ($n = 17$) | 4.4 | Axoneme − XMAP215 | [10] |
| 0.63 ± 0.11 ($n = 6$) | 2.33 ± 0.37 ($n = 11$) | 2.96 | Axoneme + XMAP215 | [10] |
| 0.44 ± 0.04 ($n = 8$) | 0.54 ± 0.08 ($n = 6$) | 0.98 | Microtubule fragments | [11] |

$n$: number of samples measured.

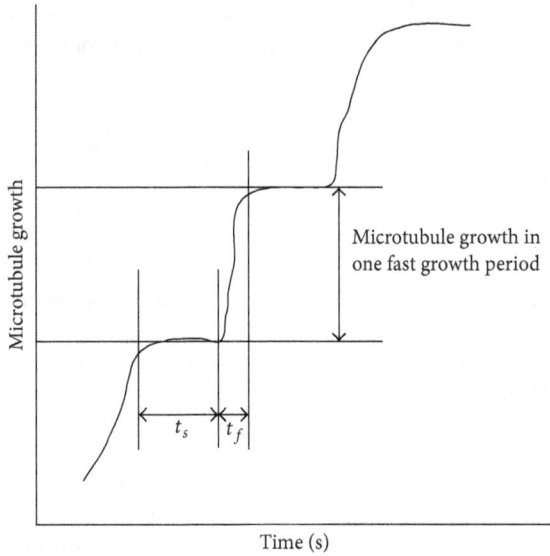

FIGURE 1: The durations of fast and slow growth periods of microtubule assembly and the length of microtubule growth in a fast growth period. The duration measurements were conducted with data in [10, 11], while the length measurements were only conducted with data in [11] since information on such lengths is available in [10].

site and the duration of the fast growth period. Then (3) is turned into

$$J_{-\text{XMAP215}} = -D \left[ \frac{(C_0 - C_{c-\text{XMAP215}})}{L_{-\text{XMAP215}}} \right]$$
$$= \frac{\left[ 39/\left( 6.022 \times 10^{23} \right) \right]}{\left( at_{f-\text{XMAP215}} \right)}, \tag{4}$$

$$J_{+\text{XMAP215}} = -D \left[ \frac{(C_0 - C_{c+\text{XMAP215}})}{L_{+\text{XMAP215}}} \right]$$
$$= \frac{\left[ 78/\left( 6.022 \times 10^{23} \right) \right]}{\left( at_{f+\text{XMAP215}} \right)}, \tag{5}$$

where −XMAP215 and +XMAP215 denote the two experiments in [10] respectively, and $a$ is the area parameter around the microtubule assembly point and is intrinsic to the

geometric form of the microtubule assembly point. $6.022 \times 10^{23}$ is the Avogadro's Constant.

The microtubule assembly conditions without and with XMAP215 are assumed to be comparable to the previously described conditions for microtubule assembly without and with microtubule-associated proteins, respectively. $C_{c-\text{XMAP215}}$ and $C_{c+\text{XMAP215}}$ are thus estimated to be $4 \, \mu M$ [12] and $2 \, \mu M$ [13, 14], respectively. Using the $t_f$ values in Table 1, letting the average $C_0 = (5 + 20)/2 = 12.5 \, \mu M$ [10, online supplementary information], and dividing (4) by (5), the following is obtained

$$\frac{\left[ (12.5 - 4)/L_{-\text{XMAP215}} \right]}{\left[ (12.5 - 2)/L_{+\text{XMAP215}} \right]} = \frac{(39 \times 0.63)}{(78 \times 0.55)}. \tag{6}$$

Consolidating (6), then

$$\frac{L_{+\text{XMAP215}}}{L_{-\text{XMAP215}}} = 0.71, \quad \text{or} \quad \frac{L_{+\text{xmap215}}^2}{L_{-\text{xmap215}}^2} = 0.5. \tag{7}$$

It is presumed that the average diffusion distance $x$ is proportional to the tubulin concentration gradient length $L$; then

$$\frac{x_{+\text{XMAP215}}^2}{x_{-\text{XMAP215}}^2} = \frac{L_{+\text{XMAP215}}^2}{L_{-\text{XMAP215}}^2} = 0.5. \tag{8}$$

From (2) and (8), the following is obtained

$$\frac{t_{d+\text{XMAP215}}}{t_{d-\text{XMAP215}}} = \frac{x_{+\text{XMAP215}}^2}{x_{-\text{XMAP215}}^2} = 0.5. \tag{9}$$

The experimental result from $t_d$ values in Table 1 is

$$\frac{t_{d+\text{XMAP215}}}{t_{d-\text{XMAP215}}} = \frac{2.96}{4.4} = 0.67. \tag{10}$$

The calculated value of $t_{d+\text{XMAP215}}/t_{d-\text{XMAP215}}$ is quite close to that of the experimental $t_{d+\text{XMAP215}}/t_{d-\text{XMAP215}}$, supporting the assumption that the periodic slowdowns in microtubule assembly are caused by the periodic occurrences of insufficient flux of tubulin that arise from the slow diffusion rate relative to the reaction rate.

In the microtubule assembly process described in [11], isolated microtubule fragments were used to seed the microtubule growth. These seeding microtubule fragments should have retained a high level of microtubule-associated proteins. It is thus assumed that $C_c \approx 2\,\mu M$ in [11]. Because $C_0 = 5\,\mu M$ in [11], the tubulin flux at the microtubule assembly site during the fast growth period is

$$J_{\text{Schek}} = -\frac{D_{\text{Schek}}(5-2)}{L_{\text{Schek}}} = \frac{\left[39/\left(6.022 \times 10^{23}\right)\right]}{\left(a t_{f\text{Schek}}\right)}. \tag{11}$$

Dividing (4) by (11), the following is obtained

$$\frac{J_{-\text{XMAP215}}}{J_{\text{Schek}}} = \frac{\left(8.5 D_{-\text{XMAP215}}/L_{-\text{XMAP215}}\right)}{\left(3 D_{\text{Schek}}/L_{\text{Schek}}\right)}$$
$$= \frac{t_{f\text{Schek}}}{t_{f-\text{XMAP215}}}. \tag{12}$$

Using the $t_f$ values in Table 1 to consolidate (12), then

$$\frac{L_{\text{Schek}}}{L_{-\text{XMAP215}}} \approx \frac{0.28 D_{\text{Schek}}}{D_{-\text{XMAP215}}} \tag{13}$$

or

$$\frac{L_{\text{Schek}}^{2}}{L_{-\text{XMAP215}}^{2}} \approx \frac{0.078 D_{\text{Schek}}^{2}}{D_{-\text{XMAP215}}^{2}}. \tag{14}$$

From (2) and based on $L_{\text{Schek}}^{2}/L_{-\text{XMAP215}}^{2} = x_{\text{Schek}}^{2}/x_{-\text{XMAP215}}^{2}$, the following is obtained

$$\frac{t_{d\text{Schek}}}{t_{d-\text{XMAP215}}} = \frac{\left(L_{\text{Schek}}^{2} D_{-\text{XMAP215}}\right)}{\left(L_{-\text{XMAP215}}^{2} D_{\text{Schek}}\right)}. \tag{15}$$

From (14) and (15), the following is derived

$$\frac{t_{d\text{Schek}}}{t_{d-\text{XMAP215}}} \approx \frac{0.078 D_{\text{Schek}}}{D_{-\text{XMAP215}}}. \tag{16}$$

For one molecular species in different solutions at the same temperature, its $D$ values are in an inverse relationship with the viscosity of the solutions. The solution used in [10] for microtubule assembly is described as "very viscous"; the solution contained up to four times of tubulin concentration of that in [11]. In addition, the assembly solution in [10] contained 0.5–1% bovine serum albumin that was absent in the assembly solution in [11]. Therefore, the viscosity of the assembly solution in [10] is expected to be significantly higher than that in [11]; that is, $D_{\text{Schek}} > D_{-\text{XMAP215}}$. Based on the values in Table 1, the experimental $t_{d\text{Schek}}/t_{d-\text{XMAP215}}$ becomes

$$\frac{t_{d\text{Schek}}}{t_{d-\text{XMAP215}}} = \frac{0.98}{4.4} = 0.22. \tag{17}$$

Let $t_{d\text{Schek}}/t_{d-\text{XMAP215}}$ in (16) and (17) be the same; then

$$\frac{0.078 D_{\text{Schek}}}{D_{-\text{XMAP215}}} = 0.22 \tag{18}$$

or

$$D_{\text{Schek}} \approx 2.8 D_{-\text{XMAP215}}. \tag{19}$$

Equation (19) gives a plausible value of $D_{\text{Schek}}$ in relation to $D_{-\text{XMAP215}}$, considering the viscosity differences in the two microtubule assembly solutions.

The above analysis is based on the diffusion flux in one dimension. To scale it up to a three-dimensional flux, it is assumed that there are a total of $n$ paths for tubulin dimers to travel towards the microtubule assembly site. It is predicted that these paths form the 3D configuration of the tubulin flux that has a geometric focus at the microtubule assembly site and is radially symmetrical around the growing microtubule. If the tubulin concentration gradient lengths of these paths are $L_1, L_2, \ldots, L_n$, there should exist a certain quantitative relationship between $L_1$ and other $L$s; that is, $L_2 = b_2 L_1$, $L_3 = b_3 L_1, \ldots, L_n = b_n L_1$. The values of the $b$s are determined by the shape of the 3D configuration of the tubulin flux. Therefore,

$$\text{The average } L_{3\text{D}} = \frac{\left(L_1 + L_2 + L_3 + \cdots + L_n\right)}{n}$$
$$= \frac{L_1\left(1 + b_2 + b_3 + \cdots + b_n\right)}{n}. \tag{20}$$

Because the experimental apparatus in [10] is essentially the same as that in [11] (Figure 2), the 3D configurations of the tubulin fluxes in the three reaction conditions discussed in this report should be of the same shape but different sizes; that is, they have the same $b$ values with respect to the same paths. The earlier discussed $L_{-\text{XMAP215}}$, $L_{+\text{XMAP215}}$, and $L_{\text{Schek}}$ can be considered the tubulin concentration gradient lengths of the same path in the three tubulin 3D fluxes, respectively. The average $L$s of the three 3D tubulin fluxes can then be expressed as

$$\text{Average } L_{-\text{XMAP215-3D}}$$
$$= \frac{L_{-\text{XMAP215}}\left(1 + b_2 + b_3 + \cdots + b_n\right)}{n},$$

$$\text{Average } L_{+\text{XMAP215-3D}}$$
$$= \frac{L_{+\text{XMAP215}}\left(1 + b_2 + b_3 + \cdots + b_n\right)}{n}, \tag{21}$$

$$\text{Average } L_{\text{Schek-3D}}$$
$$= \frac{L_{\text{Schek}}\left(1 + b_2 + b_3 + \cdots + b_n\right)}{n}.$$

Equation (21) indicates that the ratios among $L_{-\text{XMAP215}}$, $L_{+\text{XMAP215}}$, and $L_{\text{Schek}}$ equal the ratios among the average $L_{-\text{XMAP215-3D}}$, $L_{+\text{XMAP215-3D}}$, and $L_{\text{Schek-3D}}$. The results obtained from (9) and (16), therefore, also apply to the 3D tubulin flux situation.

A further test of the proposed diffusion mechanism that underlies the microtubule assembly rhythmic behavior is that, during the $t_f$ period, whether the consumption of tubulin dimers by the microtubule assembly process can temporarily reduce the tubulin concentration around the assembly site to the level of tubulin critical concentration in the three specific conditions discussed in this report.

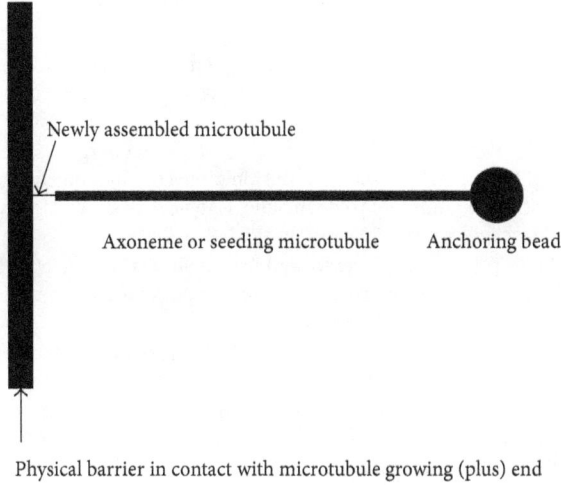

Newly assembled microtubule

Axoneme or seeding microtubule          Anchoring bead

Physical barrier in contact with microtubule growing (plus) end

FIGURE 2: Schematic representation of the experimental setup for microtubule assembly. The anchoring bead was trapped in optical tweezers. Refer to [10, 11] for more details about the experimental apparatuses and reaction conditions.

If so, it is taken as the indication that the uniform tubulin gradient near the assembly site was temporarily disrupted during $t_f$ period and it needs to be reestablished during the $t_s$ period before the next round of microtubule assembly. In other words, repetitions of a temporary disruption of the tubulin gradient followed by reestablishment of the gradient manifest into a rhythmic microtubule assembly behavior. To test the possibility of such temporary disruption in the tubulin gradient, a hemisphere with a radius $R$ of the length of the average diffusion distance during the time of $t_f$ and the assembly site as the center is deemed a relevant space in which the disruption takes place. A hemisphere is considered here because the barrier in the experimental apparatuses should block the tubulin diffusion from the left side of the barrier (Figure 2). If the tubulin concentration within the hemisphere is reduced to the tubulin critical concentration, the number of consumed tubulin dimers, $N$, is

$$N = \left(\frac{2}{3}\right)\pi R^3 \left[\frac{\left(C_{\text{edge}} - C_{\text{center}}\right)}{2}\right]\left(6.022 \times 10^{23}\right)$$
$$\approx \left(\frac{2}{3}\right)\pi R^3 \left[\frac{(R/x)\left(C_0 - C_c\right)}{2}\right]\left(6.022 \times 10^{23}\right), \tag{22}$$

where $C_{\text{edge}}$ and $C_{\text{center}}$ are the tubulin concentrations at the center and edge of the hemisphere before the disruption of the tubulin gradient, respectively, and $x$ is the average diffusion distance during the time of $t_d$.

From (2), $R = (2t_f D)^{1/2}$, and $x = (2t_d D)^{1/2}$, (22) then becomes

$$N \approx 2.96\left(6.022 \times 10^{23}\right)t_f^2 t_d^{-1/2} D^{3/2}\left(C_0 - C_c\right). \tag{23}$$

When $D = 0.07\ \mu\text{m}^2/\text{s}$ in [10] and $D = 0.07 \times 2.8\ \mu\text{m}^2/\text{s}$ in [11] (see (19)), using the $t_f$ and $t_d$ values in Table 1 and previous

$C_0$ and $C_c$ values, (23) is solved for the three experimental conditions to arrive at, respectively,

$$N_{\text{-XMAP215}} \approx 40,$$
$$N_{\text{+XMAP215}} \approx 80, \tag{24}$$
$$N_{\text{Schek}} = 91.$$

The $N_{\text{-XMAP215}}$ and $N_{\text{+XMAP215}}$ values are very close to the flux values of 39 and 78 that were estimated based on the average lengths of microtubule growth per fast growth period in the two conditions, respectively. $N_{\text{Schek}}$ is approximately two times the estimated flux value of 39 in the corresponding condition. The $R$ values in the three cases are $0.28\ \mu\text{m}$, $0.30\ \mu\text{m}$, and $0.42\ \mu\text{m}$, respectively, and the corresponding $x$ values are $0.78\ \mu\text{m}$, $0.63\ \mu\text{m}$, and $0.62\ \mu\text{m}$. These values seem to be within reasonable ranges for the experimental conditions, which suggest that, within the $t_f$ period, the microtubule assembly process is capable of depleting the tubulin dimers to a concentration at or near the critical concentration of microtubule assembly. The depicted $D$ values that produce the desired $N$ values are smaller than the $D$ value of tubulin dimers in the cytoplasm in sea urchin [5], but they may be close to the actual $D$ values in the *in vitro* experimental conditions. The $N_{\text{Schek}}$ value is somewhat higher than the desired result, but it can be potentially attributed to imprecise estimates of the other parameters in (23). In conclusion, the calculations demonstrate that it is at least plausible that the slow diffusion rates relative to the microtubule assembly rates can generate the pattern of alternating fast and slow growth periods of microtubule assembly.

## 4. Discussion

In biological systems, the diffusion coefficients of large molecules are small due to the large sizes and variable shapes of the molecules, the high viscosity of the fluid, and other more complex molecular interactions. It is not difficult to envision that the proposed mechanism based on the microtubule assembly process can be generally applicable to heterogeneous reactions in biological systems. Many molecular processes in biological systems are known to be rhythmic. For example, levels of secondary messenger molecules oscillate in cells [15], ion channels generate rhythmic electrical activities in neurons and cardiac cells [16], and single immobilized enzyme molecule exhibits rhythmic catalysis [17]. Whether and how slow diffusion rates relative to reaction rates play a role in generating these rhythmic behaviors remain to be investigated. It will be also very interesting to explore whether and how multiple rhythmic reactions at the individual reaction sites integrate into rhythmic behaviors and predictable patterns at a higher (e.g., system) level.

In the studies of the catalytic kinetics of immobilized single enzyme molecules, it was revealed that the catalytic reactions occurred in periods of high activity alternating with periods of low activity and the "waiting times" (in reference to the durations of the low activity periods by the investigators) fall into a broad time scale of milliseconds to seconds [17, 18]. It has been hypothesized that an enzyme molecule should have thousands of conformational states so that the waiting time can vary from milliseconds to seconds [17, 18]. It is argued here that the rhythmic nature of single enzyme catalysis with a broad range of waiting time can be simply explained by the same mechanism proposed for microtubule assembly in this study. The enzyme-substrate complex can be safely assumed to be a dynamic union, which undergoes separation-and-reunion cycles mainly due to the thermal dynamic movement of the substrate molecule since the enzyme molecules in the studies were immobilized. This dynamic process is expected to have a broad time scale that depends on the variable distance of the separated substrate molecule to the enzyme molecule; the waiting time range should be in proportion to the square of the distance range. Also importantly, the diffusion mechanism predicts that the waiting time should be reduced along with the increase of the substrate concentration, whereas it is difficult to fathom that the number of the conformational states should be reduced by the increase of the substrate concentration. Indeed, English et al. [17] found that the waiting time was decreased along with the increase of the substrate concentration.

Because nothing is known about the critical tubulin concentration for microtubule disassembly, this study is limited to the microtubule assembly process. However, microtubule disassembly also appears to alternate between fast and slow periods and overall occurs faster than microtubule assembly [10, 11]. Therefore, the degradation of a biological structure may also follow a similar mechanism as proposed for the synthesis of a biological structure, although the kinetics may be different from that of synthesis.

The length of *in vitro* microtubule growth in one fast growth period is much less than one unit (approximately 350 nm) of microtubule elongation observed in the *in vivo* spindle elongation process [8, 9]. It suggests that either the *in vivo* fast growth period of microtubule assembly is much longer than the *in vitro* fast growth period or there is another level of pause in assembly after a fixed number of fast and slow growth periods similar in duration to those observed *in vitro*. To determine which of the two scenarios occurs *in vivo*, it requires studies of microtubule assembly in cells at sufficient time and length resolutions.

## 5. Conclusion

This study of two *in vitro* microtubule assembly cases demonstrates that small diffusion coefficients of a reactant can lead to rhythmic behavior of the reaction in a heterogeneous reaction system.

## Conflict of Interests

The author declares that there is no conflict of interests regarding the publication of this paper.

## Acknowledgment

The author acknowledges financial support from the Oklahoma Center for the Advancement of Science and Technology.

## References

[1] M. Stenberg, L. Stiblert, and H. Nygren, "External diffusion in solid-phase immunoassays," *Journal of Theoretical Biology*, vol. 120, no. 2, pp. 129–140, 1986.

[2] J. L. van Roon, M. M. Arntz, A. I. Kallenberg et al., "A multicomponent reaction-diffusion model of a heterogeneously distributed immobilized enzyme," *Applied Microbiology and Biotechnology*, vol. 72, no. 2, pp. 263–278, 2006.

[3] M. E. Young, P. A. Carroad, and R. L. Bell, "Estimation of diffusion coefficients of proteins," *Biotechnology and Bioengineering*, vol. 22, no. 5, pp. 947–955, 1980.

[4] A. M. Mastro, M. A. Babich, W. D. Taylor, and A. D. Keith, "Diffusion of a small molecule in the cytoplasm of mammalian cells," *Proceedings of the National Academy of Sciences of the United States of America*, vol. 81, no. 11, pp. 3414–3418, 1984.

[5] E. D. Salmon, W. M. Saxton, and R. J. Leslie, "Diffusion coefficient of fluorescein-labeled tubulin in the cytoplasm of embryonic cells of a sea urchin: video image analysis of fluorescence redistribution after photobleaching," *Journal of Cell Biology*, vol. 99, no. 6, pp. 2157–2164, 1984.

[6] A. S. Verkman, "Solute and macromolecule diffusion in cellular aqueous compartments," *Trends in Biochemical Sciences*, vol. 27, no. 1, pp. 27–33, 2002.

[7] A. Nenninger, G. Mastroianni, and C. W. Mullineaux, "Size dependence of protein diffusion in the cytoplasm of *Escherichia coli*," *Journal of Bacteriology*, vol. 192, no. 18, pp. 4535–4540, 2010.

[8] M. Yang and H. Ma, "Male meiotic spindle lengths in normal and mutant Arabidopsis cells," *Plant Physiology*, vol. 126, no. 2, pp. 622–630, 2001.

[9] M. Yang and Y. Wang, "A model for discrete spindle elongation," *Cell Cycle*, vol. 10, no. 3, pp. 549–550, 2011.

[10] J. W. Kerssemakers, E. L. Munteanu, L. Laan, T. L. Noetzel, M. E. Janson, and M. Dogterom, "Assembly dynamics of microtubules at molecular resolution," *Nature*, vol. 442, no. 7103, pp. 709–712, 2006.

[11] H. T. Schek III, M. K. Gardner, J. Cheng, D. J. Odde, and A. J. Hunt, "Microtubule assembly dynamics at the nanoscale," *Current Biology*, vol. 17, no. 17, pp. 1445–1455, 2007.

[12] T. Mitchinson and M. Kirschner, "Dynamic instability of microtubule growth," *Nature*, vol. 312, no. 5991, pp. 237–242, 1984.

[13] J. B. Olmsted, J. M. Marcum, K. A. Johnson, C. Allen, and G. G. Borisy, "Microtuble assembly: some possible regulatory mechanisms," *Journal of Supramolecular Structure*, vol. 2, no. 2–4, pp. 429–450, 1974.

[14] F. Gaskin, C. R. Cantor, and M. L. Shelanski, "Turbidimetric studies of the in vitro assembly and disassembly of porcine neurotubules," *Journal of Molecular Biology*, vol. 89, no. 4, pp. 737–755, 1974.

[15] D. Willoughby and D. M. F. Cooper, "$Ca^{2+}$ stimulation of adenylyl cyclase generates dynamic oscillations in cyclic AMP," *Journal of Cell Science*, vol. 119, no. 5, pp. 828–836, 2006.

[16] S. Thon, R. Schmauder, and K. Benndorf, "Elementary functional properties of single HCN2 channels," *Biophysical Journal*, vol. 105, no. 7, pp. 1581–1589, 2013.

[17] B. P. English, W. Min, A. M. van Oijen et al., "Ever-fluctuating single enzyme molecules: Michaelis-Menten equation revisited," *Nature Chemical Biology*, vol. 2, no. 2, pp. 87–94, 2006.

[18] H. Yang, G. Luo, P. Karnchanaphanurach et al., "Protein conformational dynamics probed by single-molecule electron transfer," *Science*, vol. 302, no. 5643, pp. 262–266, 2003.

# 18

# Water-Protein Interactions: The Secret of Protein Dynamics

**Silvia Martini,**[1,2] **Claudia Bonechi,**[1,2] **Alberto Foletti,**[3] **and Claudio Rossi**[1,2]

[1] *Department of Biotechnology, Chemistry and Pharmacy, University of Siena, Via Aldo Moro 2, 53100 Siena, Italy*
[2] *Centre for Colloid and Surface Science (CSGI), University of Florence, Via della Lastruccia 3, 50019 Sesto Fiorentino, Italy*
[3] *Laboratory of Applied Mathematics and Physics, Department of Innovative Technologies-DTI,*
*University of Applied Sciences of Southern Switzerland (SUPSI), Manno, CH 6928, Switzerland*

Correspondence should be addressed to Claudio Rossi; claudio.rossi@unisi.it

Academic Editors: J. Lipfert and G. Tresset

Water-protein interactions help to maintain flexible conformation conditions which are required for multifunctional protein recognition processes. The intimate relationship between the protein surface and hydration water can be analyzed by studying experimental water properties measured in protein systems in solution. In particular, proteins in solution modify the structure and the dynamics of the bulk water at the solute-solvent interface. The ordering effects of proteins on hydration water are extended for several angstroms. In this paper we propose a method for analyzing the dynamical properties of the water molecules present in the hydration shells of proteins. The approach is based on the analysis of the effects of protein-solvent interactions on water protons NMR relaxation parameters. NMR relaxation parameters, especially the nonselective ($R_1^{NS}$) and selective ($R_1^{SE}$) spin-lattice relaxation rates of water protons, are useful for investigating the solvent dynamics at the macromolecule-solvent interfaces as well as the perturbation effects caused by the water-macromolecule interactions on the solvent dynamical properties. In this paper we demonstrate that Nuclear Magnetic Resonance Spectroscopy can be used to determine the dynamical contributions of proteins to the water molecules belonging to their hydration shells.

## 1. Introduction

Water-protein interactions play an important role in driving the protein organization at the water interface [1–4]. Water-protein interactions help to maintain flexible conformation conditions which are required for multifunctional protein recognition processes. The intimate relationship between the protein surface and hydration water can be analyzed by studying experimental water properties measured in protein systems in solution. In particular, proteins in solution modify the structure and the dynamics of the bulk water at the solute-solvent interface. The ordering effects of proteins are extended for several angstroms. This process results in a protein hydration shell in which water molecules have restricted dynamics with respect to the bulk water. The extent of interaction can be monitored studying the solvent parameters mostly affected by the presence of a large, slowly reorienting biomacromolecule [5–12]. NMR relaxation parameters, especially the nonselective ($R_1^{NS}$) and selective ($R_1^{SE}$) spin-lattice relaxation rates of water protons, are useful for investigating the solvent dynamics at the macromolecule-solvent interfaces as well as the perturbation effects caused by the water-macromolecule interactions on the solvent dynamical properties [13–25]. In this paper we demonstrate that Nuclear Magnetic Resonance Spectroscopy can be used to determine the dynamical contribution of the biomacromolecules to the water molecules belonging to their hydration shells. In a globular protein solution, three different water environments are present, that is, the buried water molecules (which are integrant part of the protein structure and cannot be removed even during protein crystallization) [3, 4], the water hydration shell around the protein, and the bulk water. The present investigation analyzes the dynamical properties of the water molecules present in the hydration shell around a protein system. Water proton relaxation rates have been used to investigate different systems and phenomena, and theoretical interpretations of the experimental results have been proposed [26–30]. Both

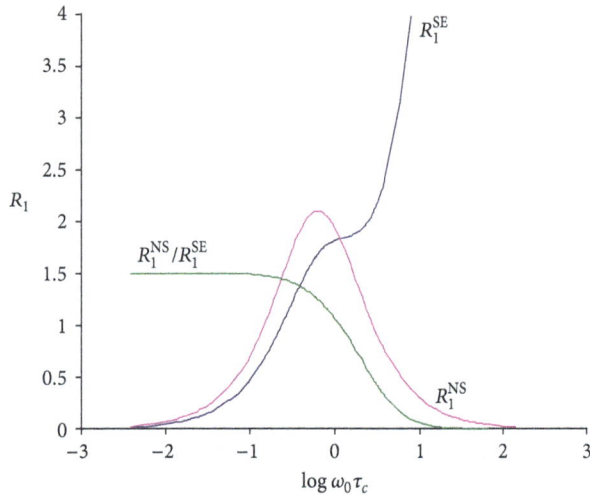

FIGURE 1: Dependence of selective and nonselective spin-lattice relaxation rates of the motion parameter $\omega_0\tau_c$.

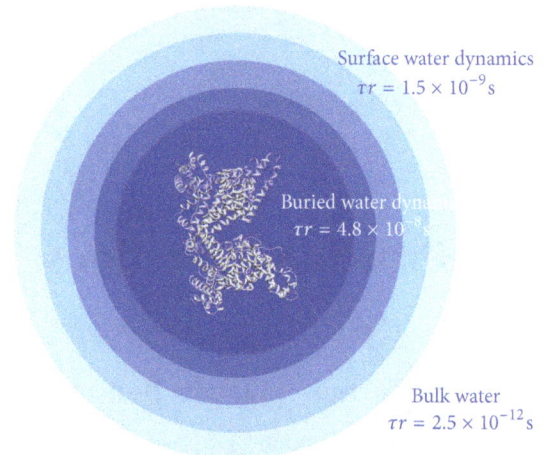

FIGURE 2: Effect of the ordering effect of proteins on water. Three water environments defined by their dynamical properties can be observed: bulk, surface, and buried water environments.

the water proton spin-lattice relaxation rates $R_1^{NS}$ and $R_1^{SE}$ in solution are analyzed considering all possible sources of dipolar contributions arising from proton environments. From this analysis an equation for the calculation of ordering effect induced by the macromolecule on the hydration water was derived. In particular the average water rotational correlation time which characterizes water protons dynamics in the protein hydration shell was calculated. This information was then used for the calculation of the dimension of the long range ordering effect caused by the protein molecules on the hydration water.

## 2. Theory

Dipolar nonselective $R_1^{NS}$ and selective $R_1^{SE}$ spin-lattice relaxation rates have the following expressions [31–36]:

$$R_1^{NS} = \sum \rho_{ij} + \sum \sigma_{ij},$$
$$R_1^{SE} = \sum \rho_{ij},$$

(1)

where $\rho_{ij}$ is the direct self-relaxation rate and $\sigma_{ij}$ the "cross-relaxation" rates.

For any $i$, $j$ dipolar coupling $R_1^{NS}$ and $R_1^{SE}$ assume the explicit form:

$$R_1^{NS} = \frac{3}{10} \frac{\gamma_H^4 \hbar^2}{r_{ij}^6} \left[ \frac{4\tau_c}{1 + 4\omega_H^2\tau_c^2} + \frac{\tau_c}{1 + \omega_H^2\tau_c^2} \right],$$
$$R_1^{SE} = \frac{1}{10} \frac{\gamma_H^4 \hbar^2}{r_{ij}^6} \left[ \frac{3\tau_c}{1 + \omega_H^2\tau_c^2} + \frac{6\tau_c}{1 + 4\omega_H^2\tau_c^2} + \tau_c \right],$$

(2)

where $\hbar$ is the reduced Plank's constant, $\omega_H$ is the proton magnetogyric ratio and Larmor frequency, respectively, $r_{ij}$ is the internuclear distance, and $\tau_c$ is the effective correlation time which modulates the $i$-$j$ magnetic interaction.

The dependence of selective and nonselective spin-lattice relaxation rates of the motion parameter $\omega_0\tau_c$ is reported in Figure 1.

In pure water, the water nonselective $wR_1^{NS}$ and selective $wR_1^{SE}$ spin-lattice relaxation rates are

$$wR_1^{NS} = \sum \rho_{ww} + \sum \sigma_{ww},$$
$$wR_1^{SE} = \sum \rho_{ww} + \sum \sigma_{ww},$$

(3)

where $\rho_{ww}$ and $\sigma_{ww}$ are the water direct and cross-relaxation rate contributions which result from water proton-proton intra- and intermolecular interactions.

In pure water both $wR_1^{NS}$ and $wR_1^{SE}$ assume the same value as the cross-relaxation term $\sigma_{ww}$ affects the selective and nonselective measurements equally.

In binary system (water-protein and/or polymer) we assume the distribution of water molecules as schematically represented by the model showed in Figure 2. Water molecules can be classified into three different categories according to their dynamical properties: (i) bulk water with a typical reorientational correlation time of the order of picoseconds; (ii) water present at the macromolecular surface which exhibits a partially restricted reorientational motion; (iii) buried water molecules. The dynamical properties of the water molecules in these conditions can be well represented by a distribution of correlation time values. These molecules are in fast chemical exchange with the microenvironments present at the protein surface and with the bulk water molecules. These long lived water molecules show dynamics which are mostly determined by the slow reorientation motion of the macromolecule with $\tau_c$ values typically of the order of $10^{-8}$ seconds. These molecules exhibit slow chemical exchange rate in the NMR time scale with the waters present at the macromolecular surface. The contribution of these water molecules to the observed spin-lattice relaxation rates is negligible due to their very low molar fraction.

Relaxometric studies have been used to determine the number and the dominant reorientational correlation time which is involved in the relaxation of water molecules buried in the macromolecular structure [37]. Nevertheless relaxometric experiments cannot monitor the dominant fluctuations which are involved in the relaxation of the water molecules present at the macromolecular surface. In fact this environment is characterized by water molecules which exhibit a distribution of the $\tau_c$ values and display fast chemical exchange with other waters of the same environment or with the bulk molecules. These are in fact the appropriate conditions for applying the selective and nonselective water spin-lattice relaxation methodologies.

In water-protein binary systems, under fast chemical exchange conditions between the free (bulk) and bound water, the changes observed in water spin-lattice relaxation rates with respect to the bulk water reflect the presence of water molecules with restricted dynamical reorientation. In these conditions nonselective ($wR_1^{NS}$) and selective ($wR_1^{SE}$) water spin-lattice relaxation rates assume different values as a consequence of a negative protein-water cross-relaxation contribution to $wR_1^{NS}$ and $wR_1^{SE}$. They are defined as

$$wR_{1\,exp} = \chi_b R_{1b} + \chi_f R_{1f}, \tag{4}$$

where $wR_{1\,exp}$ is the experimental relaxation rate of water in the presence of the protein, $R_{1b}$ and $R_{1f}$ are the water relaxation rates of the pure bound and free environments, and $\chi_b$ and $\chi_f$ are the molar fraction of water in bound and bulk conditions.

$\chi_f$ of the free water molar fraction is assumed to be $\chi_f = 1 - \chi_b \cong 1$.

At the bound site, in the presence of $D_2O > 95\%$, the residual water protons show a relaxation which is mainly dominated by the dipolar interactions with the nonexchangeable protein protons. Water-water interactions (both inter- and intra-) have a sufficient low frequency to be neglected so that

$$wR_{1\,exp}^{NS} = wR_1^{NS} + \chi_b \left( \sum \rho_{wp} + \sum \sigma_{wp} \right),$$
$$wR_{1\,exp}^{SE} = wR_1^{SE} + \chi_b \left( \sum \rho_{wp} \right). \tag{5}$$

Then the protein contribution to the water relaxation rates, $\Delta R_1$, can be calculated as

$$\Delta R_1^{NS} = wR_{1\,exp}^{NS} - wR_1^{NS} = \chi_b \left( \sum \rho_{wp} + \sum \sigma_{wp} \right) = \chi_b R_{1b}^{NS},$$
$$\Delta R_1^{SE} = wR_{1\,exp}^{SE} - wR_1^{SE} = \chi_b \left( \sum \rho_{wp} \right) = \chi_b R_{1b}^{SE}, \tag{6}$$

where $R_{1b}^{NS}$ and $R_{1b}^{SE}$ are the relaxation rates of the water molecules present in the bound conditions.

Considering the dependence of the $R_1^{NS}/R_1^{SE}$ ratio on $\tau_c$ (see (2)), $\Delta R_1^{NS}/\Delta R_1^{SE}$ ratio allows the calculation of the $\tau_c$ value resulting from the average contribution of the distribution of motions that characterizes the water dynamics at the macromolecular surface. Equation (2) holds their own validity when a single correlation time value is replaced by a distribution function which considers all different fast exchanging microenvironments.

In fact

$$\frac{\Delta R_1^{NS}}{\Delta R_1^{SE}} = \frac{\chi_b R_{1b}^{NS}}{\chi_b R_{1b}^{SE}} = \frac{R_{1b}^{NS}}{R_{1b}^{SE}}$$
$$= \frac{12\tau_{c1}/\left(1 + 4\omega_H^2 \tau_{c1}^2\right) + 3\tau_{c1}/\left(1 + \omega_H^2 \tau_{c1}^2\right)}{6\tau_{c1}/\left(1 + 4\omega_H^2 \tau_{c1}^2\right) + 3\tau_{c1}/\left(1 + \omega_H^2 \tau_{c1}^2\right) + \tau_{c1}}, \tag{7}$$

where $\tau_{c1}$ represents a distribution function which considers all individual dynamics which modulate the relaxation. The calculated $\tau_c$ value may be not directly related to a physical meaning as the presence at the macromolecular surface of a specific dynamics defined by this value is not demonstrated. Nevertheless this experimentally determined parameter represents the average value which affects the dipolar water-protein interactions at the macromolecular surface. This parameter assumes a value which has to be in between the protein $\tau_c$ reorientational motion ($\sim 10^{-8}$ s) and the solvent free tumbling reorientation ($\sim 10^{-12}$ s).

## 3. Materials and Methods

[1]H-NMR spectra were obtained on a Bruker AMX 400 spectrometer operating at 400 MHz. Spin-lattice relaxation rates were measured using the $(180°\text{-}\tau\text{-}90°\text{-}t)_n$ sequence. The $\tau$ values used for the selective and nonselective experiments were 0.01, 0.02, 0.04, 0.06, 0.08, 0.1, 0.2, 0.4, 0.8, 1, 1.5, 2, 3, 4, 5, 7, and 10 seconds. The 180° selective inversion of the proton spin population was obtained with a selective perturbation pulse, generated by the decoupler channel. The selective spin-lattice relaxation rates were calculated using the initial slope approximation and subsequent three-parameter exponential regression analysis of the longitudinal recovery curves. The maximum experimental error in the relaxation rate measurements was 5%.

Human albumin (molecular weight 66200 Dalton) was purchased from Sigma Chemical Co. All the solutions were obtained using $D_2O$ with a minimum content of deuterium of 99.9%.

## 4. Results and Discussion

The theory presented in the previous section is supported by the experimental results obtained on human albumin system.

Water selective and nonselective spin-lattice relaxation rates as a function of protein concentrations are reported in Table 1.

The proteins contribution to the water selective $\Delta R_1^{SE}$ and nonselective $\Delta R_1^{NS}$ relaxation for human albumin systems is shown in Figure 3. In this figure the fitting of the experimental results is also shown. As required by the theory, the calculated straight lines pass through the origin in the system under study. As shown in Figure 3, water selective spin-lattice relaxation rates assume a larger value with respect to the water nonselective spin-lattice relaxation rates, whose results are affected by the negative protein-water cross-relaxation contributions.

The ratio calculated from the proteins contribution to the water nonselective and selective relaxation rates, $\Delta R_1^{NS}/\Delta R_1^{SE}$,

TABLE 1: Water non-selective and selective proton spin-lattice relaxation times as a function of the human albumin content at 298 K. In the same table the protein contribution to the selective and non-selective proton spin-lattice relaxation rates $\Delta R_1^{SE}$ and $\Delta R_1^{NS}$ is also reported.

| Albumin concentration mol/L | Albumin concentration mg/mL | $T_1^{NS}$ s | $T_1^{SE}$ s | $R_1^{NS}$ s$^{-1}$ | $R_1^{SE}$ s$^{-1}$ | $\Delta R_1^{NS}$ s$^{-1}$ | $\Delta R_1^{SE}$ s$^{-1}$ |
|---|---|---|---|---|---|---|---|
| 0 | 0 | 10.10 | 10.30 | 0.099 | 0.097 | 0 | 0 |
| $1.6 \times 10^{-5}$ | 1.0 | 8.30 | 6.45 | 0.120 | 0.155 | 0.021 | 0.058 |
| $3.2 \times 10^{-5}$ | 2.0 | 7.10 | 4.55 | 0.141 | 0.220 | 0.042 | 0.123 |
| $4.8 \times 10^{-5}$ | 3.0 | 6.15 | 3.60 | 0.163 | 0.278 | 0.064 | 0.181 |
| $6.5 \times 10^{-5}$ | 4.0 | 5.50 | 2.90 | 0.182 | 0.345 | 0.083 | 0.248 |
| $7.3 \times 10^{-5}$ | 4.5 | 5.10 | 2.70 | 0.196 | 0.370 | 0.097 | 0.273 |
| $8.1 \times 10^{-5}$ | 5.0 | 4.70 | 2.45 | 0.213 | 0.408 | 0.114 | 0.311 |
| $8.9 \times 10^{-5}$ | 5.5 | 4.45 | 2.30 | 0.225 | 0.435 | 0.126 | 0.338 |
| $9.7 \times 10^{-5}$ | 6.0 | 4.15 | 2.10 | 0.241 | 0.476 | 0.142 | 0.379 |
| $1.3 \times 10^{-4}$ | 8.0 | 3.55 | 1.66 | 0.282 | 0.602 | 0.183 | 0.505 |
| $1.6 \times 10^{-4}$ | 10.0 | 3.10 | 1.40 | 0.323 | 0.714 | 0.202 | 0.559 |

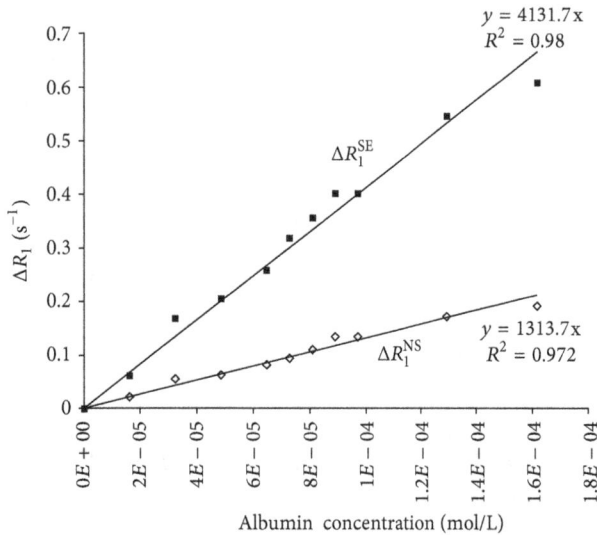

FIGURE 3: Nonselective and selective proton spin-lattice relaxation rates $\Delta R_1^{SE}$ and $\Delta R_1^{NS}$ as a function of the human albumin concentration.

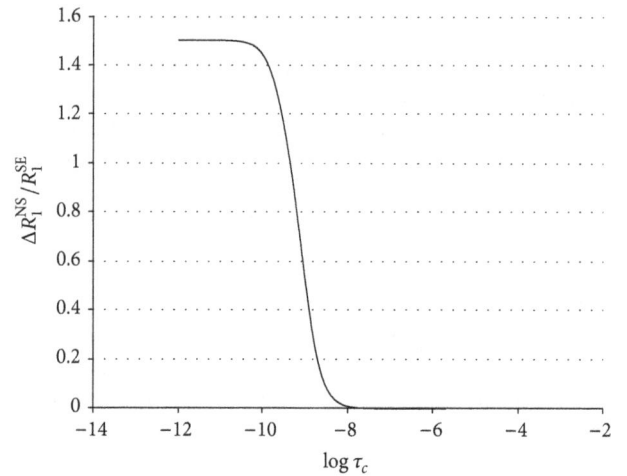

FIGURE 4: Computed values of $\Delta R_1^{NS}/\Delta R_1^{SE}$ ratio as a function of $\tau_c$ at a proton frequency of 400 MHz.

assumes a value of 0.36. The behavior of the $\Delta R_1^{NS}/\Delta R_1^{SE}$ ratio as a function of $\tau_c$ is reported in Figure 4.

Using the previously computed $\Delta R_1^{NS}/\Delta R_1^{SE}$ ratio of 0.36, an average reorientational correlation time of $1.5 \times 10^{-9}$ s was calculated for the water molecules in the protein hydration shell. In Figure 1 a summary of the water environment typical of protein systems in the case of human albumin is shown: bulk, buried, and hydration water. In the same figure the rotational correlation time values typical of each water environments are reported. The average water hydration correlation time previously computed was used to calculate the ordering effects of the protein on water molecules in the hydration shells at different distance from the protein surface. Assuming a spherical shape with a diameter of 70 Å, the volume of ten hydration spheres around human albumin was

calculated. The number of water molecules in each hydration sphere was computed as well as the number of the total water molecules contained in the first ten hydration spheres. Assuming an exponential decay of the water correlation time from its value at the protein surface to the bulk conditions, the following equation was developed:

$$\tau_{c(1,2,\ldots,10)} = a + be^{-kd}, \qquad (8)$$

where $\tau_{c(1,2,\ldots,10)}$ are the calculated correlation time values of the water molecules present in the first tenth hydration shell, $a = 2.5 \times 10^{-12}$ s is the bulk water rotational $\tau_c$, $b = 4.8 \times 10^{-8}$ s is the buried water rotational $\tau_c$, $d$ is the hydration shell distance from the protein surface assumed here to range from 1 to 10 Å, $k$ is a constant which defines how strong the ordering effect of the protein on the water molecules. In Figure 4 the computed correlation times (calculated from equation (7)) of the water molecules in each of the first tenth hydration shells

log distance

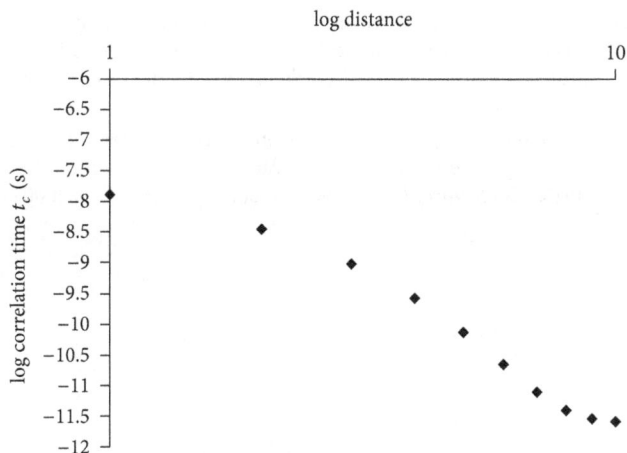

FIGURE 5: Computed values of the reorientational water correlation times typical of water molecules in the first tenth hydration shells around human albumin. Correlation time was calculated using equation (8) with $a = 2.5 \times 10^{-12}$, $b = 4.8 \times 10^{-8}$, and $k = 1.3$. The average correlation time over the ten shells was calculated using the equation $\tau_{c\,\text{average}} = \sum_{i=1}^{10} \chi_i \tau_{ci} = 1.5 \times 10^{-9}$ s.

as a function of the distance $d$ are reported. The convergence between the experimental average reorientational correlation times of the water molecules in the protein hydration shells of $1.5 \times 10^{-9}$ s with the value computed on the basis of (8) was obtained for a $k$ equal to 1.3 ($\text{Å}^{-1}$). The long range ordering effect of the protein on the hydration water is extended at least to 8 Å (Figure 5).

## 5. Conclusions

In diluted protein solutions, the bulk water proton relaxation shared the contributions from the water molecules in the protein hydration shell. These water molecules differ from the bulk water, mainly because of their correlation times, which is are short for bulk water and longer for the protein hydration waters. In slow motion conditions ($\omega_0 \tau_c \gg 1$, typical of the slow tumbling of protein molecules), these contributions are different: large and positive to $wR_1^{SE}$ and negligible or absent to $wR_1^{NS}$. This process makes $wR_1^{SE}$ larger than $wR_1^{NS}$ as showed in Figure 3. The analysis of both the selective and nonselective water spin-lattice relaxation rates allowed the calculation of the average effective correlation time for the water molecules at the water-protein interface. Moreover, using the assumption of an exponential decay of the rotational correlation time of the hydration water from its value at the protein surface to the bulk conditions, the long range ordering effect of the protein surface on the surrounded water molecules was calculated.

## Acknowledgment

This work was partially supported by LIR srl.

## References

[1] B. Halle, "Protein hydration dynamics in solution: a critical survey," *Philosophical Transactions of the Royal Society B*, vol. 359, pp. 1207–1224, 2004.

[2] S. Bone and R. Pethig, "Dielectric studies of protein hydration and hydration-induced flexibility," *Journal of Molecular Biology*, vol. 181, no. 2, pp. 323–326, 1985.

[3] K. Takano, Y. Yamagata, and K. Yutani, "Buried water molecules contribute to the conformational stability of a protein," *Protein Engineering*, vol. 16, no. 1, pp. 5–9, 2003.

[4] S. Fisher and C. S. Verma, "Binding of buried structural water increases the flexibility of proteins," *Proceedings of the National Academy of Sciences of the United States of America*, vol. 96, no. 17, p. 9613, 1999.

[5] V. P. Denisov and B. Halle, "Protein hydration dynamics in aqueous solution," *Faraday Discussions*, vol. 103, pp. 227–244, 1996.

[6] M. Vorob'ev, "Monitoring of water ordering in aqueous protein systems," *Food Hydrocolloids*, vol. 21, no. 2, pp. 309–312, 2007.

[7] N. Serge and S. N. Timasheff, "Protein-solvent preferential interactions, protein hydration, and the modulation of biochemical reactions by solvent components," *Proceedings of the National Academy of Sciences of the United States of America*, vol. 99, no. 15, pp. 9721–9726, 2002.

[8] G. Zaccai, "The effect of water on protein dynamics," *Proceedings of the National Academy of Sciences of the United States of America*, vol. 359, no. 1448, pp. 1269–1275, 2004.

[9] N. Bhattacharjee and P. Biswas, "Structure of hydration water in proteins: a comparison of molecular dynamics simulations and database analysis," *Biophysical Chemistry*, vol. 158, no. 1, pp. 73–80, 2011.

[10] M. Belotti, A. Martinelli, R. Gianferri, and E. Brosio, "A proton NMR relaxation study of water dynamics in bovine serum albumin nanoparticles," *Physical Chemistry Chemical Physics*, vol. 12, no. 2, pp. 516–522, 2010.

[11] M. Jasnin, A. Stadler, M. Tehei, and G. Zaccai, "Specific cellular water dynamics observed *in vivo* by neutron scattering and NMR," *Physical Chemistry Chemical Physics*, vol. 12, no. 35, pp. 10154–10160, 2010.

[12] D. I. Svergun, S. Richard, M. H. J. Koch, Z. Sayers, S. Kuprin, and G. Zaccai, "Protein hydration in solution: experimental observation by x-ray and neutron scattering," *Proceedings of the National Academy of Sciences of the United States of America*, vol. 95, no. 5, pp. 2267–2272, 1998.

[13] R. G. Bryant, "The dynamics of water-protein interactions," *Annual Review of Biophysics and Biomolecular Structure*, vol. 25, pp. 29–53, 1996.

[14] N. Niccolai, L. Pogliani, and C. Rossi, "Water proton selective nuclear relaxation: a proof to investigate surface characteristics of biological materials," *Chemical Physics Letters*, vol. 110, no. 3, pp. 294–297, 1984.

[15] C. Rossi, A. Donati, and M. R. Sansoni, "Nuclear magnetic resonance as a tool for the identification of specific DNA—ligand interaction," *Chemical Physics Letters*, vol. 189, no. 3, pp. 278–280, 1992.

[16] C. Bonechi, A. Donati, M. P. Picchi, C. Rossi, and E. Tiezzi, "DNA-ligand interaction detected by proton selective and nonselective spin-lattice relaxation rate analysis," *Colloids and Surfaces A*, vol. 115, pp. 89–95, 1996.

[17] C. Rossi, S. Bastianoni, C. Bonechi, G. Corbini, P. Corti, and A. Donati, "Ligand-protein recognition studies as determined by

nuclear relaxation analysis," *Chemical Physics Letters*, vol. 310, no. 5-6, pp. 495–500, 1999.

[18] N. Niccolai, C. Rossi, and E. Tiezzi, "Water proton spin-lattice relaxation behaviour in heterogeneous biological systems," *Chemical Physics Letters*, vol. 96, no. 2, pp. 154–156, 1983.

[19] G. Otting, E. Liepinsh, and K. Wuthrich, "Protein hydration in aqueous solution," *Science*, vol. 254, no. 5034, pp. 974–980, 1991.

[20] C. Rossi, A. Donati, C. Bonechi et al., "Nuclear relaxation studies in ligand-macromolecule affinity index determinations," *Chemical Physics Letters*, vol. 264, no. 1-2, pp. 205–209, 1997.

[21] S. Martini, C. Bonechi, and C. Rossi, "Interaction of quercetin and Its conjugate quercetin 3-*O*-*β*-d-glucopyranoside with albumin as determined by NMR relaxation data," *Journal of Natural Products*, vol. 71, no. 2, pp. 175–178, 2008.

[22] S. Martini, C. Bonechi, M. Casolaro, G. Corbini, and C. Rossi, "Drug-protein recognition processes investigated by NMR relaxation data: a study on corticosteroid-albumin interactions," *Biochemical Pharmacology*, vol. 71, no. 6, pp. 858–864, 2006.

[23] S. Martini, C. Bonechi, and C. Rossi, "Interaction between vine pesticides and bovine serum albumin studied by nuclear spin relaxation data," *Journal of Agricultural and Food Chemistry*, vol. 58, no. 19, pp. 10705–10709, 2010.

[24] S. Martini, M. Consumi, C. Bonechi, C. Rossi, and A. Magnani, "Fibrinogen–catecholamine interaction as observed by NMR and fourier transform infrared spectroscopy," *Biomacromolecules*, vol. 8, no. 9, pp. 2689–2696, 2007.

[25] S. Martini, C. Bonechi, G. Corbini, and C. Rossi, "Determination of the modified "affinity index" of small ligands and macromolecular receptors from NMR spin-lattice relaxation data," *Chemical Physics Letters*, vol. 447, no. 1–3, pp. 147–153, 2007.

[26] R. Damadian, "Tumor detection by nuclear magnetic resonance," *Science*, vol. 171, no. 3976, pp. 1151–1153, 1971.

[27] V. R. Mathur-De, "The NMR studies of water in biological systems," *Progress in Biophysics and Molecular Biology*, vol. 35, pp. 103–134, 1980.

[28] D. E. Woessner, "Nuclear magnetic relaxation and structure in aqueous heterogenous systems," *Molecular Physics*, vol. 34, no. 4, pp. 899–920, 1977.

[29] S. H. Koenig, R. D. Brown III, and R. Ugolini, "A unified view of relaxation in protein solutions and tissue, including hydration and magnetization transfer," *Magnetic Resonance in Medicine*, vol. 29, no. 1, pp. 77–83, 1993.

[30] R. G. Bryant and W. M. Shirley, "Dynamical deductions from nuclear magnetic resonance relaxation measurements at the water-protein interface," *Biophysical Journal*, vol. 32, no. 1, pp. 3–16, 1980.

[31] F. Bloch, "Generalized theory of relaxation," *Physical Review*, vol. 105, no. 4, pp. 1206–1222, 1957.

[32] I. Solomon, "Relaxation processes in a system of two spins," *Physical Review*, vol. 99, no. 2, pp. 559–565, 1955.

[33] J. H. Noggle and R. E. Shirmer, *The Nuclear Overhauser Effect*, Academic, New York, NY, USA, 1971.

[34] C. Rossi, S. Bastianoni, C. Bonechi, G. Corbini, P. Corti, and A. Donati, "Ligand-protein recognition studies as determined by nuclear relaxation analysis," *Chemical Physics Letters*, vol. 310, no. 5-6, pp. 495–500, 1999.

[35] D. Neuhaus and M. Williamson, *The Nuclear Overhauser Effect in Structural and Conformational Analysis*, VCH Publisher, New York, NY, USA, 1989.

[36] R. Freeman, H. D. W. Hill, B. L. Tomlinson, and L. D. Hall, "Dipolar contribution to NMR spin-lattice relaxation of protons," *Journal of Chemical Physics*, vol. 61, no. 11, pp. 4466–4473, 1974.

[37] S. Kiihne and R. G. Bryant, "Protein-bound water molecule counting by resolution of (1)H spin-lattice relaxation mechanisms," *Biophysical Journal*, vol. 78, no. 4, pp. 2163–2169, 2000.

# 19

# Biophysical Insights into Cancer Transformation and Treatment

**Jiří Pokorný,**[1] **Alberto Foletti,**[2,3] **Jitka Kobilková,**[4] **Anna Jandová,**[1] **Jan Vrba,**[5] **Jan Vrba Jr.,**[6] **Martina Nedbalová,**[7] **Aleš Čoček,**[8] **Andrea Danani,**[3] **and Jack A. Tuszyński**[9]

[1] *Institute of Photonics and Electronics, Academy of Sciences of the Czech Republic, AS CR, Chaberská 57,*
*182 51 Prague 8-Kobylisy, Czech Republic*
[2] *Institute of Translational Pharmacology, National Research Council-CNR, Via Fosso del Cavaliere 100, 00133 Rome, Italy*
[3] *University of Applied Sciences of Southern Switzerland-SUPSI, Department of Innovative Technologies,*
*Galleria 2, 6928 Manno, Switzerland*
[4] *1st Faculty of Medicine, Charles University in Prague, Department of Obstetrics and Gynaecology, Apolinářská 18,*
*128 00 Prague 2, Czech Republic*
[5] *Faculty of Electrical Engineering, Czech Technical University in Prague, Technická 2, 166 27 Prague 6, Czech Republic*
[6] *Faculty of Biomedical Engineering, Czech Technical University in Kladno, Sitná Square 3105, 272 01 Kladno, Czech Republic*
[7] *1st Faculty of Medicine, Charles University in Prague, Institute of Physiology, Albertov 5, 128 00 Prague 2, Czech Republic*
[8] *3rd Faculty of Medicine, Charles University in Prague, Department of Otorhinolaryngology, Ruská 87,*
*100 00 Prague 10, Czech Republic*
[9] *Department of Physics, University of Alberta, Edmonton, AB, Canada T6G 2J7*

Correspondence should be addressed to Jiří Pokorný; pokorny@ufe.cz

Academic Editors: A. Kukol, B. Schneider, and L. Strasak

Biological systems are hierarchically self-organized complex structures characterized by nonlinear interactions. Biochemical energy is transformed into work of physical forces required for various biological functions. We postulate that energy transduction depends on endogenous electrodynamic fields generated by microtubules. Microtubules and mitochondria colocalize in cells with microtubules providing tracks for mitochondrial movement. Besides energy transformation, mitochondria form a spatially distributed proton charge layer and a resultant strong static electric field, which causes water ordering in the surrounding cytosol. These effects create conditions for generation of coherent electrodynamic field. The metabolic energy transduction pathways are strongly affected in cancers. Mitochondrial dysfunction in cancer cells (Warburg effect) or in fibroblasts associated with cancer cells (reverse Warburg effect) results in decreased or increased power of the generated electromagnetic field, respectively, and shifted and rebuilt frequency spectra. Disturbed electrodynamic interaction forces between cancer and healthy cells may favor local invasion and metastasis. A therapeutic strategy of targeting dysfunctional mitochondria for restoration of their physiological functions makes it possible to switch on the natural apoptotic pathway blocked in cancer transformed cells. Experience with dichloroacetate in cancer treatment and reestablishment of the healthy state may help in the development of novel effective drugs aimed at the mitochondrial function.

## 1. Introduction

Partial suppression of oxidative metabolism in cancers was first discovered by Warburg et al. [1]. He proved its origin in the diminished activity of mitochondria [2]. This was an extraordinary insight that took half a century for the scientific community to be fully appreciated. It is worth noting that Warburg understood biological systems as highly organized structures at the time when little was known about the internal organization of a living cell. This led him to an intuitive conclusion that disturbances of oxidative metabolism are an essential part of cancer initiation and progression. However, at the time of Warburg's discovery, this point of view was not accepted by the scientific community regardless of its genuine significance. Contemporary comprehension of biological systems is connected with the idea of complexity (for a description of complex systems, the reader is referred, for instance, to Cohen and Havlin [3], and a simple model is developed in [4]). Biological systems are examples of complex systems composed of a large number of nonlinearly interacting elements organized into hierarchical structures [5]. These systems of interconnected entities exhibit emergent phenomena where the whole possesses properties not present in their individual parts. The organized ensemble of elements, therefore, creates new features and forms of activity. Biological systems are open since they exchange mass, energy, and information with their environments. They are dissipative structures. The whole complex of any biological system is a result of self-organization [6] under nonequilibrium thermodynamic conditions (far from thermodynamic equilibrium). Adaptability is another feature of biological systems based on interactions with their surroundings. Environmental influences lead to constant modifications of internal structure and patterns of activity. Interaction with the surroundings is not only passive but also displays active action. Branching of reaction and activity pathways is a general property of biological systems. Consequently, the knowledge regarding the composition and activity of only one component of biological systems is of limited value due to parallel interconnections.

Biological systems display a central control and steering which is provided by brain activity in mammals. The brain receives information from individual parts of the hierarchical system, processes it, and reacts to it by sending controlling signals. Body communication systems with information channels are an indispensable part of the brain's control-and-command function. However, information transfer in biological systems has been up to the present time analyzed as a transfer of quality, or an order of entities [7]. Reduction to a quantitative basis has not been performed so far, and, therefore, assessment of the amount of information or channel capacity has not been possible. Action potential propagating along nerve fibres as an information transfer medium seems to have insufficient capacity to provide a required communication function. Internal cooperation and coherent activity in mammalian species require high capacity information transfer between the central control unit—the brain—and the periphery—the organs. Photon information transfer was argued to be essential in biology in general [8]. Importantly, recent results show the ability of cancer cells to cause injury in distant healthy tissues by a physical mechanism of information transfer [9]. The changes in the tissue are possibly triggered by biophotons emitted by cancer cells. Biophotons may propagate through the soft tissues, inside nerve fibers, and inside or along other structures used as conduits in the body. Notably, biophoton transfer along the nerves of rats has been demonstrated experimentally [10]. Disturbances of the information transferred to and from the brain in a pathological cancer state have not been demonstrated yet.

Mitochondrial dysfunction in cancer cells was considered as an unimportant effect at Warburg's time and even long after his death. Biological research was primarily based on the examination of morphology, composition, chemical reactions, and information transfer by mass elements. Physical processes in biological systems were not accepted as an essential part of living activity. Subsequently, Fröhlich proposed that coherent electrical polar oscillations and the generation of electromagnetic fields play important roles in

living cells [11–15], and their disturbances occur in cancer cells [16]. Similarly to Warburg, Fröhlich was ahead of his time. Structures generating the electromagnetic field were not discovered at that time and nanotechnological measurement methods were not in existence yet. However, experimental support for Fröhlich's ideas was being gradually accumulated. Measurements performed on living cells disclosed electric and electromagnetic oscillations. Dielectrophoretic forces of the cellular oscillating electric field cause attraction of dielectric particles which depends on their permittivity [17]. Further measurements were performed by Hölzel and Lamprecht [18] and Hölzel [19] who proved electric origin of the forces acting on dielectric particles. The generation of cellular electromagnetic fields, however, was not ascribed to microtubules discovered by Amos and Klug [20] regardless of the extensive microtubule research at the time. However, both experimental and theoretical research of the cellular electromagnetic activity gradually pointed to microtubules as major sources of electromagnetic interactions [21, 22].

Very likely the most interesting phenomenon connected with the microtubule research is water ordering taking place in living cells. Layers of water without solutes observed around microtubules were called clear zones [23]. Formation of clear zones was assumed to depend on the negative electrostatic charge at the microtubule surface [24]. Ling formulated a theory of the ordering of water molecules in the electrostatic field of the surface charges at the interface [25]. The clear (exclusion) zones were proved to be layers of ordered water [26–28]. Interfacial water ordering may be formed up to a distance of about 0.1 mm from the charged surface. Ions are excluded from the ordered layer due to its strong electric field, thermal fluctuations are diminished as follows from the measurements in the range of wavelengths of 3.8–4.6 μm, and UV absorbance at 270 nm which is increased. The ordered layer of water molecules resembles a gel. Ordered water layers are formed around mitochondria [29], which follows from the experimental results published by Tyner et al. [30].

The physical structure of water was analyzed by Preparata [31], Del Giudice et al. [32], and Del Giudice and Tedeschi [33] on the basis of the quantum electrodynamic theory. The liquid water is a mixture of two phases of water: ordered water forming coherent domains and gas-like water (bulk water). The clear (exclusion) zones display macroscopic separation of these two phases of water caused by a strong electric field.

Elastic oscillations of the yeast cell membrane in the acoustic range below 2 kHz were measured by Pelling et al. [34, 35], and elastic and electric oscillations were compared [36, 37]. Microtubule polymerization in cells may be disrupted by external electromagnetic field in the frequency range 0.1–0.3 MHz [38, 39]. Electric oscillations at cellular membrane of yeast and alga cells in the frequency range 1.5–52 MHz were measured [18, 19]. The high values of the electrodynamic activity of synchronized yeast cells in the M phase coincide with the periods of arrangement of the microtubules into a mitotic spindle, during metaphase, and anaphase A and B [22]. Damping of external electromagnetic field caused by cancer tissue at the frequency 465 MHz and the first harmonic was experimentally determined by Vedruccio and Meessen [40]. Oscillations in microtubules may be damped in cancer cells by water with decreased level of ordering [41]. Cancer cells exhibit a less-ordered structure [42]. Interactions between cells mediated by cellular electromagnetic fields in the red and near-infrared range were observed by Albrecht-Buehler [43–45].

Electromagnetic resonant frequencies of microtubules were measured by Sahu et al. [46] in the range of 10–30 MHz and 100–200 MHz. The resonant frequencies were disclosed by measurement of DC conductivity after application of oscillating signal of corresponding frequency and from transmittance and reflectance of microtubule without and with compensation of parasitic reactances of contacts in the frequency range from 1 kHz to 20 GHz. Transmission of the oscillating signals is independent of the length of microtubule. At the resonant frequencies, a sharp increase of DC conductivity was observed. At the particular frequencies, the transmittance is large and the microtubule resistance is much less than 0.04 Ω. Microtubule oscillators have a high quality factor. The peaks of resonance are not observed after release of water from the microtubule cavity. It should be mentioned that microtubule is also a multilevel memory. Electric current can store and erase 500 discrete bits in a single microtubule [47].

This paper contains an overview of the cancer initiation process as a pathological state of a complex biological system. The cancer transformation in the complex system contains biochemical-genetic links on the one hand and biophysical links on the other hand. The most important mechanisms involved in these processes concern cooperation of mitochondria and microtubules in the generation of the cellular electromagnetic field and production of force effects. We believe that gaining a biophysical understanding of the complexity of cancer processes may significantly contribute to improved cancer diagnostics and treatment.

## 2. Mitochondria Form Conditions for Microtubule Oscillations

Production of ATP and GTP (adenosine and guanosine triphosphate) and triggering apoptosis are not the sole functions of mitochondria. The role of mitochondria in a cell is rather complex. Mitochondria and microtubules form a unique cooperating system in the cell [41, 48, 49]. Mitochondria alter the medium around them by the mechanism of proton transfer. Energy of pyruvate and fatty acids is used for pumping protons into the intermembrane space and in this way it is transformed into electrochemical proton gradient energy. From the intermembrane space, protons diffuse into cytosol through the outer membrane pores which are freely permeable to molecules whose relative molecular mass is 5,000 daltons or less. A layer of ordered water and a strong static electric field are formed around each properly working mitochondrion. Measurement of the intensity of the static electric field was performed by solid fluorescent particles of 30 nm in diameter [30]. At the outer mitochondrial membrane, the greatest intensity of the electric field (of about 3.5 MV/m) was measured. In the vicinity of a single mitochondrion, intensity of the electric field decreases nearly linearly as a function of distance. Even at a distance of 2 μm from a mitochondrion, significant values of the electric field

were measured (about 540 kV/m). This dependence may correspond to an ordered layer of water around a mitochondrion [29]. Water ordering is a phenomenon involving a change of the water structure from a viscosity liquid to a quasielastic gel affecting inner cellular processes, in particular providing low damping of the cytoskeleton vibration system. More than 20% of the cellular volume is occupied by mitochondria and the ordered water fills up most of the rest. Cytosol, cytoskeleton, and biological molecules are exposed to a strong electric field. Production of ATP utilizes the electrochemical proton gradient across the inner membrane. ATP is produced with efficiency higher than 40%. The rest of the nonutilized energy (nearly 60%) is liberated from mitochondria as heat, photons (emission of UV photons was detected too), and chemical energy not exploited for ATP and GTP production.

Localization of mitochondria depends on loci of energy consumption. In the interphase, mitochondria are concentrated around microtubules which are the organizing structures of the cytoskeleton. It is well known that microtubules consume energy in the form of GTP molecules required for polymerization by assembly competent tubulin dimers. In the M phase, the distribution of mitochondria is not precisely known. Microtubules are composed of tubulin heterodimers which carry large electric dipoles. Their oscillations generate an electromagnetic field (the near field whose great energy has characteristic features of the electric field is called electrodynamic field or the virtual photon field; at a greater distance from the source the electromagnetic character is dominating). The strong static electric field around mitochondria can shift microtubule oscillations into a highly nonlinear region.

The electrodynamic field generated by microtubules in the cellular cytoskeleton during different phases of the cell cycle has to be excited from the cellular energy sources. Energy supply is an essential condition for oscillations and generation of the electrodynamic field. Microtubules are dynamic cylindrically shaped polymers that display a so-called dynamic instability process with periods of growth interspersed with rapid shortening called a catastrophe. Energy supply to microtubules is provided by hydrolysis of GTP to GDP (guanosine diphosphate) in the ß tubulin unit of a heterodimer after the heterodimer is polymerized into a microtubule. Motor proteins transporting "cargo" move along microtubules using ATP molecules as energy sources. A part of the energy of motion is transferred to the microtubules (but the motor proteins might also cause disturbances of coherence and damping). In the interphase the greatest energy supply to microtubules is very likely provided by nonutilized energy liberated from mitochondria. In the M phase energy is supplied by treadmilling—polymerization from one end and depolymerization from the other end of microtubules in the mitotic spindle.

We argue that biological cellular activity depends on the generated electrodynamic field. Its role in the directional transport of mass particles and electrons [50, 51], organization of living matter [52], interactions between systems [53], and information transfer [8] was extensively analyzed and described. These works represent a new contribution to our understanding of the biological activity of living cells.

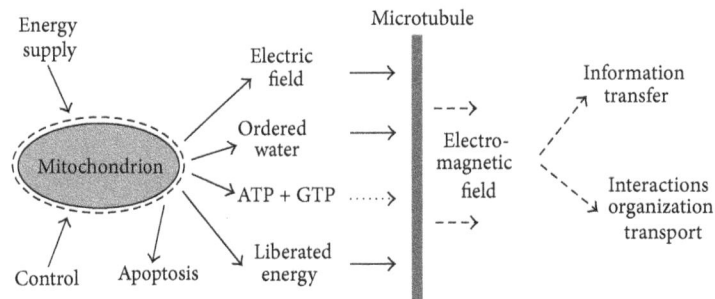

FIGURE 1: Biophysical mechanisms of biological activity of living cells depend on cooperation of mitochondria and microtubules. Mitochondrial function in healthy cells depends on transfer of protons from the matrix space into intermembrane space and to cytosol. Proton transfer is connected with formation of a strong static electric field and high level of water ordering in the mitochondrial neighborhood. Consequently, microtubule oscillations are strongly nonlinear and their damping is low. Microtubule oscillations are excited by supply of energy produced by mitochondria. Microtubules are electrically polar structures whose oscillations generate electrodynamic field which may participate in organization, transport of molecules and particles, interactions, and information transfer. Mitochondria function is disturbed in cancers. Inhibition of the pyruvate pathway in mitochondrion [54] results in partial suppression of proton transfer from the matrix space (nevertheless, diminished proton transfer may be caused also by other disturbances, for instance, in the citric acid cycle). Mitochondrial dysfunction causes lowering of static electric field and water ordering. Cancer cells with blocked pyruvate pathway (i.e., glycolytic phenotype cells) form a large group of cancers. The other large group of cancers has dysfunctional mitochondria in fibroblasts associated with cancer cells.

Figure 1 shows a schematic picture of the mitochondrion and microtubule activity and their cooperation. Mitochondria form conditions for coherent excitation of microtubules by energy supply, low damping, and shift of oscillations into a highly nonlinear region.

## 3. Biophysical Processes Disturbed by Cancer

Warburg assumed that partial suppression of the oxidative production of ATP and its replacement by fermentative (glycolytic) processes diminishes functional (and possibly structural) order in the cell. He commented on it stating that "the adenosine triphosphate synthesized by respiration therefore involves more structure than adenosine triphosphate synthesized by fermentation" [2]. Mitochondrial dysfunction disturbs all consequent physical processes and biological activity dependent on mitochondria. In healthy cells the oxidative energy production may be up to 100 times greater than the fermentative one (for instance, in kidney and liver cells). In cancer cells, only about one half of the ATP cell production is provided by mitochondrial supply. One type of mitochondrial dysfunction (called the glycolytic phenotype) is caused by inhibition of the pyruvate pathway by PDK—pyruvate dehydrogenase kinase [54]. Mitochondrial dysfunction was found in many types of cancer [55, 56]. In this connection, we stress the following facts: (a) a diverse group of information channels and oncogenes results in mitochondrial dysfunction with increased glycolysis and resistance to apoptosis, (b) the majority of carcinomas have so-called hyperpolarization of the mitochondrial inner membrane, and (c) most solid tumors exhibit an increased glucose uptake. These properties prompted Michelakis et al. to advocate targeting mitochondria in cancer treatments which may be effective in a large number of diverse malignant tumors, in particular using DCA (dichloroacetate) [57].

Potential of the inner membrane is an essential parameter for assessment of mitochondrial function. The potential is measured by uptake and retention of positively charged fluorescent dye, such as Rhodamine 123. Large uptake and retention is termed hyperpolarization. However, the uptake and retention may depend also on distribution of ions (for instance, $K^+$) in the cell, production of lactate, and water ordering level and need not strictly correspond to the real mitochondrial inner membrane potential. The lack of mitochondrial hyperpolarization in certain types of malignant tumors, including oat cells lung cancer, lymphomas, neuroblastomas, sarcomas, and some other cancers [57, 58], suggests either a modified glycolytic phenotype or existence of another type(s) of mitochondrial defects and apoptosis blocking. By an electrically neutral exchange of protons and potassium ions, the pH gradient decreases and the membrane potential increases [58]. Defects in the mitochondrial respiratory enzyme complexes and electron carriers in the mitochondrial inner membrane might also diminish the proton transfer resulting in mitochondrial dysfunction in cancer cells. However, another deviation develops in cancerous tissue. Mitochondrial dysfunction is formed in fibroblasts associated with the cancer cell with fully active mitochondria—the reverse Warburg effect [59–62]. Energy rich metabolites (lactate, glutamine, etc.) are transported from the fibroblasts to the cancer cell. The state of enhanced mitochondrial energy production and activity may correspond to the lack of hyperpolarization. Therefore, the method of uptake and retention of a fluorescent dye can distinguish two different cancer mechanisms.

Dependence of life processes on the real mitochondrial membrane potential may suggest its possible promotion of both life and death [63]. Basic processes of life are affected by potential disturbances caused by insufficient energy supply from pyruvate or fatty acids. Activity of pyruvate dehydrogenase (PDH) enzymes is regulated by PDH kinases (PDK-1-PDK-4). Mitochondrial dysfunction in the glycolytic phenotype cancer cell is caused by blocking the pyruvate pathway by the PDH kinases (see Figure 2). Hyperpolarization is

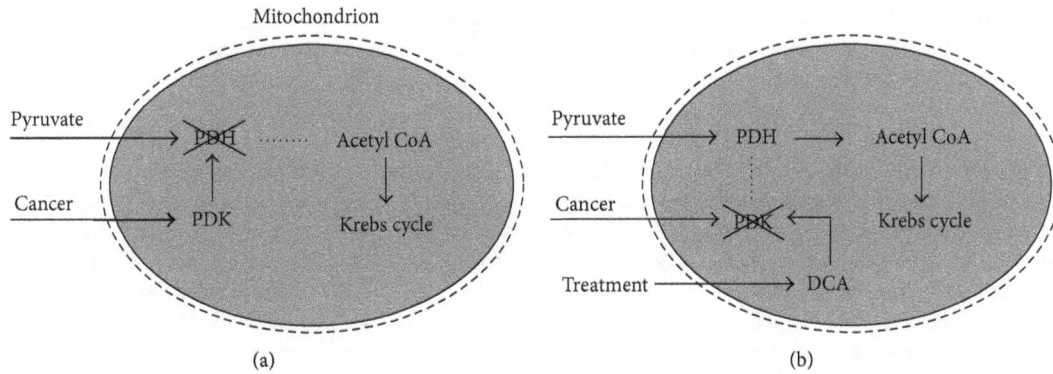

FIGURE 2: Glycolytic phenotype cancer cell may be treated by DCA (up to now the only known drug capable of restoring normal function of cancer cells). (a) The pyruvate pathway into the mitochondrial matrix space is blocked. PDH (pyruvate dehydrogenase) enzymes in the mitochondrial matrix (the grey area) phosphorylated by kinase PDK are dysfunctional, and pyruvate is not transferred to be broken down into the two-carbon acetyl groups on acetyl CoA (Coenzyme A). (b) DCA inhibits activity of PDK. Function of PDH enzymes is restored, and the pyruvate pathway in mitochondria is open (after [54]). The cell needs sufficient amount of oxygen for normal function.

accompanied by a low level of water ordering, diminution of the intensity of the static electric field around a mitochondrion, decrease of the nonutilized energy efflux, and low expression of the $K^+$ channels. Importantly, DCA disturbs PDK-1, -2, and -4 [64] and in this way restores a normal mitochondrial activity resulting in normal cell function or switching on apoptosis of aberrant cells. Hyperpolarization is always associated with increased resistance to apoptosis [54]. A need for developing better PDK inhibitors than DCA was suggested [65]. It should also be mentioned that DCA action is based on attacking PDK and not on the mechanism of its production which could lead to the development of different pharmacological agents in the future.

Mitochondrial function is controlled by chemical and genetic signaling. But the altered mitochondrial function changes physical conditions in the cell affecting microtubule oscillations. As a final result, physical processes in the cell are altered, in particular mechanisms dependent on the electromagnetic field. Organization, transport, interactions, and information transfer are examples of such processes which are liable to be strongly disturbed. Consequently, the whole complex of the system activity exhibits disturbed behavior.

## 4. Cytoskeleton Filaments

Actin filaments, microtubules, and intermediate filaments form a three-dimensional network providing a mechanical integrity of a cell. This network is collectively referred to as a cytoskeleton. Actin and tubulin proteins in their respective filaments bind a large number of different proteins, for example, ARP and MAP proteins, to enable participation in different functions in the cells. Microtubules form highly dynamic structures organizing the cell and generating electrodynamic fields around them. Cellular mechanical properties, dynamical behavior through the cell cycle including transport, and biological activity may strongly depend on the cytoskeleton organization and also on the generated electrodynamic field, in particular on its intensity, frequency spectrum, coherence, and spatial distribution pattern. The space pattern of the

generated field is determined by geometrical arrangement of microtubules and other cytoskeleton structures. Cytoskeleton disturbances are presumably induced along the pathway of cancer cell development before malignant properties are fully established. Mechanical properties of healthy and cancer cells of the same tissue (investigated under action of external forces) are significantly different [66, 67]. Deformability of different cells of human origin was measured. Human nontumorigenic epithelial breast cells (MCF-10), nonmetastatic adenocarcinoma cells (MCF-7), and increased metastatic potential cells (modMCF-7) have different deformabilities 10%, 20%, and 30%, respectively [68]. Mechanical properties of human pancreatic cells (Pac-1) are altered after application of SPC (sphingosylphosphorylcholine) that plays a critical role in the metastatic invasion of gastrointestinal cancers. The keratin network shrinks around the nucleus, elasticity of the cell is reduced, and energy dissipated by mechanical deformation increased [69–71]. These effects might be caused by diminished electrodynamic interactions that are long range in comparison with chemical bond-making forces and biophysical contact interactions (a generated electromagnetic field may mediate interactions at a distance greater than 0.1 micrometer).

Some morphological changes used for cytological and histological evaluation of cancer development may result from cytoskeleton defects. For instance, in the cytological pictures, the keratin network shrinkage may be characterized by wrinkling of the nuclear membrane and disturbances of chromatin regular distribution—coarse chromatin clumping. Mitochondrial dysfunction may result in a lower intensity, lack of coherence, and a spatially diffused pattern of the electrodynamic field generated by microtubules in cancer cells of glycolytic phenotype in comparison with normal cells. Interaction forces between cells depend on the power and coherence of the generated electrodynamic field and due to the microtubule spatial organization on its spatial distribution pattern too. Interaction forces between cancer cells may be smaller than those between normal cells or between a normal and a cancer cell. For instance, cancer cells might be attracted by the normal cells around the tumor and pulled into healthy tissue. This force effect may constitute an essential part of the local invasion of the healthy tissue by malignant cancer cells [53].

Shrinkage of phosphorylated keratin filaments around the nucleus in response to SPC treatment precedes metastatic processes [69–71]. Due to the cytoskeleton disorganization, the space pattern of the generated electrodynamic field may be damaged to such an extent that the cancer cell can release itself from interactions with surrounding cells, liberate, and make metastases in distant organs. This process is well described in the cancer research literature and referred to as the epithelial-to-mesenchymal transition. However, we propose that it can be connected with a further decrease of the electromagnetic field intensity, level of coherence, and nonlinear properties of microtubules and a disturbance of the frequency spectrum. These mechanisms may be closely connected with the extracellular matrix defects which are known to be associated with the initiation of cancer.

Enslavement of cells in a tissue is assumed to depend on the generated electrodynamic field. The cells are identical and have the same conditions for generation of the electromagnetic field. Without essential changes the cell cannot evade enslavement and start independent activity in the body. On top of this, the changed cell has to escape from the supervision by the immune system. The region of tolerance of the electromagnetic field changes should be determined in order to elucidate this effect.

## 5. Treatment of Mitochondria in Cancer Patients

DCA is the first known drug capable of restoring normal healthy activity of a large variety of cancer cells. Regardless of this fact, its application to cancer treatment has been rare. Very likely the greatest clinical experience has been accumulated in the Medicor Cancer Centres in Canada where patients with advanced stages of cancer are treated by DCA; casual reports are available ([72]; http://www.medicorcancer .com/). A wide spectrum of various tumors has been treated. For instance, metastatic renal, lung, and ovarian carcinoma, mesothelioma, glioblastoma, and melanoma with brain metastasis were reported. The minimum and the maximum doses were 10 and 25–50 mg/kg/day, respectively. The courses were continual or cyclic (1–3 weeks on followed by 1 week off). Duration of the treatment should be at least one month. Doses are limited by severity of the side effects. For doses equal to or greater than about 20 mg/kg/day, the response is developed within 2–4 weeks. For smaller doses the response is weaker or delayed. A positive response was reported to have been experienced by 60%–70% of treated patients.

Side effects are claimed to be mild and reversible in the majority of cases ([72]; http://www.medicorcancer.com/). Description of the side effects is also in an overview on mitochondrial targeting for cancer treatment [73]. The side effects are dose and age dependent. In some experimental animals dichloroacetic acid may induce liver cancer by long-term exposure at doses of 100–1000 mg/kg/day. However, in humans a short-term DCA administration appears to

be relatively nontoxic. Exposure to low doses of about 25 mg/kg/day for several months did not reveal adverse side effects [74]. The side effects observed so far are of neurological and gastrointestinal origin. Neurological side effects concern peripheral neuropathy, sedation, fatigue, confusion, hallucination, memory problems, hand tremor, and gait disturbances. Gastrointestinal side effects include heartburn, nausea, vomiting, and indigestion. The most dangerous process involved is the tumor lysogenic syndrome. If a large number of cells is decomposed by apoptosis in a short time, then a sudden release of the dead cell material into the blood stream may cause abnormal heart rhythms and kidney failure.

DCA has antitumor effects, relatively low toxicity (however, dependent on the dose and period of administration), and low cost. In the positively responded patients, DCA offers a palliative effect. In these "palliative" patients, DCA appears to demonstrably improve the quality of life. However, it can also transform advanced stages of cancer from a fatal disease to a chronic disease treatable with simple medications. Adjuvant DCA can occasionally cure stage 4 cancer. Cancer treatment provided in the Medicor Cancer Centres is based on a combination of DCA treatment with chemotherapy, radiotherapy, and surgery. The most effective combination of drugs for each individual tumor is found by the chemosensitiveness test. Medicor Cancer Centres announced that they had treated about 1300 patients. All these patients were previously treated by standard methods and capabilities of conventional cancer therapies which came to an end—the treatment was either ineffective or could not continue. Most of the patients were in advanced stages of the disease. As expected, DCA appears to be more effective in healthier patients as opposed to patients with a very advanced disease. Therefore, response to DCA in patients with an initial disease stage may be more promising.

## 6. Discussion

Excitation of electromagnetic fields in living cells is one of the essential parts of the biophysical processes taking place at subcellular levels. Its generation in living cells was claimed to be impossible due to water viscosity damping [75] or insufficient energy sources for excitation of oscillations in the cell [76, 77]. However, the former authors neglected to consider water ordering and the latter authors did not take into consideration high quality factor of biological oscillators and the key nonlinear properties of the cellular system.

Cellular electromagnetic fields are generated by microtubules. Cooperation of microtubules with mitochondria plays an essential role in living cells leading to the establishment of a functional level of biophysical processes. Generation of electromagnetic fields by microtubules in living cells crucially depends on the function of mitochondria. Mitochondria are regulated by chemical-genetic signaling, but besides triggering apoptosis their activity is mainly connected with physical mechanisms. Mitochondrial function cannot be reduced to energy conversion into ATP and GTP. Transfer of protons from the matrix space into cytosol creates strong static electric fields around mitochondria with consequences that include nonlinear effects on microtubules and water ordering in the cytosol. Mitochondria perform an essential role in cell organization and cell activity in general. Their dysfunction disturbs biophysical processes. This is the case in the vast majority of cancers. At a certain stage of cancer development, mitochondrial dysfunction is formed and affects numerous properties of cells including spatial organization and functional order. Chemical, genetic, and physical mechanisms are mutually coupled.

Diversity of cancer origin agents also led to a hypothesis that mitochondrial dysfunction is a primary cause of cancer and biochemical and genetic deviations develop as consequent events [78]. This hypothesis has not been proved yet and some inconsistency with experimental results may be found. In cervical cancer cells formation of the mitochondrial dysfunction is observed in the time period of the development from precancerous lesions to cancer cells (measured by the immune system response to LDH virus antigen and specific tumor antigen—Jandová et al. [79]), that is, after biochemical and genetic changes. Mitochondrial dysfunction is a result of chemical-genetic defects (on the other hand dysfunction of mitochondria might be caused by a specific agent directly without any previous changes in the biochemical-genetic region). Mitochondria are the boundary entities between chemical-genetic and biophysical processes. Mitochondrial dysfunction disturbs essential biophysical processes in living cells [41, 48, 49]. However, it should be kept in mind that there is an exceptional cancer type characterized by normally functioning mitochondria. Pavlides et al. [59] describe a type of cancer, where energy rich metabolites are transported to cancer cells for utilization in efficient mitochondrial production. These cancer cells with highly active mitochondria display increased malignity which may correspond to enhanced electrodynamic excitation.

Disturbances of biophysical processes might be also-caused by defects in the link to the generation of the electromagnetic fields. For instance, asbestos carcinogenicity is hypothetically explained as having a capability to lower the electrodynamic activity in cells by forming transmission fibers which short-circuit distant parts of the cell with different levels of the electromagnetic field [80]. If it is so, asbestos transformation pathway begins beyond the mitochondrial link and disturbs directly the electrodynamic activity of the cell. Asbestos carcinogenicity has been explained on the basis of several different mechanisms including oxidative stress, chromosome tangling, or adsorption of specific proteins and carcinogenic molecules which may contain iron atoms (Toyokuni, [81]). Toyokuni also explained the carcinogenicity of iron as an effect of reactive oxygen species—ROS [82]. But metal particles might disturb the electromagnetic activity of the cell by a mechanism similar to the short-circuits created by asbestos. The iron carcinogenicity needs not be only a result of ROS but also of the short-circuit effects. Therefore, this raises a question whether treatment based on drugs increasing electric conductivity might cause adverse effects in the afflicted normal cells.

Nevertheless, mitochondrial dysfunction seems to be the most common defect in cancers disturbing their biophysical and consequently the biological behavior. Discovery of electromagnetic activity in living cells may improve our understanding of biological activity and its disordering by cancer. Microtubule oscillation frequencies are one of the fundamental parameters required to be determined in this connection. Nanotechnological sensors and amplifiers may be used for measurement of electrodynamic activity of healthy and cancer cells in the frequency range from about 1 MHz to 1 GHz to determine physical differences (nevertheless, electrodynamic activity in kHz range is reported too). The resonant frequency may depend on excitation due to nonlinear nature of oscillations in microtubules. If the frequency is determined by the secondary structure of tubulin, then proteins should be able to oscillate at resonant frequencies and electrically polar protein molecules generate electrodynamic fields. Electrodynamic fields are also generated by rotation and rotation-vibrational motion of electrically polar molecules and organized structures. Based on these considerations long-range interactions between individual proteins are likely to not only exist but to play important roles in living cells. Therefore, drugs interacting attractively with a convenient target can be synthesized. This supports the idea of preparation of carrier or helper particles that would transport molecules of chemotherapeutic drugs and direct them at the predetermined targets. Efficacy of treatment would be enhanced and side effects diminished.

A cancer transformation pathway is formed by a complex microevolution, multistep, and multibranched process. Essential life mechanisms are misused and gradually altered by cancer. The complex biological system is deformed. Adaptability of cancer cells to environmental changes and diverse cellular stresses is high, arguably higher than in normal cells. Any "one-point" treatment may be overcome by altered cancer mechanism. Adaptability and heterogeneity of cancer processes is an obstacle in their efficient treatment. Therefore, the treatment should be complex, targeting essential links along the cancer transformation pathway and reproducible in recurrent cases. Dysfunctional mitochondria are such an essential link that cannot be bypassed by altered cancer mechanism. Moreover, differences between healthy and cancer cells used for treatment seem to be very often of quantitative type. Treatment of malignant tumors based on the therapeutic strategy of killing the tumor cells almost always negatively affects the healthy cells, in particular those that proliferate rapidly, such as epithelial cells, blood cells, and the immune system. The negative side effects may be individual and vary from negligible to serious. For instance, the differences in increased fermentative ATP production levels may belong to a quantitative type. Cancer therapy based on inhibiting fermentative energy production may considerably damage healthy cells too.

Cancer treatment should target the processes and structures that exhibit the most significant deviations from a normal physiological state. Dysfunction of mitochondria is a very remarkable difference between a healthy and a cancer cell of the glycolytic phenotype. As a result of mitochondrial dysfunction (due to diminished static electric field and water ordering), the endogenous electrodynamic field generated in cancer cells has a decreased intensity, coherence, disturbed frequency spectrum, and spatial pattern. Except for some cases mentioned earlier, mitochondrial dysfunction represents the greatest functional differences in comparison with healthy cells [41, 48, 49]. Restoration of normal mitochondrial function reestablishes conditions for normal physical processes and unlocks the apoptotic pathway. If the cell is too aberrant, for instance, by disorganization of the cytoskeleton or the DNA structure, mitochondria can send a signal to start the preprogrammed cell death (apoptosis). Targeting mitochondria very likely acts on the region of essential differences between healthy and cancer cells. It may be assumed that an effective therapeutic strategy of cancer treatment should aim at a restoration of normal mitochondrial function. In the case of the reverse Warburg effect cancer cells are highly excited due to supply of energy rich metabolites. The normal mitochondrial function should be restored in associated fibroblast and the transfer of energy rich metabolites to cancer cells cut off.

Defects causing mitochondrial dysfunction may form the main target of cancer therapy. Inhibition of the pyruvate pathway into mitochondria is a well-known defect. Exchange of protons for potassium ions in the transfer across the inner membrane may disturb mitochondrial function too [58]. But other types of defects may also exist, for example, inhibition of the electron pathway of the oxidation cycle in the inner mitochondrial membrane. The only known effective drug for restoration of the pyruvate pathway and normal mitochondrial function in a large group of cancers is DCA. This drug inhibits some PDK (-1, -2, -4) blocking pyruvate transfer and its utilization in the mitochondrial matrix. Some other drugs (such as vitamin E analogs [83, 84]) targeted at mitochondria kill cells through destabilization of mitochondria and induction of apoptosis. Nevertheless, the killing process needs not be limited only to cancer cells and could also inflict negative effects on healthy cells.

A considerable amount of preclinical evidence of DCA effects *in vitro* and *in vivo* has already been accumulated. Supporting experience in human cancer treatment is substantial too. The treatment effects in Medicor Research Centres were mainly palliative. However, it should be noted that mainly advanced stage cancer patients were treated there. Clinical trials with less advanced cancer patients should be started to bring further confirmation of positive effects in DCA cancer treatment. Observations of tumor reaction and development should be performed together with examinations of patient states based on laboratory tests, measurements, clinical findings, and analysis of specific symptoms after DCA application. Besides, many essential issues of DCA application have remained unknown, in particular, whether the treated cells had sufficient amount of oxygen to start normal mitochondrial function. Important points may also concern the type of chemical reaction provided by DCA, its target loci at the PDK's and other structures, and possible changes caused by DCA in its target molecules or structures. Some compounds containing sodium, chlorine, and oxygen elements might be more effective than DCA, especially those whose conformation of oxygen atoms is similar to that in DCA (such as chlorine dioxide). It is known that sodium chloride induces necrosis of ovarian carcinoma cells and hypochlorous acid enhances immunogenicity by activation of tumor-specific cytotoxic T cells [85]. Clinical trials with DCA

and examination of DCA reactions in the cell may determine specific requirements for future drug development and open a way for a new strategy in cancer treatment, restoring the normal function of mitochondria, the whole cell, and unlocking apoptosis, depriving cancer cells of their immortality which is the main proliferative advantage over normal cells.

## 7. Conclusions

Coherent electromagnetic fields generated by microtubules in living cells represent a new and outstanding issue in present-day cell biology. Effectiveness of microtubule oscillations depends on mitochondrial function. Besides ATP and GTP production and liberation of nonutilized energy mitochondria form important boundary links between chemical-genetic and physical processes in living cells. They set up conditions for physical mechanisms in living cells. Establishment of a strong static electric field and formation of a layer of ordered water around mitochondria belong to essential conditions for the generation of coherent electrodynamic field by microtubules. The electrodynamic field can provide directional transport in the cell, facilitate organization of structures and organelles, and affect interactions in the cell and between cells, including information transfer.

Inhibition of the pyruvate transfer into mitochondrial matrix causes mitochondrial dysfunction in a large group of cancers. Other defects (for instance, in the citric acid cycle) diminish proton transfer from the matrix space too. Dysfunction of mitochondria results in decreased static electric field, diminished water ordering around them, and lowered energy supply. Consequently, electrodynamic field generated by microtubules is characterized by low power, diminished coherence, and altered frequency spectrum. The space pattern of the field may be disordered too. Biological functions dependent on generated electrodynamic fields are very likely to be disturbed as a result.

However, there is another group of cancers. Mitochondrial dysfunction is not developed in a cancer cell but in the associated fibroblasts. The supply of energy rich metabolites from associated fibroblasts results in a strong aggressiveness of cancer cells. This cancer type deserves special analysis concerning increased electrodynamic activity.

We propose that essential and specific differences between healthy and cancer cells should be exploited for the development of a new generation of cancer therapies. The conventional cancer therapeutic strategy is based on cancer cell killing with associated collateral damage to healthy tissue. The main effort is aimed at finding sufficiently specific property to kill the cancer cells and to limit damage to healthy cells. However, the healthy cells may be damaged too and treatment of recurrent cancers remains unsolved.

Mitochondrial dysfunction may be a specific and essential difference between healthy and cancer tissues. One cause of the dysfunction is known—inhibition of the pyruvate pathway into mitochondrial matrix. Treatment by DCA of cancers with the Warburg effect and by DCA in combination with glycolysis inhibitors of cancers with the reverse Warburg effect can stimulate synthesis of new drugs restoring physiological role of mitochondria and normal function of various cancer cells.

## Acknowledgment

This study was supported by Grant no. P102/11/0649 of the Czech Science Foundation GA CR.

## References

[1] O. Warburg, K. Posener, and E. Negelein, "Über den stoffwechsel der carcinomzelle," *Biochem Z*, vol. 152, pp. 309–344, 1924.

[2] O. Warburg, "On the origin of cancer cells," *Science*, vol. 123, no. 3191, pp. 309–314, 1956.

[3] R. Cohen and S. Havlin, *Complex Networks: Structure, Robustness and Function*, Cambridge University Press, 2010.

[4] A. Hübler, C. Stephenson, D. Lyon, and R. Swindeman, "Fabrication and programming of large physically evolving networks," *Complexity*, vol. 16, no. 5, pp. 7–8, 2011.

[5] J. Kwapień and S. Drożdż, "Physical approach to complex systems," *Physics Reports*, vol. 15, no. 3-4, pp. 115–226, 2012.

[6] G. Nicolis and I. Prigogine, *Self-Organization in Nonequilibrium Systems*, John Willey & Sons, New York, NY, USA, 1977.

[7] P. Hogeweg, "The roots of bioinformatics in theoretical biology," *PLoS Computational Biology*, vol. 7, no. 3, Article ID e1002021, 2011.

[8] J. Pokorný, T. Martan, and A. Foletti, "High capacity optical channels for bioinformation transfer: acupuncture meridians," *Journal of Acupuncture and Meridian Studies*, vol. 5, no. 1, pp. 34–41, 2012.

[9] H. K. Roy, H. Subramanian, D. Damania et al., "Optical detection of buccal epithelial nanoarchitectural alterations in patients harboring lung cancer: implications for screening," *Cancer Research*, vol. 70, no. 20, pp. 7748–7754, 2010.

[10] Y. Sun, C. Wang, and J. Dai, "Biophotons as neural communication signals demonstrated by in situ biophoton autography," *Photochemical and Photobiological Sciences*, vol. 9, no. 3, pp. 315–322, 2010.

[11] H. Fröhlich, "Bose condensation of strongly excited longitudinal electric modes," *Physics Letters A*, vol. 26, no. 9, pp. 402–403, 1968.

[12] H. Fröhlich, "Long-range coherence and energy storage in biological systems," *International Journal of Quantum Chemistry*, vol. 2, no. 5, pp. 641–649, 1968.

[13] H. Fröhlich, "Quantum mechanical concepts in biology," in *Theoretical Physics and Biology*, M. Marois, Ed., pp. 13–22, North Holland, 1969, (Proceedings of the 1st International Conference on. Theoretical physics Biology, Versailles, France , 1967).

[14] H. Fröhlich, "Collective behaviour of non-linearly coupled oscillating fields (with applications to biological systems)," *Journal of Collective Phenom*, vol. 1, pp. 101–109, 1973.

[15] H. Fröhlich, "The biological effects of microwaves and related questions," in *Advances in Electronics and Electron Physics*, vol. 53, pp. 85–152, 1980.

[16] H. Fröhlich, "Coherent electric vibrations in biological systems and cancer problem," *IEEE Transactions MTT*, vol. 26, pp. 613–617, 1978.

[17] H. A. Pohl, "Oscillating fields about growing cells," *International Journal of Quantum Chemistry*, vol. 7, pp. 411–431, 1980.

[18] R. Holzel and I. Lamprecht, "Electromagnetic fields around biological cells," *Neural Network World*, vol. 4, no. 3, pp. 327–337, 1994.

[19] R. Hölzel, "Electric activity of non-excitable biological cells at radio frequencies," *Electro- and Magnetobiology*, vol. 20, pp. 1–13, 2001.

[20] L. A. Amos and A. Klug, "Arrangement of subunits in flagellar microtubules," *Journal of Cell Science*, vol. 14, no. 3, pp. 523–549, 1974.

[21] J. A. Tuszyński, S. Hameroff, M. V. Satari Ć, B. Trpisová, and M. L. A. Nip, "Ferroelectric behavior in microtubule dipole lattices: implications for information processing, signaling and Assembly/Disassembly," *Journal of Theoretical Biology*, vol. 174, no. 4, pp. 371–380, 1995.

[22] J. Pokorný, J. Hašek, F. Jelínek, J. Šaroch, and B. Palán, "Electromagnetic activity of yeast cells in the M phase," *Electro- and Magnetobiology*, vol. 20, pp. 371–396, 2001.

[23] L. A. Amos, "Structure of microtubules," in *Microtubules*, K. Roberts and J. S. Hyam, Eds., pp. 1–64, Academic Press, London, UK, 1979.

[24] H. Stebbings and C. Hunt, "The nature of the clear zone around microtubules," *Cell and Tissue Research*, vol. 227, no. 3, pp. 609–617, 1982.

[25] G. N. Ling, "A new theoretical foundation for the polarized-oriented multilayer theory of cell water and for inanimate systems demonstrating long-range dynamic structuring of water molecules," *Physiological Chemistry and Physics and Medical NMR*, vol. 35, no. 2, pp. 91–130, 2003.

[26] J. Zheng and G. H. Pollack, "Long-range forces extending from polymer-gel surfaces," *Physical Review E*, vol. 68, no. 3, part 1, Article ID 031408, 1 page, 2003.

[27] G. Pollack, I. Cameron, and D. Wheatley, *Water and the Cell*, Springer, Dordrecht, The Netherlands, 2006.

[28] J. Zheng, W. Chin, E. Khijniak, E. Khijniak Jr., and G. H. Pollack, "Surfaces and interfacial water: evidence that hydrophilic surfaces have long-range impact," *Advances in Colloid and Interface Science*, vol. 127, no. 1, pp. 19–27, 2006.

[29] J. Pokorný, "Physical aspects of biological activity and cancer," *AIP Advances*, vol. 2, no. 1, Article ID 011207, 11 pages, 2012.

[30] K. M. Tyner, R. Kopelman, and M. A. Philbert, "'Nanosized voltmeter' enables cellular-wide electric field mapping," *Biophysical Journal*, vol. 93, no. 4, pp. 1163–1174, 2007.

[31] G. Preparata, *QED Coherence in Matter*, World Scientific, Hong Kong, China, 1995.

[32] E. Del Giudice, V. Elia, and A. Tedeschi, "The role of water in the living organisms," *Neural Network World*, vol. 19, no. 4, pp. 355–360, 2009.

[33] E. Del Giudice and A. Tedeschi, "Water and autocatalysis in living matter," *Electromagnetic Biology and Medicine*, vol. 28, no. 1, pp. 46–52, 2009.

[34] A. E. Pelling, S. Sehati, E. B. Gralla, J. S. Valentine, and J. K. Gimzewski, "Local nanomechanical motion of the cell wall of *Saccharomyces cerevisiae*," *Science*, vol. 305, no. 5687, pp. 1147–1150, 2004.

[35] A. E. Pelling, S. Sehati, E. B. Gralla, and J. K. Gimzewski, "Time dependence of the frequency and amplitude of the local nanomechanical motion of yeast," *Nanomedicine*, vol. 1, no. 2, pp. 178–183, 2005.

[36] J. Pokorný, J. Hašek, J. Vaniš, and F. Jelínek, "Biophysical aspects of cancer—electromagnetic mechanism," *Indian Journal of Experimental Biology*, vol. 46, pp. 310–321, 2008.

[37] F. Jelínek, M. Cifra, J. Pokorný et al., "Measurement of electrical oscillations and mechanical vibrations of yeast cells membrane around 1 kHz," *Electromagnetic Biology and Medicine*, vol. 28, pp. 223–232, 2009.

[38] E. D. Kirson, Z. Gurvich, R. Schneiderman et al., "Disruption of cancer cell replication by alternating electric fields," *Cancer Research*, vol. 64, no. 9, pp. 3288–3295, 2004.

[39] E. D. Kirson, V. Dbalý, F. Tovaryš et al., "Alternating electric fields arrest cell proliferation in animal tumor models and human brain tumors," *Proceedings of the National Academy of Sciences of the United States of America*, vol. 104, no. 24, pp. 10152–10157, 2007.

[40] C. Vedruccio and A. Meessen, "EM cancer detection by means of non-linear resonance interaction," in *Proceedings of the Progress in Electromagnetics Research Symposium (PIERS '04)*, pp. 909–912, March 2004.

[41] J. Pokorný, C. Vedruccio, M. Cifra, and O. Kučera, "Cancer physics: diagnostics based on damped cellular elastoelectrical vibrations in microtubules," *European Biophysics Journal*, vol. 40, no. 6, pp. 747–759, 2011.

[42] R. Damadian, "Tumor detection by nuclear magnetic resonance," *Science*, vol. 171, no. 3976, pp. 1151–1153, 1971.

[43] G. Albrecht-Buehler, "Surface extensions of 3T3 cells towards distant infrared light sources," *Journal of Cell Biology*, vol. 114, no. 3, pp. 493–502, 1991.

[44] G. Albrecht-Buehler, "Rudimentary form of cellular 'vision'," *Proceedings of the National Academy of Sciences of the United States of America*, vol. 89, no. 17, pp. 8288–8292, 1992.

[45] G. Albrecht-Buehler, "A long-range attraction between aggregating 3T3 cells mediated by near-infrared light scattering," *Proceedings of the National Academy of Sciences of the United States of America*, vol. 102, no. 14, pp. 5050–5055, 2005.

[46] S. Sahu, S. Ghosh, B. Ghosh et al., "Atomic water channel controlling remarkable properties of a single brain microtubule: correlating single protein to its supramolecular assembly," *Biosensors and Bioelectronics*, vol. 47, pp. 141–148, 2013.

[47] S. Sahu, S. Ghosh, K. Hirata, D. Fujita, and A. Bandyopadhyay, "Multi-level memory-switching properties of a single brain microtubule," *Applied Physics Letters*, vol. 102, no. 12, Article ID 123701, 4 pages, 2013.

[48] J. Pokorný, "Biophysical cancer transformation pathway," *Electromagnetic Biology and Medicine*, vol. 28, pp. 105–123, 2009.

[49] J. Pokorný, "Fröhlich's coherent vibrations in healthy and cancer cells," *Neural Network World*, vol. 19, pp. 369–378, 2009.

[50] J. Pokorný, "Endogenous electromagnetic forces in living cells: implications for transfer of reaction components," *Electro- and Magnetobiology*, vol. 20, no. 1, pp. 59–73, 2001.

[51] J. Pokorný, J. Hašek, and F. Jelínek, "Electromagnetic field of microtubules: effects on transfer of mass particles and electrons," *Journal of Biological Physics*, vol. 31, pp. 401–514, 2005.

[52] J. Pokorný, J. Hašek, and F. Jelínek, "Endogenous electric field and organization of living matter," *Electromagnetic Biology and Medicine*, vol. 24, pp. 185–197, 2005.

[53] J. Pokorný, "The role of Fröhlich's coherent excitations in cancer transformation of cells," in *Herbert Fröhlich, FRS: A Physicist Ahead of his Time*, G. J. Hyland and P. Rowlands, Eds., pp. 177–207, The University of Liverpool, Liverpool, UK, 2006.

[54] S. Bonnet, S. L. Archer, J. Allalunis-Turner et al., "A mitochondria-$K^+$ channel axis is suppressed in cancer and its normalization promotes apoptosis and inhibits cancer growth," *Cancer Cell*, vol. 11, no. 1, pp. 37–51, 2007.

[55] J. S. Carew and P. Huang, "Mitochondrial defects in cancer," *Molecular Cancer*, vol. 1, article 9, 2002.

[56] J. M. Cuezva, M. Krajewska, M. López de Heredia et al., "The bioenergetic signature of cancer: a marker of tumor progression," *Cancer Research*, vol. 62, no. 22, pp. 6674–6681, 2002.

[57] E. D. Michelakis, L. Webster, and J. R. Mackey, "Dichloroacetate (DCA) as a potential metabolic-targeting therapy for cancer," *British Journal of Cancer*, vol. 99, no. 7, pp. 989–994, 2008.

[58] L. B. Chen, "Mitochondrial membrane potential in living cells," *Annual Review of Cell Biology*, vol. 4, pp. 155–181, 1988.

[59] S. Pavlides, D. Whitaker-Menezes, R. Castello-Cros et al., "The reverse Warburg effect: aerobic glycolysis in cancer associated fibroblasts and the tumor stroma," *Cell Cycle*, vol. 8, no. 23, pp. 3984–4001, 2009.

[60] G. Bonuccelli, D. Whitaker-Menezes, R. Castello-Cros et al., "The reverse Warburg effect: glycolysis inhibitors prevent the tumor promoting effects of caveolin-1 deficient cancer associated fibroblasts," *Cell Cycle*, vol. 9, no. 10, pp. 1960–1971, 2010.

[61] U. E. Martinez-Outschoorn, R. M. Balliet, D. B. Rivadeneira et al., "Oxidative stress in cancer associated fibroblasts drives tumor-stroma co-evolution: a new paradigm for understanding tumor metabolism, the field effect and genomic instability in cancer cells," *Cell Cycle*, vol. 9, no. 16, pp. 3256–3276, 2010.

[62] M. P. Lisanti, U. E. Martinez-Outschoorn, B. Chiavarina et al., "Understanding the "lethal" drivers of tumor-stroma co-evolution: emerging role(s) for hypoxia, oxidative stress and autophagy/mitophagy in the tumor micro-environment," *Cancer Biology and Therapy*, vol. 10, no. 6, pp. 537–542, 2010.

[63] E. D. Michelakis, "Mitochondrial medicine: a new era in medicine opens new windows and brings new challenges," *Circulation*, vol. 117, no. 19, pp. 2431–2434, 2008.

[64] R. C. Sun, M. Fadia, J. E. Dahlstrom, C. R. Parish, P. G. Board, and A. C. Blackburn, "Reversal of the glycolytic phenotype by dichloroacetate inhibits metastatic breast cancer cell growth *in vitro* and *in vivo*," *Breast Cancer Research and Treatment*, vol. 120, no. 1, pp. 253–260, 2010.

[65] T. McFate, A. Mohyeldin, H. Lu et al., "Pyruvate dehydrogenase complex activity controls metabolic and malignant phenotype in cancer cells," *Journal of Biological Chemistry*, vol. 283, no. 33, pp. 22700–22708, 2008.

[66] S. E. Cross, Y. Jin, J. Rao, and J. K. Gimzewski, "Nanomechanical analysis of cells from cancer patients," *Nature Nanotechnology*, vol. 2, no. 12, pp. 780–783, 2007.

[67] G. Y. H. Lee and C. T. Lim, "Biomechanics approaches to studying human diseases," *Trends in Biotechnology*, vol. 25, no. 3, pp. 111–118, 2007.

[68] J. Guck, S. Schinkinger, B. Lincoln et al., "Optical deformability as an inherent cell marker for testing malignant transformation and metastatic competence," *Biophysical Journal*, vol. 88, no. 5, pp. 3689–3698, 2005.

[69] M. Beil, A. Micoulet, G. Von Wichert et al., "Sphingosylphosphorylcholine regulates keratin network architecture and viscoelastic properties of human cancer cells," *Nature Cell Biology*, vol. 5, no. 9, pp. 803–811, 2003.

[70] S. Suresh, J. Spatz, J. P. Mills et al., "Connections between single-cell biomechanics and human disease states: gastrointestinal cancer and malaria," *Acta Biomaterialia*, vol. 1, no. 1, pp. 15–30, 2005.

[71] S. Suresh, "Biomechanics and biophysics of cancer cells," *Acta Materialia*, vol. 55, no. 12, pp. 3989–4014, 2007.

[72] A. Khan, "Use of oral dichloroacetate for palliation of leg pain arising from metastatic poorly differentiated carcinoma: a case report," *Journal of Palliative Medicine*, vol. 14, no. 8, pp. 973–977, 2011.

[73] J. Pokorný, M. Cifra, A. Jandová et al., "Targeting mitochondria for cancer treatment," *European Journal of Cancer*, vol. 17, pp. 23–36, 2012.

[74] P. W. Stacpoole, G. N. Henderson, Z. Yan, and M. O. James, "Clinical pharmacology and toxicology of dichloroacetate," *Environmental Health Perspectives*, vol. 106, no. 4, pp. 989–994, 1998.

[75] K. R. Foster and J. W. Baish, "Viscous damping of vibrations in microtubules," *Journal of Biological Physics*, vol. 26, no. 4, pp. 255–260, 2000.

[76] L. K. McKemmish, J. R. Reimers, R. H. McKenzie, A. E. Mark, and N. S. Hush, "Penrose-Hameroff orchestrated objective-reduction proposal for human consciousness is not biologically feasible," *Physical Review E*, vol. 80, no. 2, part 1, Article ID 021912, 2009.

[77] J. R. Reimers, L. K. McKemmish, R. H. McKenzie, A. E. Mark, and N. S. Hush, "Weak, strong, and coherent regimes of Fröhlich condensation and their applications to terahertz medicine and quantum consciousness," *Proceedings of the National Academy of Sciences of the United States of America*, vol. 106, no. 11, pp. 4219–4224, 2009.

[78] T. N. Seyfried and L. M. Shelton, "Cancer as a metabolic disease," *Nutrition and Metabolism*, vol. 7, article 7, 2010.

[79] A. Jandová, J. Pokorný, J. Kobilková et al., "Cell-mediated immunity in cervical cancer evolution," *Electromagnetic Biology and Medicine*, vol. 28, pp. 1–14, 2009.

[80] R. R. Traill, "Asbestos as 'toxic short-circuit' optic-fibre for UV within the cell-net: likely roles and hazards for secret UV and IR metabolism," in *Proceedings of the 9th International Fröhlich's Symposium*, vol. 329 of *Journal of Physics: Conference Series*, p. 012017, 2011.

[81] S. Toyokuni, "Mechanisms of asbestos-induced carcinogenesis," *Nagoya Journal of Medical Science*, vol. 71, no. 1-2, pp. 1–10, 2009.

[82] S. Toyokuni, "Iron-induced carcinogenesis: the role of redox regulation," *Free Radical Biology and Medicine*, vol. 20, no. 4, pp. 553–566, 1996.

[83] J. Neuzil, M. Tomasetti, Y. Zhao et al., "Vitamin E analogs, a novel group of "mitocans," as anticancer agents: the importance of being redox-silent," *Molecular Pharmacology*, vol. 71, no. 5, pp. 1185–1199, 2007.

[84] K. Valis, L. Prochazka, E. Boura et al., "Hippo/Mst1 stimulates transcription of the proapoptotic mediator *NOXA* in a FoxO1-dependent manner," *Cancer Research*, vol. 71, no. 3, pp. 946–954, 2011.

[85] Ch. L.-L. Chiang, J. A. Ledermann, A. N. Rad, D. R. Katz, and B. M. Chain, "Hypochlorous acid enhances immunogenicity and uptake of allogeneic ovarian tumor cells by dendritic cells to cross-prime tumor-specific T cells," *Cancer Immunology, Immunotherapy*, vol. 55, no. 11, pp. 1384–1395, 2006.

# *In Vivo* Healthy Knee Kinematics during Dynamic Full Flexion

**Satoshi Hamai,**[1,2] **Taka-aki Moro-oka,**[1,2] **Nicholas J. Dunbar,**[1] **Hiromasa Miura,**[3] **Yukihide Iwamoto,**[2] **and Scott A. Banks**[1]

[1] *Department of Mechanical and Aerospace Engineering, University of Florida, 231 MAE-A Building, P.O. Box 116250, Gainesville, FL 32611-6250, USA*

[2] *Department of Orthopaedic Surgery, Faculty of Medical Sciences, Kyushu University, 3-1-1 Maidashi, Higashi-ku, Fukuoka 812-8582, Japan*

[3] *Department of Orthopaedic Surgery, Faculty of Medicine, Ehime University, 10-13 Dogo-himata, Matsuyama, Ehime 790-8577, Japan*

Correspondence should be addressed to Satoshi Hamai; hamachan@ortho.med.kyushu-u.ac.jp

Academic Editor: José M. Vilar

Healthy knee kinematics during dynamic full flexion were evaluated using 3D-to-2D model registration techniques. Continuous knee motions were recorded during full flexion in a lunge from 85° to 150°. Medial and lateral tibiofemoral contacts and femoral internal-external and varus-valgus rotations were analyzed as a function of knee flexion angle. The medial tibiofemoral contact translated anteroposteriorly, but remained on the center of the medial compartment. On the other hand, the lateral tibiofemoral contact translated posteriorly to the edge of the tibial surface at 150° flexion. The femur exhibited external and valgus rotation relative to the tibia over the entire activity and reached 30° external and 5° valgus rotations at 150° flexion. Kinematics' data during dynamic full flexion may provide important insight as to the designing of high-flexion total knee prostheses.

## 1. Introduction

3D-to-2D model registration techniques [1–4] and MRI-based methods [5–9] are used for assessing 3D knee kinematics without implants. Although deep knee flexion is an important function for many activities of daily living, few studies [4, 7, 9] have explored the *in vivo* knee kinematics beyond 140° of flexion. Kinematic analysis of healthy knees during dynamic full flexion is one key to designing for full flexion after total knee arthroplasty.

The purpose of this study was to observe healthy kinematics during full flexion in a lunge using 3D-to-2D model registration techniques. We sought to answer a specific question: How does the femur translate and rotate relative to the tibia during dynamic full flexion?

## 2. Materials and Methods

Five healthy male subjects, averaging 29 years (28–30), 172 cm (169–177), and 68 kg (55–80), gave informed consent to participate in this Institutional Review Board approved study. Continuous knee motions were recorded using a flat panel detector (Hitachi, Clavis, Tokyo, Japan: 3 frames/sec, $0.20 \times 0.20$ mm/pixel resolution) during full flexion in a lunge with their foot placed on a 25-cm step.

Cortical bone edges were segmented from CT images (Aquilion, Toshiba, Tochigi, Japan) using commercial software (SliceOmatic, Tomovision, Montreal, CA, USA), and these point clouds were converted into polygonal surface models (Geomagic Studio, Raindrop Geomagic, NC, USA). Bone model-embedded Cartesian coordinate systems for each femur and tibia were aligned with the cylindrical axis described by Eckhoff et al. [10]. The 3D position and orientation of the femur and tibia/fibula were determined using 3D-to-2D CT model-to-flat panel image registration techniques [3, 4] (Figure 1(a)). Medial and lateral tibiofemoral contacts were computed as the geometric center of the region having less than 6 mm tibiofemoral separation [3, 4] (Figure 1(b)). Femoral internal-external and varus-valgus rotations relative to the tibia were analyzed as a function of knee flexion angle. The center of axial rotation was determined from the femoral

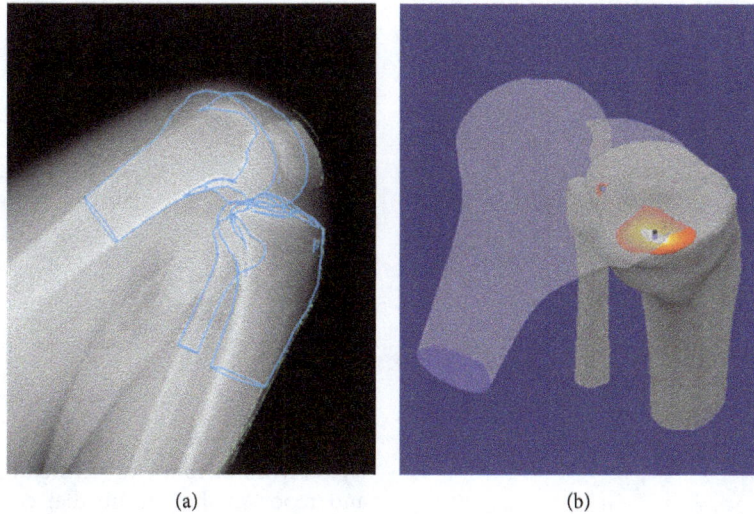

FIGURE 1: 3D-to-2D CT model-to-flat panel image registration techniques were used to determine the *in vivo* healthy knee kinematics (a). Medial and lateral tibiofemoral contacts were computed as the geometric center of the region having less than 6 mm tibiofemoral separation (b).

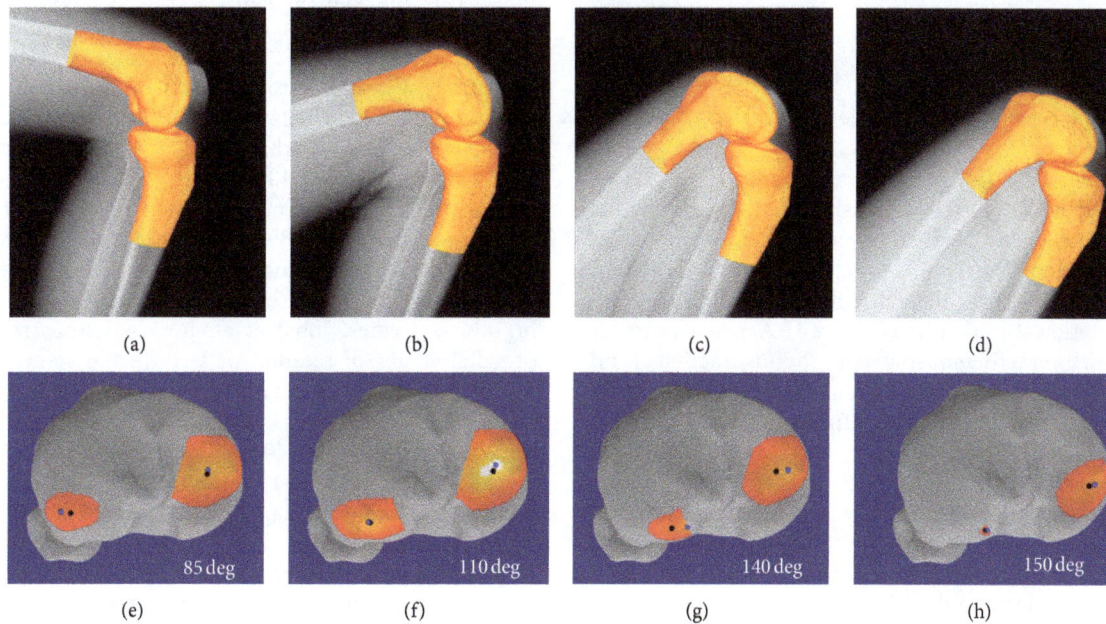

FIGURE 2: Matching of bone models to x-ray images: (a) 85°, (b) 110°, (c) 140°, and (d) 150° flexion. Medial and lateral tibiofemoral contacts were computed during weight-bearing dynamic knee flexion (e, f, g, and h). Black points on the tibial surfaces in the lower stand mean geometric center of the contact regions.

flexion/extension axis [11, 12]. The mediolateral location of the center of rotation was normalized to the dimensions of each tibial plateau, and expressed as a percentage of the tibial width, −50% (lateral) to +50% (medial). Spline interpolation with 10° flexion increments was used to create average kinematics for the group. The best-case accuracy of this matching method was 0.53 mm for in-plane translation, 1.6 mm for out-of-plane translation, and 0.54° for rotations in a previous study [3].

## 3. Results and Discussion

The medial tibiofemoral contact translated anteroposteriorly, but remained centered in the medial compartment (Figure 2). The lateral tibiofemoral contact translated posteriorly to the edge of the tibial surface at 150° flexion. From 85° to 150° flexion, the medial and lateral tibiofemoral contacts demonstrated 2 mm and 8 mm posterior translations on the tibial surface, respectively (Figure 3). The lateral femoral

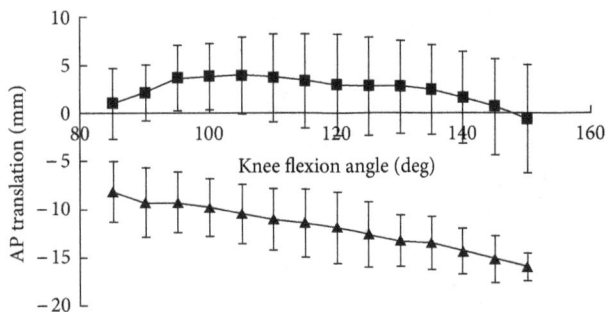

FIGURE 3: Anteroposterior (AP) translations of the medial (■) and lateral (▲) tibiofemoral contacts on the tibial surfaces.

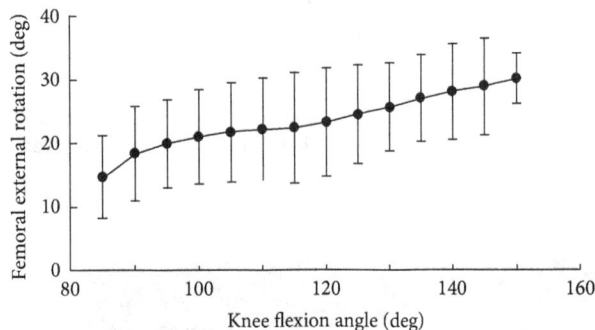

FIGURE 5: Femoral valgus rotation relative to the tibia.

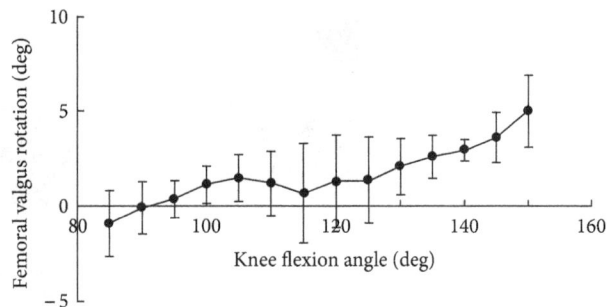

FIGURE 4: Femoral external rotation relative to the tibia.

condyle had larger posterior translations than the medial femoral condyle and femoral external rotation increased with knee flexion. The femur exhibited 15° external rotation relative to the tibia over the entire activity and reached 30° external rotation at 150° flexion (Figure 4). A medial center of rotation (+24%) was demonstrated for the lunge activity from 85° to 150° flexion. The femoral valgus rotation relative to the tibia was demonstrated over the entire activity and reached 5° at 150° flexion (Figure 5).

In this study, 3D-to-2D model registration techniques have been used to measure healthy knee kinematics during dynamic full flexion. The medial contact remained centered on the tibial surface. On the contrary, the lateral contact translated posteriorly to the edge of the tibial surface at full flexion. The femur externally rotated relative to tibia around a medially positioned axis from 85° to 150° flexion and reached 30° external and 5° valgus rotation relative to the tibia at 150° flexion.

Anteroposterior translations of the medial and lateral tibiofemoral contacts during deep knee flexion are generally consistent with previous 3D-to-2D model registration [3] and MR analyses [7, 9]. Nakagawa et al. [7] reported that the lateral condyle had the larger translation and lies posterior to the tibia at 162°. The lateral meniscus also moves with the lateral condyle, assuming a key role in distributing tibiofemoral compressive force [8]. Moro-oka et al. [3] reported that the medial contact translated 3 mm anteriorly and the lateral contact translated 8 mm posteriorly during kneeling from 100° to 150°. On the other hand, Pinskerova et al. [9] used the centers of the posterior circular portions of the medial and

lateral femoral condyles (termed the Flexion Facet Centers) and reported that the medial condyle moved back 8 mm and the lateral 5 mm from 120° to 160°. Variations in the anteroposterior translations could be explained by different anatomic coordinate systems, activity, and foot position [3, 5, 13].

The average center of rotation provides a simple metric to describe the general pattern of knee motion over an entire activity [1]. This metric permits intuitive comparisons between activities. In this study, the center of axial rotation for lunging was approximately located at the center of the medial compartment, +24%. This is consistent with reports from other papers, which examined nonambulatory activities. Yamaguchi et al. [12] reported that average center of rotation for squatting from extension to 110° was located in the medial tibia, +23%. Komistek et al. [1] compared kinematics in healthy knees during deep knee bends and rising from a chair and showed a medial center of rotation for both activities. Johal et al. [6] also similarly showed a medial center of rotation pattern during several sagittal plane activities. Moro-oka et al. [3] reported femoral external rotation increased with knee flexion for stair from 0° to 70°, squat from 0° to 140°, and kneel from 100° to 150°. On the other hand, recent studies using motion capture and dynamic stereo X-ray imaging have reported greater medial than lateral translations during walking and running and suggest that there is a lateral center for tibial axial rotations during weight-bearing activities near extension [14–16].

The femur exhibited 5° valgus rotation at 150° flexion. Since the medial and lateral tibial surfaces have different sagittal slopes, valgus rotation can result from the condyles moving anteroposteriorly while remaining in contact with these surfaces. In addition to the asymmetric geometry of the tibial surfaces, trapezoidal joint laxity in flexion is also considered to be necessary for deep knee flexion. In the healthy knee, the lateral tibiofemoral joint gap is significantly lax [17]. Posterior displacement with posterior stress was larger in the lateral compartment at 90° and 135° flexion [18]. Although achieving normal stability and kinematics in replaced knees is challenging, these kinematic data should be beneficial for designing high-flexion total knee prostheses.

This analysis has several drawbacks. First, the current approach to analyzing joint contact ignores the menisci, which are invisible on X-ray but obviously affect joint contact and load distribution. There is not currently an X-ray-based technique that will overcome this limitation, but 3D-to-2D registration techniques have the capability to reveal continuous dynamic *in vivo* kinematics. Second, the study included only five knees, but this cohort is similar to previous fluoroscopic studies that have analyzed four or five healthy knees [1, 2] and is consistent with minimizing X-ray exposure to healthy individuals while still obtaining important information. Finally, this study did not include the assessment of healthy patellofemoral kinematics, which also is important for designing knee replacements for natural flexion kinematics.

## 4. Conclusions

The lateral tibiofemoral contact translated posteriorly to the edge of the tibial surface at full flexion with 30° external and 5° valgus rotation relative to the tibia. The medial tibiofemoral contact remained centered on the tibial surface. The results of this study were generally consistent with reports from previous studies. Healthy knee kinematic data during full flexion may provide important insight for designing high-flexion total knee prostheses.

## Acknowledgment

The first author is supported by a Postdoctoral Fellowship from The Uehara Memorial Foundation.

## References

[1] R. D. Komistek, D. A. Dennis, and M. Mahfouz, "*In vivo* fluoroscopic analysis of the normal human knee," *Clinical Orthopaedics and Related Research*, no. 410, pp. 69–81, 2003.

[2] G. Li, L. E. DeFrate, E. P. Sang, T. J. Gill, and H. E. Rubash, "*In vivo* articular cartilage contact kinematics of the knee: an investigation using dual-orthogonal fluoroscopy and magnetic resonance image-based computer models," *American Journal of Sports Medicine*, vol. 33, no. 1, pp. 102–107, 2005.

[3] T. A. Moro-oka, S. Hamai, H. Miura et al., "Can magnetic resonance imaging-derived bone models be used for accurate motion measurement with single-plane three-dimensional shape registration?" *Journal of Orthopaedic Research*, vol. 25, no. 7, pp. 867–872, 2007.

[4] T. A. Moro-Oka, S. Hamai, H. Miura et al., "Dynamic activity dependence of *in vivo* normal knee kinematics," *Journal of Orthopaedic Research*, vol. 26, no. 4, pp. 428–434, 2008.

[5] P. F. Hill, V. Vedi, A. Williams, H. Iwaki, V. Pinskerova, and M. A. R. Freeman, "Tibiofemoral movement 2: the loaded and unloaded living knee studied by MRI," *Journal of Bone and Joint Surgery B*, vol. 82, no. 8, pp. 1196–1198, 2000.

[6] P. Johal, A. Williams, P. Wragg, D. Hunt, and W. Gedroyc, "Tibio-femoral movement in the living knee. A study of weight bearing and non-weight bearing knee kinematics using "interventional" MRI," *Journal of Biomechanics*, vol. 38, no. 2, pp. 269–276, 2005.

[7] S. Nakagawa, Y. Kadoya, S. Todo et al., "Tibiofemoral movement 3: full flexion in the living knee studied by MRI," *Journal of Bone and Joint Surgery B*, vol. 82, no. 8, pp. 1199–1200, 2000.

[8] J. Yao, S. L. Lancianese, K. R. Hovinga, J. Lee, and A. L. Lerner, "Magnetic resonance image analysis of meniscal translation and tibio-menisco-femoral contact in deep knee flexion," *Journal of Orthopaedic Research*, vol. 26, no. 5, pp. 673–684, 2008.

[9] V. Pinskerova, K. M. Samuelson, J. Stammers, K. Maruthainar, A. Sosna, and M. A. R. Freeman, "The knee in full flexion: an anatomical study," *Journal of Bone and Joint Surgery B*, vol. 91, no. 6, pp. 830–834, 2009.

[10] D. G. Eckhoff, T. F. Dwyer, J. M. Bach, V. M. Spitzer, and K. D. Reinig, "Three-dimensional morphology of the distal part of the femur viewed in virtual reality," *Journal of Bone and Joint Surgery A*, vol. 83, no. 2, pp. 43–50, 2001.

[11] S. A. Banks and W. A. Hodge, "2003 Hap Paul Award Paper of the International Society for Technology in Arthroplasty: design and activity dependence of kinematics in fixed and mobile-bearing knee arthroplasties," *Journal of Arthroplasty*, vol. 19, no. 7, pp. 809–816, 2004.

[12] S. Yamaguchi, K. Gamada, T. Sasho, H. Kato, M. Sonoda, and S. A. Banks, "*In vivo* kinematics of anterior cruciate ligament deficient knees during pivot and squat activities," *Clinical Biomechanics*, vol. 24, no. 1, pp. 71–76, 2009.

[13] S. Mu, T. Moro-Oka, P. Johal, S. Hamai, M. A. R. Freeman, and S. A. Banks, "Comparison of static and dynamic knee kinematics during squatting," *Clinical Biomechanics*, vol. 26, no. 1, pp. 106–108, 2011.

[14] S. Koo and T. P. Andriacchi, "The knee joint center of rotation is predominantly on the lateral side during normal walking," *Journal of Biomechanics*, vol. 41, no. 6, pp. 1269–1273, 2008.

[15] M. Kozanek, A. Hosseini, F. Liu et al., "Tibiofemoral kinematics and condylar motion during the stance phase of gait," *Journal of Biomechanics*, vol. 42, no. 12, pp. 1877–1884, 2009.

[16] Y. Hoshino and S. Tashman, "Internal tibial rotation during *in vivo*, dynamic activity induces greater sliding of tibio-femoral joint contact on the medial compartment," *Knee Surgery, Sports Traumatology, Arthroscopy*, vol. 20, no. 7, pp. 1268–1275, 2012.

[17] Y. Tokuhara, Y. Kadoya, S. Nakagawa, A. Kobayashi, and K. Takaoka, "The flexion gap in normal knees. An MRI study," *Journal of Bone and Joint Surgery B*, vol. 86, no. 8, pp. 1133–1136, 2004.

[18] S. Fukagawa, S. Matsuda, Y. Tashiro, M. Hashizume, and Y. Iwamoto, "Posterior displacement of the tibia increases in deep flexion of the knee," *Clinical Orthopaedics and Related Research*, vol. 468, no. 4, pp. 1107–1114, 2010.

# Permissions

The contributors of this book come from diverse backgrounds, making this book a truly international effort. This book will bring forth new frontiers with its revolutionizing research information and detailed analysis of the nascent developments around the world.

We would like to thank all the contributing authors for lending their expertise to make the book truly unique. They have played a crucial role in the development of this book. Without their invaluable contributions this book wouldn't have been possible. They have made vital efforts to compile up to date information on the varied aspects of this subject to make this book a valuable addition to the collection of many professionals and students.

This book was conceptualized with the vision of imparting up-to-date information and advanced data in this field. To ensure the same, a matchless editorial board was set up. Every individual on the board went through rigorous rounds of assessment to prove their worth. After which they invested a large part of their time researching and compiling the most relevant data for our readers. Conferences and sessions were held from time to time between the editorial board and the contributing authors to present the data in the most comprehensible form. The editorial team has worked tirelessly to provide valuable and valid information to help people across the globe.

Every chapter published in this book has been scrutinized by our experts. Their significance has been extensively debated. The topics covered herein carry significant findings which will fuel the growth of the discipline. They may even be implemented as practical applications or may be referred to as a beginning point for another development. Chapters in this book were first published by Hindawi Publishing Corporation; hereby published with permission under the Creative Commons Attribution License or equivalent.

The editorial board has been involved in producing this book since its inception. They have spent rigorous hours researching and exploring the diverse topics which have resulted in the successful publishing of this book. They have passed on their knowledge of decades through this book. To expedite this challenging task, the publisher supported the team at every step. A small team of assistant editors was also appointed to further simplify the editing procedure and attain best results for the readers.

Our editorial team has been hand-picked from every corner of the world. Their multi-ethnicity adds dynamic inputs to the discussions which result in innovative outcomes. These outcomes are then further discussed with the researchers and contributors who give their valuable feedback and opinion regarding the same. The feedback is then collaborated with the researches and they are edited in a comprehensive manner to aid the understanding of the subject.

Apart from the editorial board, the designing team has also invested a significant amount of their time in understanding the subject and creating the most relevant covers. They scrutinized every image to scout for the most suitable representation of the subject and create an appropriate cover for the book.

The publishing team has been involved in this book since its early stages. They were actively engaged in every process, be it collecting the data, connecting with the contributors or procuring relevant information. The team has been an ardent support to the editorial, designing and production team. Their endless efforts to recruit the best for this project, has resulted in the accomplishment of this book. They are a veteran in the field of academics and their pool of knowledge is as vast as their experience in printing. Their expertise and guidance has proved useful at every step. Their uncompromising quality standards have made this book an exceptional effort. Their encouragement from time to time has been an inspiration for everyone.

The publisher and the editorial board hope that this book will prove to be a valuable piece of knowledge for researchers, students, practitioners and scholars across the globe.

# List of Contributors

**Milda Bilinauskaite**
Department of Mechanical Engineering, Kaunas University of Technology, LT-44029 Kaunas, Lithuania
Mechatronics Centre for Research, Studies and Information, Kaunas University of Technology, LT-44029 Kaunas, Lithuania
Department of Mechanical Engineering, University of Tras-os-Montes and Alto Douro, 5001-801 Vila Real, Portugal
Centre of Research in Sports, Health and Human Development, CIDESD, 5001-801 Vila Real, Portugal

**Vishveshwar Rajendra Mantha**
Department of Mechanical Engineering, University of Tras-os-Montes and Alto Douro, 5001-801 Vila Real, Portugal
Centre of Research in Sports, Health and Human Development, CIDESD, 5001-801 Vila Real, Portugal
Department of Sport Sciences, Exercise and Health, University of Tras-os-Montes and Alto Douro, 5001-801 Vila Real, Portugal

**Abel Ilah Rouboa**
Department of Mechanical Engineering, University of Tras-os-Montes and Alto Douro, 5001-801 Vila Real, Portugal
Department of Mechanical Engineering and Applied Mechanics, University of Pennsylvania, Philadelphia, PA 19104, USA

**Pranas Ziliukas**
Department of Mechanical Engineering, Kaunas University of Technology, LT-44029 Kaunas, Lithuania

**Antonio Jose Silva**
Centre of Research in Sports, Health and Human Development, CIDESD, 5001-801 Vila Real, Portugal
Department of Sport Sciences, Exercise and Health, University of Tras-os-Montes and Alto Douro, 5001-801 Vila Real, Portugal

**Alberto Ranavolo, Alessio Silvetti, Sergio Iavicoli and Francesco Draicchio**
Department of Occupational Medicine, INAIL, Via Fontana Candida 1, Monte Porzio Catone, 00040 Rome, Italy

**Lorenzo M. Donini**
Department of Experimental Medicine, Medical Physiopathology, Food Science and Endocrinology Section, Food Science and Human Nutrition Research Unit, Sapienza University of Rome, Ple Aldo Moro 5, 00185 Rome, Italy
Villa delle Querce Clinical Rehabilitation Institute, Unit of Metabolic and Nutritional Rehabilitation, Via delle Vigne 19, Nemi, 00040 Rome, Italy

**Silvia Mari**
Fondazione Don Gnocchi, 20148 Milan, Italy

**Mariano Serrao**
Rehabilitation Centre, Policlinico Italia, Piazza del Campidano 6, 00162 Rome, Italy
Department of Medical and Surgical Science and Biotechnologies, Sapienza University of Rome, Via Faggiana 34, 40100 Latina, Italy

**Edda Cava, Rosa Asprino and Alessandro Pinto**
Department of Experimental Medicine, Medical Physiopathology, Food Science and Endocrinology Section, Food Science and Human Nutrition Research Unit, Sapienza University of Rome, Ple Aldo Moro 5, 00185 Rome, Italy

**Nihal S. El-Bialy and Monira M. Rageh**
Biophysics Department, Faculty of Science, Cairo University, Al Gammaa Street, Giza 12613, Egypt

**Hansang Kim**
Department of Mechanical and Automotive Engineering, Gachon University, Seongnam-Si, Gyeonggi-do 461-701, Republic of Korea

**Lawrence Yoo and Andrew Shin**
Department of Ophthalmology, Jules Stein Eye Institute, University of California, Los Angeles, CA 90095-7002, USA
Department of Mechanical Engineering, University of California, Los Angeles, CA, USA

**Joseph L. Demer**
Department of Ophthalmology, Jules Stein Eye Institute, University of California, Los Angeles, CA 90095-7002, USA
Biomedical Engineering Interdepartmental Program, University of California, Los Angeles, CA, USA
Neuroscience Interdepartmental Program, University of California, Los Angeles, CA, USA
Department of Neurology, University of California, Los Angeles, CA, USA

**Giedrius Gorianovas, Albertas Skurvydas, Vytautas Streckis, Marius Brazaitis, and Sigitas Kamandulis**
Department of Applied Biology and Physiotherapy, Research Centre for Fundamental and Clinical Movement Sciences, Lithuanian Sports University, Sporto 6, 4422 Kaunas, Lithuania

**Malachy P. McHugh**
Nicholas Institute of Sports Medicine and Athletic Trauma, Lenox Hill Hospital, 130 East 77th Street, New York, NY 10075, USA

**Rafael R. G. Maciel, Adriele A. de Almeida, Odin G. C. Godinho, Filipe D. S. Gorza, Graciela C. Pedro, Tarquin F. Trescher, Josmary R. Silva, and Nara C. de Souza**
Grupo de Materiais Nanoestruturados, Campus Universit´ario do Araguaia, Universidade Federal de Mato Grosso, 78600-000 Barra do Garc¸as, MT, Brazil

**Nupur Hajela**
Electrophysiological Analysis of Gait and Posture Laboratory, Sensory Motor Performance Program, Rehabilitation Institute of Chicago, 345 East Superior Street, Chicago, IL 60611, USA
Department of Physical Medicine and Rehabilitation, Northwestern University Feinberg School of Medicine, Chicago, IL 60611, USA

**Chaithanya K. Mummidisetty, Andrew C. Smith**
Electrophysiological Analysis of Gait and Posture Laboratory, Sensory Motor Performance Program, Rehabilitation Institute of Chicago, 345 East Superior Street, Chicago, IL 60611, USA

**Maria Knikou**
Electrophysiological Analysis of Gait and Posture Laboratory, Sensory Motor Performance Program, Rehabilitation Institute of Chicago, 345 East Superior Street, Chicago, IL 60611, USA
Department of Physical Medicine and Rehabilitation, Northwestern University Feinberg School of Medicine, Chicago, IL 60611, USA
Department of Physical Therapy and the Graduate Center, The City University of New York, Staten Island, NY 10314, USA

**I. V. Ogneva**
State Research Center of Russian Federation Institute of Biomedical Problems, Russian Academy of Sciences, 76-a, Khoroshevskoyoe shosse, Moscow 123007, Russia

**Jumat Salimon, Nadia Salih, and Bashar Mudhaffar Abdullah**
School of Chemical Sciences and Food Technology, Faculty of Science and Technology, Universiti Kebangsaan Malaysia, Bangi, 43600 Selangor, Malaysia

**Elena Ibarz, Yolanda Más, José Cegoñino and Sergio Puértolas**
Department of Mechanical Engineering, University of Zaragoza, 50018 Zaragoza, Spain

**Antonio Herrera and Javier Rodríguez-Vela**
Department of Surgery, University of Zaragoza, 50009 Zaragoza, Spain
Department of Orthopaedic Surgery and Traumatology, Miguel Servet University Hospital, 50009 Zaragoza, Spain
Aragón Health Sciences Institute, 50009 Zaragoza, Spain

**Luis Gracia**
Department of Mechanical Engineering, University of Zaragoza, 50018 Zaragoza, Spain
Engineering and Architecture School, University of Zaragoza, María de Luna 3, 50018 Zaragoza, Spain

**Monira M. Rageh, Reem H. EL-Gebaly, and Nihal S. El-Bialy**
Department of Biophysics, Faculty of Science, Cairo University, Giza 12013, Egypt

**Júlia Maria D'Andréa Greve, Guilherme Carlos Brech and Angelica Castilho Alonso**
Laboratory of Movement Studies (LEM), Institute of Orthopedics and Traumatology (IOT), Hospital das Clínicas (HC), School of Medicine, University of São Paulo, 05403-010 São Paulo, SP, Brazil

**Mutlu Cuğ**
Physical Education and Sports Department, Cumhuriyet University, 58140 Sivas, Turkey

**Deniz Dülgeroğlu**
Diskapi Yildirim Beyazit Education and Research Hospital, Physical Medicine and Rehabilitation Clinic, Ankara, Turkey

**Pedro Figueiredo**
Centre of Research, Education, Innovation and Intervention in Sport, Faculty of Sport, University of Porto, Rua Dr. Pl´acido Costa 91, 4200-450 Porto, Portugal
Higher Education Institute of Maia (ISMAI), Avenida Carlos Oliveira Campos, 4475-690 Maia, Portugal

**David R. Pendergast**
Center for Research and Education in Special Environments, Department of Physiology and Biophysics, University at Buffalo, 3435 Main Street, Buffalo, NY 14214, USA

**João Paulo Vilas-Boas and Ricardo J. Fernandes**
Centre of Research, Education, Innovation and Intervention in Sport, Faculty of Sport, University of Porto, Rua Dr. Pl´acido Costa 91, 4200-450 Porto, Portugal
Porto Biomechanics Laboratory, University of Porto, Rua Dr. Pl´acido Costa 91, 4200-450 Porto, Portugal

**Joseph Vorro**
Department of Family Medicine, College of Osteopathic Medicine, Michigan State University, East Lansing, MI 48824, USA

**Tamara R. Bush**
Department of Mechanical Engineering, College of Engineering, Michigan State University, East Lansing, MI 48824, USA

**Brad Rutledge**
BiomechanicsDivision, MEA Forensic Engineers & Scientists, Laguna Hills, CA 92653, USA

**Mingfei Li**
Department of Mathematical Sciences and Center for Quantitative Analysis, Bentley University,Waltham, MA 02452, USA

**Jieshi Ma, Canhua Xu, Meng Dai, Fusheng You, Xuetao Shi, Xiuzhen Dong and Feng Fu**
Department of Biomedical Engineering, Fourth Military Medical University, Xi'an 710032, China

**Lauryl E. Campbell, Jennifer Nelson, Elizabeth Gibbons, Allan M. Judd, and John D. Bell**
Department of Physiology and Developmental Biology, BrighamYoung University, Provo, UT 84601, USA

**Ming Yang**
Department of Botany, Oklahoma State University, 301 Physical Sciences, Stillwater, OK 74078, USA

**Silvia Martini, Claudia Bonechi and Claudio Rossi**
Department of Biotechnology, Chemistry and Pharmacy, University of Siena, Via Aldo Moro 2, 53100 Siena, Italy
Centre for Colloid and Surface Science (CSGI), University of Florence, Via della Lastruccia 3, 50019 Sesto Fiorentino, Italy

**Alberto Foletti**
Laboratory of Applied Mathematics and Physics, Department of Innovative Technologies-DTI, University of Applied Sciences of Southern Switzerland (SUPSI), Manno, CH 6928, Switzerland

**Jilí Pokorný, and Anna Jandová**
Institute of Photonics and Electronics, Academy of Sciences of the Czech Republic, AS CR, Chabersk´a 57, 182 51 Prague 8-Kobylisy, Czech Republic

**Alberto Foletti**
Institute of Translational Pharmacology, National Research Council-CNR, Via Fosso del Cavaliere 100, 00133 Rome, Italy
University of Applied Sciences of Southern Switzerland-SUPSI, Department of Innovative Technologies, Galleria 2, 6928 Manno, Switzerland

**Jitka Kobilková**
1st Faculty of Medicine, Charles University in Prague, Department of Obstetrics and Gynaecology, Apolin´aˇrsk´a 18, 128 00 Prague 2, Czech Republic

**Jan Vrba**
Faculty of Electrical Engineering, Czech Technical University in Prague, Technick´a 2, 166 27 Prague 6, Czech Republic

**Jan Vrba Jr.**
Faculty of Biomedical Engineering, Czech Technical University in Kladno, Sitn´a Square 3105, 272 01 Kladno, Czech Republic

**Martina Nedbalová**
1st Faculty of Medicine, Charles University in Prague, Institute of Physiology, Albertov 5, 128 00 Prague 2, Czech Republic

**Aleš Holek**
3rd Faculty of Medicine, Charles University in Prague, Department of Otorhinolaryngology, Rusk´a 87, 100 00 Prague 10, Czech Republic

**Andrea Danani**
University of Applied Sciences of Southern Switzerland-SUPSI, Department of Innovative Technologies, Galleria 2, 6928 Manno, Switzerland

**Jack A. TuszyNski**
Department of Physics, University of Alberta, Edmonton, AB, Canada T6G 2J7

**Satoshi Hamai and Taka-aki Moro-oka**
Department of Mechanical and Aerospace Engineering, University of Florida, 231 MAE-A Building, P.O. Box 116250, Gainesville, FL 32611-6250, USA
Department of Orthopaedic Surgery, Faculty of Medical Sciences, Kyushu University, 3-1-1 Maidashi, Higashi-ku, Fukuoka 812-8582, Japan

**Nicholas J. Dunbar and Scott A. Banks**
Department of Mechanical and Aerospace Engineering, University of Florida, 231 MAE-A Building, P.O. Box 116250, Gainesville, FL 32611-6250, USA

**Hiromasa Miura**
Department of Orthopaedic Surgery, Faculty of Medicine, Ehime University, 10-13 Dogo-himata, Matsuyama, Ehime 790-8577, Japan

**Yukihide Iwamoto**
Department of Orthopaedic Surgery, Faculty of Medical Sciences, Kyushu University, 3-1-1 Maidashi, Higashi-ku, Fukuoka 812-8582, Japan

www.ingramcontent.com/pod-product-compliance
Lightning Source LLC
Chambersburg PA
CBHW050449200326

41458CB00014B/5119